Microbiology Monographs

Volume 34

Series Editor

Alexander Steinbüchel
Münster, Germany

The Springer book series ***Microbiology Monographs*** presents carefully refereed volumes on selected microbiological topics. Microbiology is a still rapidly expanding field with significant impact on many areas of basic and applied science. The growth in knowledge of microbial physiology, cell structure, biotechnological capabilities and other aspects of microorganisms is increasing dramatically even in the face of the breakthroughs that already have been made.

Reflecting these most recent achievements, the series' wide scope encompasses such topics as inclusions in prokaryotes, predatory prokaryotes, magnetoreception and magnetosomes in bacteria, uncultivable microorganisms, microbial endosymbionts, bacterial resistance, extremophilic microorganisms, analyses of genome sequences and structures, microorganisms as cell factories for chemicals and fuels, metabolic engineering, gene transfer and expression systems and distinct physiological groups of bacteria.

The volume editors are well-known experts in their particular fields, and each volume offers 10 to 20 comprehensive review articles covering all relevant aspects of the topic in focus.

More information about this series at http://www.springer.com/series/7171

Dieter Jendrossek
Editor

Bacterial Organelles and Organelle-like Inclusions

Second Edition

Springer

Editor
Dieter Jendrossek
Institute of Microbiology
University of Stuttgart
Stuttgart, Baden-Württemberg, Germany

ISSN 1862-5576 ISSN 1862-5584 (electronic)
Microbiology Monographs
ISBN 978-3-030-60172-0 ISBN 978-3-030-60173-7 (eBook)
https://doi.org/10.1007/978-3-030-60173-7

This Springer imprint is published by the registered company Springer Nature Switzerland AG.
The registered company address is: Gewerbestrasse 11, 6330 Cham, Switzerland

Preface

When we studied microbiology in the seventies and eighties of the last century, the knowledge on the presence and composition of compartmented structures in prokaryotes was poor. It was assumed that the metabolic activities of prokaryotic cells with average diameters in the lower micrometer range are sufficiently explained simply by diffusion of the metabolites between macromolecules (DNA and proteins) and other supramolecular structures (ribosomes). Consequently, the presence of organelles or related functional compartments was negated in prokaryotic cells at that time. It was and still is amazing for us to realize how wrong this view was.

Meanwhile, well-described intracellular compartments in prokaryotes range from supramolecular complexes such as sulfur globules or PHB granules (carbonosomes) to protein-encapsulated compartments (gas vesicles, carboxysomes) to truly membrane-enclosed organelles such as anammoxosomes or magnetosomes. Recent data on the composition of polyphosphate granules in the famous "Knallgasbakterium" *Ralstonia eutropha* (*Cupriavidus necator*) revealed the presence of (currently) up to eight proteins that can be in vivo attached to the surface of polyphosphate granules and has led to the suggestion to address these structures as polyphosphatosomes. Currently, preliminary data indicate that other insoluble structures ("inclusions") such as granulose, cyanophycin, and lipid-like inclusions probably also represent functional compartments in prokaryotic cells.

With two volumes on inclusions in prokaryotes, the book series *Microbiology Monographs* was initiated in 2006. Both volumes were edited by Jessup Shively. In this book, we have collected updates of a previous version of this *Microbiology Monographs* series mostly from Volume 1 *Inclusions in Prokaryotes*. This volume focuses on the ten most important and best investigated inclusions. These are timely updates provided by experts in the field. Another value is that these updates occur in the same volume and thereby allow a good comparison of the different inclusions.

We thank all authors. Without the author's sophisticated contributions and others' engagement it would not have been possible to compile such an interesting volume. We hope that the readers will enjoy it! We also thank Markus Späth and Bibhuti Sharma from Springer for their valuable inputs.

Stuttgart, Germany
Münster, Germany
August 2020

Dieter Jendrossek
Alexander Steinbüchel

Contents

Polyphosphate Granules and Acidocalcisomes 1
Dieter Jendrossek

Bacterial Intracellular Sulphur Globules 19
Christiane Dahl

Biosynthesis and Intracellular Organization of Magnetosomes in Magnetotactic Bacteria 53
Dirk Schüler and Frank D. Müller

Gas Vesicles of Archaea and Bacteria 71
Felicitas Pfeifer

The Anammoxosome Organelle: The Power Plant of Anaerobic Ammonium-Oxidizing (Anammox) Bacteria 107
Laura Claret Fernández, Rob Mesman, and Laura van Niftrik

Bacterial Microcompartments 125
Sabine Heinhorst and Gordon C. Cannon

The Cyanophycin Granule Peptide from Cyanobacteria 149
Björn Watzer, Friederike Klemke, and Karl Forchhammer

Storage Polysaccharides in Prokaryotes: Glycogen, Granulose, and Starch-Like Granules 177
Matthieu Colpaert, Malika Chabi, Ugo Cenci, and Christophe Colleoni

Wax Ester and Triacylglycerol Inclusions 211
Alexander Steinbüchel and Marc Wältermann

Carbonosomes 243
Dieter Jendrossek

Polyphosphate Granules and Acidocalcisomes

Dieter Jendrossek

Contents

1 Introduction 2
2 Staining and Detection of polyP 2
3 Functions of polyP 3
4 Key Enzymes of polyP Metabolism 4
5 PolyP Granules in Prokaryotes 5
6 The polyP Granule Proteome 6
7 Phosins: polyP Granule-Associated Proteins with CHAD-Domains 6
8 Other polyP Granule-Associated Proteins 8
9 Acidocalcisomes in Prokaryotes 10
References 12

Abstract Polyphosphate (polyP) is a linear polymer of phosphate residues linked by energy-rich phospho-anhydride bonds. It can be formed abiotically by heating of phosphate solutions (e.g. by vulcanism) or biologically by the action of enzymes, so-called polyphosphate kinases (PPKs). PolyP is an ubiquitous polymer present in virtually all organisms that have been looked at and therefore should be of fundamental importance. It can have an astonishing variety of functions in different species from that of a reservoir for phosphorous in many prokaryotes to a part of the blood coagulation system in mammalians. This chapter focuses only on polyP in prokaryotes where it is present in the form of granule-like inclusions of roughly 50 to about 500 nm diameter. Despite their ubiquitous occurrence in bacteria, the structures and composition of polyP granules in bacteria are only poorly understood. In this chapter I will provide an update of the various functions polyP can have in prokaryotes, will give an overview on PPK enzymes and describe some recent findings showing that polyP granules have an astonishing high number of proteins specifically bound to the surface of polyP granules (polyphosphatosomes). In addition, I will discuss the pros and cons whether acidic, membrane-enclosed, polyP and calcium-ions containing organelles (acidocalcisomes) exist in prokaryotes.

D. Jendrossek (✉)
Institute of Microbiology, University of Stuttgart, Stuttgart, Baden-Württemberg, Germany
e-mail: dieter.jendrossek@imb.uni-stuttgart.de

D. Jendrossek (ed.), *Bacterial Organelles and Organelle-like Inclusions*,
Microbiology Monographs 34, https://doi.org/10.1007/978-3-030-60173-7_1

1 Introduction

The presence of insoluble phosphate-containing inclusions in microorganisms is known for a long time. The first descriptions go back to the end of the nineteenth century (Liebermann 1888; Grimme 1902; Meyer 1904). PolyP was referred to as metaphosphate or volutin granules (present in *Spirillum volutans*, (Grimme 1902)) because of the metachromic shift of basic dyes upon binding to polyphosphate (Wiame 1947). PolyP was regularly found in yeast (Schmidt et al. 1946; Lindegren 1947; Widra 1959) and in several bacterial species such as *Aerobacter aerogenes* (Smith et al. 1954) and in particular in *Mycobacterium* and *Corynebacterium* species (Cammidge 1901; Harada 1957; Sall et al. 1958; Knaysi 1959) Later, it became evident that volutin (polyP) granules are widespread in microorganisms (for early reviews see Widra (1959); Harold (1966); Kulaev and Vagabov (1983)), meanwhile it is known that polyP is present in all kingdoms of life and apparently is present in every living organism (Kornberg et al. 1999; Kulaev and Kulakovskaya 2000; Rao et al. 2009).

2 Staining and Detection of polyP

PolyP granules can be observed in living microorganisms by light microscopy (Bringmann 1951; Smith et al. 1954). However, polyP granules are usually stained by metachromic dyes for higher sensitivity and specificity. The classical Neisser stain (Neisser 1888) (for examples, see Smith et al. (19540; Harada (1957); Sall et al. (1958); Roinestad and Yall (1970); Strohl and Larkin (1978); Pallerla et al. (2005)) uses methylene blue but was later replaced by DAPI-staining and fluorescence microscopy (Kulakova et al. 2011), for specific staining conditions and polyP quantification, see Gomes et al. (2013). DAPI-polyP complexes emit light at higher wavelength ($\approx$540 nm) than complexes of DAPI-dsDNA ($\approx$460 nm) or DAPI-ssRNA ($\approx$500 nm). Details of the basis of polyP detection in bacteria by fluorescence microscopy have been described in (Pallerla et al. 2005; Klauth et al. 2006; Aschar-Sobbi et al. 2008). Other dyes to detect polyP are tetracycline and toluidine blue. Staining with tetracycline has been used to detect polyP granules in wastewater organisms (Günther et al. 2009). Toluidine blue is frequently used to stain extracted polyP molecules of different lengths after polyacryl amide gel electrophoresis. Details of gel preparation, separation of polyP molecules of different lengths and staining with toluidine are well described in Clark et al. (1986); Clark and Wood (1987). Unfortunately, none of the above-mentioned dyes can be used to stain polyP in living and growing cells. Two non-toxic green-fluorescent dyes, JC-D7 and JCD8, have been described that can be used to stain polyP in living cells (Angelova et al. 2014). However, the specificity and fluorescence intensity of the JC-dye-polyP complexes is less efficient than for DAPI-polyP (unpublished own observation) and therefore staining with DAPI is still the method of choice for the sensitive detection of polyP. Methods for polyP quantification have been recently summarized in a recommendable method paper (Christ et al. 2020).

3 Functions of polyP

The most evident function of polyP is that of a storage compound for phosphorous and the electrostatic counterparts to the phosphate anions such as Ca^{2+} and Mg^{2+}. Binding of other cations to polyP might effectuate a detoxification of heavy metal ions (Jensen 1968; Jensen et al. 1982; Keasling 1997; Alvarez and Jerez 2004; Jasso-Chávez et al. 2019). However, several additional putative functions have been addressed to polyP granules: reservoir for energy (due to a high number of energy-rich phospho-anhydride bonds in polyP), alternative to ATP and NTPs, source of phosphate residues in phosphorylation (kinase) reactions, pathogenicity, gene expression and gene regulation, regulation of enzyme activity, stress response and adaptation to stationary phase and others (Harold 1966; Kulaev and Vagabov 1983; Albi and Serrano 2016). PolyP has a prominent role in bacteria that are part of the biological phosphate removal process (EBPR) in sewage sludge (Kortstee et al. 2000; Blackall et al. 2002; Seviour et al. 2003; Yang et al. 2017). In *E. coli*, it is well established that polyP has a prominent function in stringent response to starvation conditions (Kulaev and Vagabov 1983; Shiba et al. 1997; Rao et al. 1998; Kornberg et al. 1999; Kuroda et al. 1999; Kulaev and Kulakovskaya 2000; Kuroda and Ohtake 2000; Kuroda et al. 2001, 2006; Gray 2019).

Multiple evidence has been published that polyP is involved in and important for tolerance against various forms of stresses (UV light, solvents, reactive oxygen species, heavy metals, antibiotics, survival in stationary phase and others) (Kim et al. 2002; Fraley et al. 2007; Seufferheld et al. 2008; Varela et al. 2010; Nikel et al. 2013; Tocheva et al. 2013; Alcántara et al. 2014; Gray and Jakob 2015; Ortíz-Severín et al. 2015). PolyP granules most likely are connected with the bacterial nucleoid (Henry and Crosson 2013), and polyP seems to be important for proper control of the cell cycle (Bru et al. 2016; Racki et al. 2017; Bru et al. 2017). In humans polyP is part of the blood coagulation system (Hagemann factor).

Recently, a novel type of posttranslational modification of enzymes by polyP ("polyphosphorylation") of lysine residues was described in eukaryotic cells (Azevedo et al. 2015; Azevedo and Saiardi 2016). The authors provided evidence that polyphosphorylation can modulate the biological activity of enzymes similar as covalent modification by phosphorylation. Another important discovery of an unexpected complex function that a simple inorganic molecule such as polyP can have is the finding that polyP can function as a chaperone. PolyP is able bind to proteins and to refold unfolded proteins in an ATP-independent manner and to prevent protein aggregation (Gray et al. 2014; Yoo et al. 2018). Recent studies in human cell lines suggested that polyP interacts with preforms of α-synuclein fibrils and prevents them to be taken up by neuron cells. α-Synuclein is a key protein in neurodegenerative diseases such as Alzheimer or Parkinson (Lempart and Jakob 2019; Lempart et al. 2019) when it forms amyloid-like aggregates. These findings suggest that polyP levels play a prominent role in the velocity of the development of amyloid aggregates during neurodegenerative diseases. Another function of polyP in humans lies in the pre-activation of the Hageman factor (factor XII). Factor XII is part of the blood coagulation system and it binds to negatively charged surfaces such as polyP that is present in human platelets (Ruiz et al. 2004) and can be released from them.

The function of membrane-located low-molecular-weight polyP complexed with calcium and with low-molecular-weight polyhydroxybutyrate (PHB) is not subject of this review, and the interested reader is referred to recent reviews (Reusch 2012; Jendrossek and Pfeiffer 2014).

4 Key Enzymes of polyP Metabolism

Most of the primary research on the biosynthesis and the degradation of polyP has been made by Nobel laureate A. Kornberg in the last two decades of his life: He and his co-workers were the first to identify the polyP synthesizing enzymes, the so-called polyP kinases (PPK). The first isolated and biochemically characterized PPK was that of *E. coli* (Ahn and Kornberg 1990; Kumble et al. 1996; Tzeng and Kornberg 2000). PPK of *E. coli* (PPK_{Ec}) is the prototype of so-called type 1 PPKs (PPK1). The polypeptides of PPK1 enzymes usually are ≈80 kDa proteins and consist of four domains (N-terminal domain (N), head domain (H), and two C-terminal domains (C1 and C2)). The active site (His435 and His 592 in PPK_{Ec}) is located in the C1 domain and is part of a tunnel that accommodates the growing polyP chain during the processive elongation (Zhu et al. 2005).

Kornberg's group also identified, purified and biochemically characterized a second type of PPK (PPK2) identified in *P. aeruginosa* ($PPK2_{Pa}$) (Ishige et al. 1998; Zhang et al. 2002). PPK2s have only about half of the "size" of PPK1s (≈40 kDa) and consist of only one domain. In contrast to PPK1s, $PPK2_{Pa}$ prefers GTP (instead of ATP as in case of PPK_{Ec}) and "works" in the direction of GTP synthesis. Many other bacteria contain PPKs highly related to either $PPK1_{Ec}$ or to $PPK2_{Pa}$. Some bacteria contain both a PPK1 and a PPK2 encoding gene or can even have multiple *ppk* genes in the genome. The *R. eutropha* genome, for example, codes for seven PPKs, two of the PPK1 type (PPK1a and PPK1b) and five of the PPK2 type (PPK2a—PPK2e, for details see below). One of the *R. eutropha's* PPKs, PPK2c, has been biochemically characterized recently (Hildenbrand et al. 2019, 2020) and turned out to be highly unspecific and to work with all natural nucleotides and desoxynucleotides. Recently, PPK2s were divided into three subclasses based on whether they catalyze phosphorylation of nucleoside diphosphate, nucleoside monophosphate or both (Motomura et al. 2014). Some PPK2s are able to catalyse the formation of oligo-nucleoside-phosphates with four or even with five phosphate residues (ATP_4, ATP_5) (Mordhorst et al. 2019). Very recently, the formation of oligo-phosphorylated nucleosides up to nona-phosphates by a PPK2 protein purified from *Agrobacterium tumefaciens* was shown (Frank et al. 2020). At present, it is not known whether these oligo-nucleoside-phosphates have a specific function beyond that of ATP.

The degradation/utilization of polyP is catalyzed by polyphosphatases (PPXs). The first PPX was purified and biochemically characterized from *E. coli* by the

Kornberg lab (Akiyama et al. 1993; Keasling et al. 1993). Remarkably, the *ppx* gene and the *ppk* gene in *E. coli* are organized in an operon. Later, it turned out that *ppx* and *ppk* genes in other bacterial species are often located adjacent on the respective host chromosome but in opposite direction of transcription.

5 PolyP Granules in Prokaryotes

PolyP granules or volutin granules are present in all bacterial species that have been looked at (see Introduction for the selection of references). The diameters of polyP granules vary in most species between ≈50 and ≈200 nm, depending on medium composition and growth conditions. The presence of magnesium salts in the growth medium favoured the formation of polyP granules in several species (e.g. in *Corynebacterium glutamicum* (Pallerla et al. 2005; Klauth et al. 2006), in *Acetobacter xylinum* (Ryazanova et al. 2009), in *Agrobacterium tumefaciens* (unpublished data) or in sewage sludge organisms performing the EBPR (enhanced biological phosphate removal process) (Schönborn et al. 2001)). PolyP granules might have even larger diameters as size determination in thin sections and transmission electron microscopy can underestimate the diameter if the polyP granule was not cut in the middle. Some reports showed that polyP granules can have diameters up to half a micrometre (Ogawa and Amano 1987) or claimed sizes of up to 0.8μm in preparations of polyP granules isolated from *Desulfovibrio gigas* (Jones and Chambers 1975). In the latter case, a coalescence of polyP granules during isolation process and centrifugation steps might have led to these large dimensions.

While many working groups with interest in polyP formation in microorganisms have reported on procedures how to extract polyP from polyP-containing cells, reports that describe the isolation of intact polyP granules ("native" polyP granules) are rare. Friedberg and Avigad were able to enrich polyP granules from *Micrococcus lysodeikticus* simply by lysozyme and DNAse treatment followed by centrifugation (Friedberg and Avigad 1968). In most other reports on polyP granule isolation cell preparations, the authors used centrifugation of lysed cell preparations to spin down polyP which forms a greyish pellet at the bottom (Jones and Chambers 1975; Pallerla et al. 2005). One could imagine that gradient centrifugation techniques (e.g. with sucrose, glycerol, or Percoll) should be useful to enrich or to purify polyP granule in an as native state as possible. However, attempts in our lab to enrich polyP granules (from *R. eutropha*) by gradient centrifugation were not successful even if mutant cells were used in which the formation of other granules, PHB granules, that might contaminate the polyP granule fraction, does not occur. Therefore, we enriched polyP granules from lysed cells also mainly by centrifugation and washing steps. In addition, we combined this with a filtration step through a 0.45μm filter to remove large cell debris and thereby we could at least enrich polyP granules (Tumlirsch et al. 2015).

6 The polyP Granule Proteome

Only little information is available on proteins that are specifically bound to polyP granules in microorganism: a polyP-dependent glucokinase was reported to be associated with polyP granules in *Coryebacterium glutamicum* (Pallerla et al. 2005). In *Caulobacter crescentus* PPK was identified as a polyP-bound protein by fusion with Venus and fluorescence microscopy (Henry and Crosson 2013). The first systematic analysis of a polyP granule proteome was performed with *R. eutropha* (Tumlirsch et al. 2015; Tumlirsch 2017). Twenty-seven proteins were identified to be specifically present only in a polyP granule fraction by comparative proteome analysis of *R. eutropha* cell fractions (polyP fraction, soluble fraction, membrane fraction and PHB granule fraction; for details see Tumlirsch et al. (2015)). Among those, three proteins represented polyP kinases (PPK1a, PPK2c and PPK2d) and were confirmed as true polyP granule-associated proteins by fluorescence microscopical detection of protein fusions of these PPKs with fluorescent proteins (eYFP). Most of the remaining 24 proteins of the polyP granule fraction most likely were false-positive, as detection of constructed fusions with eYFP showed distinct locations and colocalization with polyP granules was not observed. However, three (additional) proteins were identified to be generally attached to polyP granules, thus raising the number of polyP granule-associated proteins in *R. eutropha* to six (Table 1). These are: PPK2e (in addition to PPK1a, PPK2c and PPK2d) and two proteins with CHAD-motifs (PptA and PptB, see below) (Tumlirsch 2017). All six polyP granule-associated proteins had strong basic isoelectric points of pH_{IEP} 7.8–11.5 being in agreement with the attachment to the negatively charged polyP granule. The remaining three PPKs of *R. eutropha* (PPK1b, PPK2a and PPK2b, each in form of a fusion with eYFP) were either soluble in the cytoplasm (PPK2a) or formed fluorescent foci that did, however, not colocalize with polyP granules (PPK1b, PPK2b). Interestingly, the IEP values of these non-polyP-attached PPK2 were in the acidic range (pH_{IEP} 5.8–6.7). The physiological roles of these PPKs in *R. eutropha* await clarification. *R. eutropha* has one exo-polyphosphatase (*ppx*) gene; its gene product turned out to be soluble and was not associated with polyP granules when fused to eYFP (Tumlirsch et al. 2015).

7 Phosins: polyP Granule-Associated Proteins with CHAD-Domains

The amino acid sequences of two of the specifically polyP-granule-associated proteins of *R. eutropha* H16 had a so-called CHAD-domain (conserved histidine alpha helical domain). These are H16_A0104 and H16_B1017. CHAD domains are characterized by a conserved secondary structure consisting of only alpha helices including conserved (eponymous) histidine residues (Iyer and Aravind 2002). While the H16_A0104 gene product (35.5 kDa) harbours only one defined (CHAD)

Table 1 PPKs, PPX and PolyP-associated/interacting proteins of *Ralstonia eutropha*

Protein Gene no	Molecular mass [kDa]	Domain structure/motif	IEP	Localization	Remarks
PPK1a A2436	78.2	N, H, C1, C2	7.8	polyP granule	
PPK1b B1019	79.6	N, H, C1, C2	5.9	Near cell pole	Not associated with polyP
PPK2a A0226	31.1	PPK2	6.7	Cytoplasm	Not associated with polyP
PPK2b A0997	36.2	PPK2	5.8	Near cell pole	Not associated with polyP
PPK2c A1212	42.1	PPK2	9.4	polyP granule	
PPK2d A1271	33.9	PPK2	9.6	polyP granule	
PPK2e A1979	30.7	PPK2	9.1	polyP granule	
PPX A2435	56.5	Ppx-GppH	8.8	Cytoplasm	Not associated with polyP
PptA A0104	35.5	CHAD	11.5	polyP granule	
PptB B1017	56.7	CYTH-CHAD	8.9	polyP granule	
PPI27 A0526	26.9	UPF0054	4.9	polyP granule in stat. phase	Near cell pole in exp. phase
PPI18 A3108	18.1	Ferritin	5.8	polyP granule in stat. phase	
GlnA1 A2335	52.5	Gln-synth-N Gln-synth-C	5.4	Interaction with PptA	

Abbreviations: unknown protein family (UPF), CyaB thiamine triphosphatase (CYTH), conserved histidine alpha helical domain (CHAD)

domain, the H16_B1017 gene product (56.7 kDa) consists of two domains and has a CYTH domain (CyaB, thiamine triphosphatase) domain in addition to the CHAD domain. Based on bioinformatic data, proteins with CYTH and/or CHAD domains had been previously suspected to be involved in polyP and nucleoside-phosphate metabolism (Iyer and Aravind 2002). The finding that H16_A0104 and H16_B1017 represent indeed polyP granule-associated proteins was the first experimental evidence supporting this assumption (Tumlirsch and Jendrossek 2017). Overexpression of eYFP-fusions with either H16_A0104 or with H16_B1017 resulted in a shift of the localization of formed polyP granules with attached fusion protein from the more or less mid-cell position (nucleoid region) to a localization close to the cell poles. This was also observed in stationary cells of *R. eutropha*, in which the H16_A0104 gene was expressed from a plasmid without fusion to *eyfp* indicating that it was not the eYFP moiety that was responsible for the localization of polyP granules near the cell poles. This finding suggested that the CHAD proteins where not only specifically bound to polyP granules in vivo but (somehow) also influenced the subcellular positioning of polyP granules. The two CHAD-proteins of *R. eutropha* (H16_A0104 or with H16_B1017) were therefore designated as polyP targeting proteins PptA and PptB, respectively (Tumlirsch and Jendrossek 2017). That polyP granules are not randomly localized in bacterial cells has been shown previously: in all cases a localization of polyP granules in the nucleoid region has been found as in *Caulobacter crescentus* (Henry and Crosson 2013), *Pseudomonas aeruginosa* (Racki et al. 2017) and *R. eutropha* (Beeby et al. 2012; Tumlirsch et al. 2015; Tumlirsch and Jendrossek 2017). The strong binding of eYFP-PptA to polyP granules was also observed when *eyfp-pptA* was expressed in α-proteobacteria (*Magnetospirillum gryphiswaldense*) and in γ-proteobacteria (*Pseudomonas putida*). A colocalization with polyP granules was also determined when gene fusions of CHAD-domain encoding genes of *M. gryphiswaldense* or *P. putida* were expressed in *R. eutropha* (Tumlirsch and Jendrossek 2017). These findings suggest that most likely all CHAD-containing proteins are in vivo bound to polyP granules. This assumption is supported by two recently published independent findings: Hothorn and coworkers determined the structures of CHAD proteins from a plant (*Rhizinus communis*), a bacterial (*Chlorobium tepidum)* and an archaeal species (*Sulfolobus solfataricus*) (Lorenzo-Orts et al. 2019). All three CHAD proteins revealed a central pore revetted by positively charged amino acids and were able to bind to polyP in vitro. A very similar finding was published by Werten et al. (Werten et al. 2019) who investigated the structure and the polyP binding ability of PptA of *Streptomyces chartreusis*. Taken together, PptA proteins from species of all kingdoms of life showed a similar structure and revealed a strong binding ability to polyP. Based on the analogy of small polyhydroxybutyrate (PHB)-binding proteins without (known) enzymatic function, the so-called phasins (PhaPs) (Steinbüchel et al. 1995; Mezzina and Pettinari 2016), and on similar proteins present on oil droplets in plants, oleosins (Murphy 1993), the designation as phosins for polyphosphate-associated proteins has been suggested for CHAD-proteins (Tumlirsch and Jendrossek 2017). Since polyP is ubiquitous, I expect that phosins will turn out to be present in many, if not all, polyP granule-forming organisms.

8 Other polyP Granule-Associated Proteins

Three other proteins were identified to colocalize with polyP granules in *R. eutropha* at least in some phases during growth. These are the H16_A0526, H16_A2335 and H16-A3108 gene products. The H16_A0526 protein was enriched in the polyP granule fraction but was also detected in the soluble fraction. The protein (27 kDa) has an acidic IEP of 4.8 and is annotated as a putative rRNA maturation factor. Constitutive expression of a fusion with the *eyfp* gene revealed that the A0526 protein colocalized with polyP granules only in the lag and in the stationary growth phase but was distributed in the cytoplasm during exponential growth (Tumlirsch

2017). The protein will be referred to as PPI-27 (polyP-interacting 27 kDa protein). The function of PPI27 in polyP metabolism is unknown.

The H16_A2335 protein (GlnA1) was identified in pull-down experiments with the PptA-CHAD protein as bait that had been overexpressed in *R. eutropha* wild type. When fused to eYFP, the H16_A2335 protein colocalized with polyP granules at the cell poles and/or with DsRedEC2-PptA. The colocalization of H16_A2335 with PptA was not dependent on the presence of polyP and also occurred in a polyP-deficient strain in which all seven *ppk* genes had been deleted (Δ*ppk-7* mutant). Therefore H16_A2335 presumably is not directly bound to polyP granules but indirectly to polyP via binding to PptA. This is in agreement with the identification of the H16_A2335 protein in pull-down experiments with eYFP-PptA that had been performed in the polyP-deficient Δ*ppk-7* mutant (Tumlirsch 2017). The H16_A2335 protein is one of the three glutamine synthetases present in *R. eutropha* (GlnA1). The *glnA1* gene is able to complement a gutamine synthetase mutant of *E. coli* by restoring growth on a mineral salts medium and therefore GlnA1 is a true gutamine synthetase (unpublished data). The physiological function of the interaction between PptA and GlnA1 and its involvement in polyP metabolism is not known. All attempts to find evidence for an (additional) modulation of gutamine synthetase activity of GlnA1 by the absence or presence of PptA, that would be indicative for a link between N- and P metabolism, were not successful so far (unpublished data).

The H16_A3108 protein was identified in pull-down experiments with the PptA-CHAD protein as bait. The protein (18 kDa) has an IEP of 5.9 and is annotated as a 'metalloregulation DNA-binding stress protein/starvation-inducible DNA-binding protein'. Constitutive expression of a fusion with the *eyfp* gene revealed that the A0526 protein colocalized with polyP granules only in the lag and in the late exponential and stationary growth phases. Interestingly, the eYFP-A3108 protein formed fluorescent foci at the cell pole during exponential growth and did not colocalize with polyP (Tumlirsch 2017). The protein will be referred to as PPI18 (polyP-interacting 18 kDa protein). The function of PPI18 in polyP metabolism is unknown.

In summary, six proteins are in vivo attached to polyP granules in *R. eutropha.* Two additional proteins (PPI25 and PPI18) colocalize with polyP granules in the stationary growth phase, and one protein (GlnA1) seems to be indirectly attached to polyP granules via interaction with the phosin PptA. Figure 1 and Table 1 show a cartoon of the structure of polyP granules in *R. eutropha* and provide an overview of the currently known polyP-attached proteins. We can't exclude that there are even more proteins attached to polyP granules that have not yet been identified. The presence of so many proteins with apparently different (or yet unknown) functions shows that polyP granules represent supramolecular complexes. Despite the absence of evidence for the presence of a polyP-surrounding membrane (similar to carboxysomes and carbonosomes) polyP granules apparently have organelle-like functions for which the designation as "polyphosphatosomes" has been proposed (Tumlirsch and Jendrossek 2017). I predict that investigation of isolated polyP granules in other species will turn out to have a similar complexity of polyP-granule attached proteins.

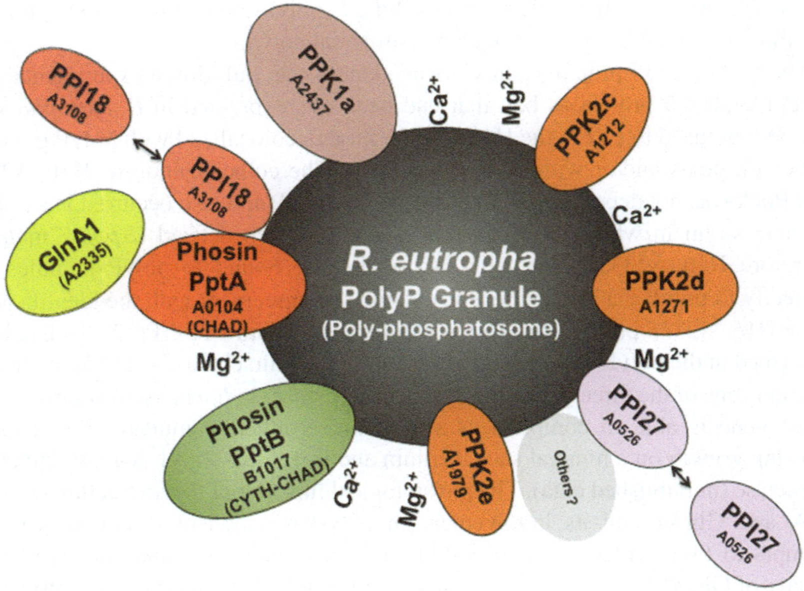

Fig. 1 Model of a polyP granule in *R. eutropha* H16. Proteins attached to polyP or to polyP-attached proteins are indicated as spheres with proposed designations. The annotation numbers of the respective gene are also indicated. Arrows indicate that some protein can detach from polyP at certain growth conditions. The putative presence of cations such as magnesium or calcium is indicated. The sizes of the protein are enlarged relative to the polyP granules for better visibility

9 Acidocalcisomes in Prokaryotes

Acidocalcisomes are acidic vacuole-like compartments rich in phosphorous (in form of polyP and pyrophosphate) and metal ions (in particular calcium). Acidocalcisomes are well-known organelles in eukaryotes from unicellular micro-organisms to humans. They are surrounded by a membrane bilayer into which several proteins (mainly transport proteins) are embedded. For reviews on acidocalcisomes and proteome analysis of eukaryotic acidocalcisomes, see Docampo et al. (2005); Docampo and Moreno (2011); Docampo et al. (2013); Huang et al. (2014); Huang and Docampo (2015); Lander et al. (2016). Characteristic for acidocalcisomes is the presence of a membrane-bound vacuolar proton translocating pyrophosphatase (V-H^+-PPase), a vacuolar proton translocating ATPase (V-H^+-ATPase) or both. Recently, proteome analysis of acidocalcisomes purified from the pathogen *Trypanosomas brucei* revealed the presence of several additional proteins that were specifically associated with acidocalcisome membranes (in particular several transporters for cations) (Huang et al. 2014; Huang and Docampo 2015). Analysis of the membrane lipids of isolated acidocalcisomes revealed the presence of mainly phosphatidyl-choline (PC) phosphatidyl-inositol

(PI), phosphatidyl-ethanolamine (PE), glycoinositol-phospholipid (GIPL) and traces of sterols (Salto et al. 2008).

In prokaryotes, membrane-enclosed acidocalcisomes with embedded V-H^+-PPase have been identified in *Rhodospirillum rubrum* and *Agrobacterium tumefaciens* (Seufferheld et al. 2003; Seufferheld et al. 2004). A bioinformatic approach (search for V-H^+-PPase sequences as a potential marker for membrane-enclosed acidocalcisomes) suggested, however, that acidocalcisomes could be wide-spread in *Eubacteria* and apparently exist also in some *Archaea* species (Seufferheld et al. 2011). Biochemical evidence for the presence of acidocalcisomes in prokaryotes other than for *R. rubrum* and *A. tumefaciens*, however, has not been published so far; a few years ago, the presence of a membrane-enclosed polyP-containing compartment was described for a marine *Beggiatoa* strain (Brock et al. 2012). However, the evidence that the polyP containing structures were enclosed by a lipid membrane layer was not fully convincing.

Despite a huge amount of literature on the ability to accumulate polyP in many prokaryotes, no solid information is available on the presence or absence of membranes around polyP granules. Images obtained by cryo-tomography did not indicate the presence of a continuous membrane layer around polyP granules in *Halothiobacillus neopolitanus* or in *Ralstonia eutropha* (Iancu et al. 2010; Beeby et al. 2012). Experiments with *R. eutropha*, in which a fusion protein of the phospholipid-specific lactC2 domain (of bovine milk adherin) and a fluorescent protein were expressed, could clearly detect the cytoplasmic membrane but gave no indication for the presence of intracellular membrane-surrounded granule-like structures (Bresan et al. 2016). In conclusion, is seems as if two types of polyP inclusions may exist in prokaryotes: (1) many/most bacteria synthesize polyP in complex with cations that are not surrounded by a membrane layer. Only the specific attachment of several proteins (e. g. PPKs, phosins and others) to polyP granules is well established by in vivo data for polyP granules of *R. eutropha* (see above) and for PPK of *Caulobacter crescentus* (Henry and Crosson 2013) and points to the presence of a proteinaceous surface layer; (2) at least in two α-proteobacterial species (*A. tumefaciens* and *R. rubrum*) polyP granules are membrane-enclosed acidic compartments and constitute true organelles. Attempts in our laboratory to isolate acidocalcisomes in intact form from *A. tumefaciens* cells or to find evidence for the existence of acidic compartments with Lysosensor dyes that are fluorescent only in a protonated form at low pH, failed. Furthermore, we could show that PPK2 of *A. tumefaciens* was bound to DAPI-stainable polyP and that PPK2 and the marker enzyme of acidocalcisomes, the vacuolar proton-translocating pyrophosphatase (HppA) did not colocalize with each other when fused to fluorescent proteins (Frank and Jendrossek, 2020). These data did not exclude the existence of acidocalcisomes but suggested that polyP in *A. tumefaciens* is present in the form of membrane-less 'ordinary' granules as in other prokaryotic species. The nature and composition of the structures designated as acidocalcisomes remain to be identified.

Acknowledgements I thank the Deutsche Forschungsgemeinschaft for support. I deeply appreciate the work of my current and previous co-workers and students in polyP metabolism: Chiara

Brück, Celina Frank, Jennie Hildenbrand, Andrea Kaufmann, Lea Kneißle, Anna Kerber, Hannes Lothholz, Simone Reinhardt, Hanna Rosigkeit, Helen Sibold, Tony Tumlirsch.

References

Ahn KH, Kornberg A (1990) Polyphosphate kinase from Escherichia coli – purification and demonstration of a phosphoenzyme intermediate. J Biol Chem 265(20):11734–11739

Akiyama M, Crooke E, Kornberg A (1993) An exopolyphosphatase of *Escherichia coli*. The enzyme and its ppx gene in a polyphosphate operon. J Biol Chem 268(1):633–639

Albi T, Serrano A (2016) Inorganic polyphosphate in the microbial world. Emerging roles for a multifaceted biopolymer. World J Microbiol Biotechnol 32(2):27. https://doi.org/10.1007/s11274-015-1983-2

Alcántara C, Blasco A, Zúñiga M, Monedero V (2014) Accumulation of polyphosphate in *Lactobacillus* spp. and its involvement in stress resistance. Appl Environ Microbiol 80:1650–1659. https://doi.org/10.1128/AEM.03997-13

Alvarez S, Jerez CA (2004) Copper ions stimulate polyphosphate degradation and phosphate efflux in *Acidithiobacillus ferrooxidans*. Appl Environ Microbiol 70:5177–5182. https://doi.org/10.1128/AEM.70.9.5177-5182.2004

Angelova PR, Agrawalla BK, Elustondo PA, Gordon J, Shiba T, Abramov AY, Chang Y-T, Pavlov EV (2014) In situ investigation of mammalian inorganic polyphosphate localization using novel selective fluorescent probes JC-D7 and JC-D8. ACS Chem Biol 9:2101–2110. https://doi.org/10.1021/cb5000696

Aschar-Sobbi R, Abramov AY, Diao C, Kargacin ME, Kargacin GJ, French RJ, Pavlov E (2008) High sensitivity, quantitative measurements of polyphosphate using a new DAPI-based approach. J Fluoresc 18:859–866. https://doi.org/10.1007/s10895-008-0315-4

Azevedo C, Saiardi A (2016) The new world of inorganic polyphosphates. Biochem Soc Trans 44:13–17. https://doi.org/10.1042/BST20150210

Azevedo C, Livermore T, Saiardi A (2015) Protein polyphosphorylation of lysine residues by inorganic polyphosphate. Mol Cell 58:71–82. https://doi.org/10.1016/j.molcel.2015.02.010

Beeby M, Cho M, Stubbe J, Jensen GJ (2012) Growth and localization of polyhydroxybutyrate granules in *Ralstonia eutropha*. J Bacteriol 194:1092–1099. https://doi.org/10.1128/JB.06125-11

Blackall LL, Crocetti G, Saunders AM, Bond PL (2002) A review and update of the microbiology of enhanced biological phosphorus removal in wastewater treatment plants. Antonie Van Leeuwenhoek 81:681–691

Bresan S, Sznajder A, Hauf W, Forchhammer K, Pfeiffer D, Jendrossek D (2016) Polyhydroxyalkanoate (PHA) granules have no phospholipids. Sci Rep 6:26612. https://doi.org/10.1038/srep26612

Bringmann G (1951) Light and electro microscopic investigations on granules cytology of *Corynebacterium diphtheriae*. Zentralbl Bakteriol Orig 156:493–502

Brock J, Rhiel E, Beutler M, Salman V, Schulz-Vogt HN (2012) Unusual polyphosphate inclusions observed in a marine Beggiatoa strain. Antonie Van Leeuwenhoek 101:347–357. https://doi.org/10.1007/s10482-011-9640-8

Bru S, Martínez-Laínez JM, Hernández-Ortega S, Quandt E, Torres-Torronteras J, Martí R, Canadell D, Ariño J, Sharma S, Jiménez J, Clotet J (2016) Polyphosphate is involved in cell cycle progression and genomic stability in *Saccharomyces cerevisiae*. Mol Microbiol 101:367–380. https://doi.org/10.1111/mmi.13396

Bru S, Samper-Martín B, Quandt E, Hernández-Ortega S, Martínez-Laínez JM, Garí E, Rafel M, Torres-Torronteras J, Martí R, Ribeiro MPC, Jiménez J, Clotet J (2017) Polyphosphate is a key factor for cell survival after DNA damage in eukaryotic cells. DNA Repair (Amst) 57:171–178. https://doi.org/10.1016/j.dnarep.2017.08.001

Cammidge PJ (1901) The value of Neisser's stain in the diagnosis of diphtheria. BMJ 2:1016–1016. https://doi.org/10.1136/bmj.2.2127.1016
Christ JJ, Willbold S, Blank LM (2020, March 17) Methods for the analysis of polyphosphate in the life sciences. Anal Chem 92(6):4167–4176. https://doi.org/10.1021/acs.analchem.9b05144
Clark JE, Wood HG (1987) Preparation of standards and determination of sizes of long-chain polyphosphates by gel electrophoresis. Anal Biochem 161:280–290
Clark JE, Beegen H, Wood HG (1986) Isolation of intact chains of polyphosphate from "*Propionibacterium shermanii*" grown on glucose or lactate. J Bacteriol 168:1212–1219
Docampo R, Moreno SNJ (2011) Acidocalcisomes. Cell Calcium 50:113–119. https://doi.org/10.1016/j.ceca.2011.05.012
Docampo R, De Souza W, Miranda K, Rohloff P, Moreno SNJ (2005) Acidocalcisomes - conserved from bacteria to man. Nat Rev Microbiol 3(3):251–261. https://doi.org/10.1038/nrmicro1097
Docampo R, Jimenez V, Lander N, Li Z-H, Niyogi S (2013) New insights into roles of acidocalcisomes and contractile vacuole complex in osmoregulation in protists. Int Rev Cell Mol Biol 305:69–113. https://doi.org/10.1016/B978-0-12-407695-2.00002-0
Fraley CD, Rashid MH, Lee SSK, Gottschalk R, Harrison J, Wood PJ, Brown MRW, Kornberg A (2007) A polyphosphate kinase 1 (ppk1) mutant of *Pseudomonas aeruginosa* exhibits multiple ultrastructural and functional defects. Proc Natl Acad Sci U S A 104:3526–3531. https://doi.org/10.1073/pnas.0609733104
Frank C, Jendrossek, D (2020) Acidocalcisomes and polyphosphate granules are different subcellular structures in *Agrobacterium tumefaciens*. Appl Environ Microbiol. 86(8). Epub: https://doi.org/10.1128/AEM.02759-19
Frank C, Teleki A, Jendrossek D (2020) Characterization of *Agrobacterium tumefaciens* PPKs reveals the formation of oligophosphorylated products up to nucleoside nona-phosphates. Appl Microbiol Biotechnol 265:11734–10. https://doi.org/10.1007/s00253-020-10891-7
Friedberg I, Avigad G (1968) Structures containing polyphosphate in *Micrococcus lysodeikticus*. J Bacteriol 96:544–553
Gomes FM, Ramos IB, Wendt C, Girard-Dias W, De Souza W, Machado EA, Miranda K (2013) New insights into the in situ microscopic visualization and quantification of inorganic polyphosphate stores by 4′,6-diamidino-2-phenylindole (DAPI)-staining. Eur J Histochem 57: e34. https://doi.org/10.4081/ejh.2013.e34
Gray MJ (2019) Inorganic polyphosphate accumulation in *Escherichia coli* is regulated by DksA but not by (p)ppGpp. J Bacteriol 201:e00664-18. https://doi.org/10.1128/JB.00664-18
Gray MJ, Jakob U (2015) Oxidative stress protection by polyphosphate – new roles for an old player. Curr Opin Microbiol 24:1–6. https://doi.org/10.1016/j.mib.2014.12.004
Gray MJ, Wholey W-Y, Wagner NO, Cremers CM, Mueller-Schickert A, Hock NT, Krieger AG, Smith EM, Bender RA, Bardwell JCA, Jakob U (2014) Polyphosphate is a primordial chaperone. Mol Cell 53:689–699. https://doi.org/10.1016/j.molcel.2014.01.012
Grimme A (1902) Die wichtigsten Methoden der Bacterienfärbung in ihrer Wirkung auf die Membran, den Protoplasten und die Einschlüsse der Bacterienzelle. Thesis, University Marburg.
Günther S, Trutnau M, Kleinsteuber S, Hause G, Bley T, Röske I, Harms H, Müller S (2009) Dynamics of polyphosphate-accumulating bacteria in wastewater treatment plant microbial communities detected via DAPI (4″,6-″diamidino-2-phenylindole) and tetracycline labeling. Appl Environ Microbiol 75:2111–2121. https://doi.org/10.1128/AEM.01540-08
Harada K (1957) Selective staining of volutin in *Corynebacterium diphtheriae*. Stain Technol 32:249–253
Harold FM (1966) Inorganic polyphosphates in biology: structure, metabolism, and function. Bacteriol Rev 30:772–794
Henry JT, Crosson S (2013) Chromosome replication and segregation govern the biogenesis and inheritance of inorganic polyphosphate granules. Mol Biol Cell 24:3177–3186. https://doi.org/10.1091/mbc.E13-04-0182

Hildenbrand JC, Reinhardt S, Jendrossek D (2019) Formation of an organic-inorganic biopolymer: polyhydroxybutyrate-polyphosphate. Biomacromolecules 20(9):3253–3260. https://doi.org/10.1021/acs.biomac.9b00208

Hildenbrand JC, Teleki A, Jendrossek D (2020) A universal polyphosphate kinase: PPK2c *of Ralstonia eutropha* accepts purine and pyrimidine nucleotides including uridine diphosphate. Appl Microbiol Biotechnol 265(5):11734–11739. https://doi.org/10.1007/s00253-020-10706-9

Huang G, Docampo R (2015) Proteomic analysis of acidocalcisomes of *Trypanosoma brucei* uncovers their role in phosphate metabolism, cation homeostasis, and calcium signaling. Commun Integr Biol 8:e1017174. https://doi.org/10.1080/19420889.2015.1017174

Huang G, Ulrich PN, Storey M, Johnson D, Tischer J, Tovar JA, Moreno SNJ, Orlando R, Docampo R (2014) Proteomic analysis of the acidocalcisome, an organelle conserved from bacteria to human cells. PLoS Pathog 10:e1004555. https://doi.org/10.1371/journal.ppat.1004555

Iancu CV, Morris DM, Dou Z, Heinhorst S, Cannon GC, Jensen GJ (2010) Organization, structure, and assembly of alpha-carboxysomes determined by electron cryotomography of intact cells. J Mol Biol 396:105–117. https://doi.org/10.1016/j.jmb.2009.11.019

Ishige K, Kameda A, Noguchi T, Shiba T (1998) The polyphosphate kinase gene of *Pseudomonas aeruginosa*. DNA Res: Int J Rapid Publ Rep Genes Genomes 5(3):157–162

Iyer LM, Aravind L (2002) The catalytic domains of thiamine triphosphatase and CyaB-like adenylyl cyclase define a novel superfamily of domains that bind organic phosphates. BMC Genomics 3:33. https://doi.org/10.1186/1471-2164-3-33

Jasso-Chávez R, Lira-Silva E, González-Sánchez K, Larios-Serrato V, Mendoza-Monzoy DL, Pérez-Villatoro F, Morett E, Vega-Segura A, Torres-Márquez ME, Zepeda-Rodríguez A, Moreno-Sánchez R (2019) Marine Archaeon *Methanosarcina acetivorans* enhances polyphosphate metabolism under persistent cadmium stress. Front Microbiol 10:2432. https://doi.org/10.3389/fmicb.2019.02432

Jendrossek D, Pfeiffer D (2014) New insights in the formation of polyhydroxyalkanoate granules (carbonosomes) and novel functions of poly(3-hydroxybutyrate). Environ Microbiol 16:2357–2373. https://doi.org/10.1111/1462-2920.12356

Jensen TE (1968) Electron microscopy of polyphosphate bodies in a blue-green alga, *Nostoc pruniforme*. Arch Mikrobiol 62:144–152

Jensen TE, Rachlin JW, Jani V, Warkentine B (1982) An X-ray energy dispersive study of cellular compartimentalization of lead and zinc in *Chlorella saccharophila* (*Chlorophyta*), *Navicula incerta* and *Nitzschia closterium* (*Bacillariophyta*). Environ Exp Bot 22:319–328

Jones HE, Chambers LA (1975) Localized intracellular polyphosphate formation by *Desulfovibrio gigas*. J Gen Microbiol 89:67–72. https://doi.org/10.1099/00221287-89-1-67

Keasling JD (1997) Regulation of intracellular toxic metals and other cations by hydrolysis of polyphosphate. Ann N Y Acad Sci 829:242–249. https://doi.org/10.1111/j.1749-6632.1997.tb48579.x

Keasling JD, Bertsch L, Kornberg A (1993) Guanosine pentaphosphate phosphohydrolase of *Escherichia coli* is a long-chain exopolyphosphatase. Proc Natl Acad Sci USA 90(15):7029–7033

Kim K-S, Rao NN, Fraley CD, Kornberg A (2002) Inorganic polyphosphate is essential for long-term survival and virulence factors in *Shigella* and *Salmonella* spp. Proc Natl Acad Sci U S A 99:7675–7680. https://doi.org/10.1073/pnas.112210499

Klauth P, Pallerla SR, Vidaurre D, Ralfs C, Wendisch VF, Schoberth SM (2006) Determination of soluble and granular inorganic polyphosphate in *Corynebacterium glutamicum*. Appl Microbiol Biotechnol 72:1099–1106. https://doi.org/10.1007/s00253-006-0562-8

Knaysi G (1959) Chemical composition of the granules of *Mycobacterium thamnopheos*, with special reference to their biological identity and the chemical nature of volutin. J Bacteriol 77:532–544

Kornberg A, Rao NN, Ault-Riche D (1999) Inorganic polyphosphate: a molecule of many functions. Annu Rev Biochem 68:89–125. https://doi.org/10.1146/annurev.biochem.68.1.89

Kortstee GJ, Appeldoorn KJ, Bonting CF, van Niel EW, van Veen HW (2000) Recent developments in the biochemistry and ecology of enhanced biological phosphorus removal. Biochemistry Mosc 65:332–340

Kulaev I, Kulakovskaya T (2000) Polyphosphate and phosphate pump. Annu Rev Microbiol 54:709–734. https://doi.org/10.1146/annurev.micro.54.1.709

Kulaev IS, Vagabov VM (1983) Polyphosphate metabolism in microorganisms. Adv Microb Physiol 24:83–171. https://doi.org/10.1016/S0065-2911(08)60385-9

Kulakova AN, Hobbs D, Smithen M, Pavlov E, Gilbert JA, Quinn JP, McGrath JW (2011) Direct quantification of inorganic polyphosphate in microbial cells using 4′-6-Diamidino-2-Phenylindole (DAPI). Environ Sci Technol 45:7799–7803. https://doi.org/10.1021/es201123r

Kumble KD, Ahn K, Kornberg A (1996) Phosphohistidyl active sites in polyphosphate kinase of *Escherichia coli*. Proc Natl Acad Sci USA 93(25):14391–14395. https://doi.org/10.1073/pnas.93.25.14391

Kuroda A, Ohtake H (2000) Molecular analysis of polyphosphate accumulation in bacteria. Biochemistry Mosc 65:304–308

Kuroda A, Tanaka S, Ikeda T, Kato J, Takiguchi N, Ohtake H (1999) Inorganic polyphosphate kinase is required to stimulate protein degradation and for adaptation to amino acid starvation in *Escherichia coli*. Proc Natl Acad Sci U S A 96:14264–14269

Kuroda A, Nomura K, Ohtomo R, Kato J, Ikeda T, Takiguchi N, Ohtake H, Kornberg A (2001) Role of inorganic polyphosphate in promoting ribosomal protein degradation by the Lon protease in *E. coli*. Science 293:705–708. https://doi.org/10.1126/science.1061315

Kuroda A, Nomura K, Takiguchi N, Kato J, Ohtake H (2006) Inorganic polyphosphate stimulates lon-mediated proteolysis of nucleoid proteins in *Escherichia coli*. Cell Mol Biol (Noisy-le-Grand) 52:23–29

Lander N, Cordeiro C, Huang G, Docampo R (2016) Polyphosphate and acidocalcisomes. Biochem Soc Trans 44:1–6. https://doi.org/10.1042/BST20150193

Lempart J, Jakob U (2019) Role of polyphosphate in Amyloidogenic processes. Cold Spring Harb Perspect Biol 11:a034041. https://doi.org/10.1101/cshperspect.a034041

Lempart J, Tse E, Lauer JA, Ivanova MI, Sutter A, Yoo N, Huettemann P, Southworth D, Jakob U (2019) Mechanistic insights into the protective roles of polyphosphate against amyloid cytotoxicity. Life Sci Alliance 2:e201900486. https://doi.org/10.26508/lsa.201900486

Liebermann L (1888) Ueber das Nucleïn der Hefe und künstliche Darstellung eines Nucleïns aus Eiweiss und Metaphosphorsäure. Ber Dtsch Chem Ges 21:598–600. https://doi.org/10.1002/cber.188802101109

Lindegren CC (1947) Function of volutin (metaphosphate) in mitosis. Nature 159:63

Lorenzo-Orts L, Hohmann U, Zhu J, Hothorn M (2019) Molecular characterization of CHAD domains as inorganic polyphosphate-binding modules. Life Sci Alliance 2:e201900385. https://doi.org/10.26508/lsa.201900385

Meyer A (1904) Orientirende Untersuchungen über Verbreitung Morphologie und Chemie des Volutins. Bot Zft 62:113–152

Mezzina MP, Pettinari MJ (2016) Phasins, multifaceted Polyhydroxyalkanoate granule-associated proteins. Appl Environ Microbiol 82:5060–5067. https://doi.org/10.1128/AEM.01161-16

Motomura K, Hirota R, Okada M, Ikeda T, Ishida T, Kuroda A (2014) A new subfamily of polyphosphate kinase 2 (class III PPK2) catalyzes both nucleoside monophosphate phosphorylation and nucleoside diphosphate phosphorylation. Appl Environ Microbiol 80(8):2602–2608. https://doi.org/10.1128/AEM.03971-13

Mordhorst S, Singh J, Mohr MKF, Hinkelmann R, Keppler M, Jessen HJ, Andexer JN (2019) Several polyphosphate kinase 2 enzymes catalyse the production of adenosine 5′-polyphosphates. Chembiochem 20:1019–1022. https://doi.org/10.1002/cbic.201800704

Murphy DJ (1993) Structure, function and biogenesis of storage lipid bodies and oleosins in plants. Prog Lipid Res 32:247–280. https://doi.org/10.1016/0163-7827(93)90009-l

Neisser A (1888) Versuche über die Sporenbildung bei Xerobacillen, Streptokokken und Choleraspirillen. Zeitschrift f Hygiene 4:165–196

Nikel PI, Chavarria M, Martinez-Garcia E, Taylor AC, de Lorenzo V (2013) Accumulation of inorganic polyphosphate enables stress endurance and catalytic vigour in *Pseudomonas putida* KT2440. Microb Cell Factories 12:50. https://doi.org/10.1186/1475-2859-12-50

Ogawa J, Amano Y (1987) Electron microprobe X-ray analysis of polyphosphate granules in *Plesiomonas shigelloides*. Microbiol Immunol 31:1121–1125. https://doi.org/10.1111/j.1348-0421.1987.tb01345.x

Ortíz-Severín J, Varas M, Bravo-Toncio C, Guiliani N, Chávez FP (2015) Multiple antibiotic susceptibility of polyphosphate kinase mutants (*ppk1* and *ppk2*) from *Pseudomonas aeruginosa* PAO1 as revealed by global phenotypic analysis. Biol Res 48:22. https://doi.org/10.1186/s40659-015-0012-0

Pallerla SR, Knebel S, Polen T, Klauth P, Hollender J, Wendisch VF, Schoberth SM (2005) Formation of volutin granules in *Corynebacterium glutamicum*. FEMS Microbiol Lett 243:133–140. https://doi.org/10.1016/j.femsle.2004.11.047

Racki LR, Tocheva EI, Dieterle MG, Sullivan MC, Jensen GJ, Newman DK (2017) Polyphosphate granule biogenesis is temporally and functionally tied to cell cycle exit during starvation in *Pseudomonas aeruginosa*. Proc Natl Acad Sci U S A 114:201615575–E2449. https://doi.org/10.1073/pnas.1615575114

Rao NN, Liu S, Kornberg A (1998) Inorganic polyphosphate in *Escherichia coli*: the phosphate regulon and the stringent response. J Bacteriol 180:2186–2193

Rao NN, Gómez-García MR, Kornberg A (2009) Inorganic polyphosphate: essential for growth and survival. Annu Rev Biochem 78:605–647. https://doi.org/10.1146/annurev.biochem.77.083007.093039

Reusch RN (2012) Physiological importance of poly-(R)-3-hydroxybutyrates. Chem Biodivers 9:2343–2366. https://doi.org/10.1002/cbdv.201200278

Roinestad FA, Yall I (1970) Volutin granules in *Zoogloea ramigera*. Appl Microbiol 19:973–979

Ruiz FA, Lea CR, Oldfield E, Docampo R (2004) Human platelet dense granules contain polyphosphate and are similar to acidocalcisomes of bacteria and unicellular eukaryotes. J Biol Chem 279:44250–44257. https://doi.org/10.1074/jbc.M406261200

Ryazanova LP, Suzina NE, Kulakovskaya TV, Kulaev IS (2009) Phosphate accumulation of *Acetobacter xylinum*. Arch Microbiol 191:467–471. https://doi.org/10.1007/s00203-009-0470-2

Sall T, Mudd S, Takagi A (1958) Polyphosphate accumulation and utilization as related to synchronized cell division of *Corynebacterium diphtheriae*. J Bacteriol 76:640–645

Salto ML, Kuhlenschmidt T, Kuhlenschmidt M, de Lederkremer RM, Docampo R (2008) Phospholipid and glycolipid composition of acidocalcisomes of Trypanosoma cruzi. Mol Biochem Parasitol 158:120–130. https://doi.org/10.1016/j.molbiopara.2007.12.001

Schmidt G, Hecht L, Thannhauser SJ (1946) The enzymatic formation and the accumulation of large amounts of a metaphosphate in bakers' yeast under certain conditions. J Biol Chem 166:775

Schönborn C, Bauer HD, Röske I (2001) Stability of enhanced biological phosphorus removal and composition of polyphosphate granules. Water Res 35:3190–3196. https://doi.org/10.1016/s0043-1354(01)00025-2

Seufferheld M, Vieira M, Ruiz FA, Rodrigues CO, Moreno S, Docampo R (2003) Identification of organelles in bacteria similar to acidocalcisomes of unicellular eukaryotes. J Biol Chem 278:29971–29978. https://doi.org/10.1074/jbc.M304548200

Seufferheld M, Lea CR, Vieira M, Oldfield E, Docampo R (2004) The H+-pyrophosphatase of *Rhodospirillum rubrum* is predominantly located in polyphosphate-rich acidocalcisomes. J Biol Chem 279:51193–51202. https://doi.org/10.1074/jbc.M406099200

Seufferheld MJ, Alvarez HM, Farias ME (2008) Role of polyphosphates in microbial adaptation to extreme environments. Appl Environ Microbiol 74:5867–5874. https://doi.org/10.1128/AEM.00501-08

Seufferheld MJ, Kim KM, Whitfield J, Valerio A, Caetano-Anolles G (2011) Evolution of vacuolar proton pyrophosphatase domains and volutin granules: clues into the early evolutionary origin of the acidocalcisome. Biol Direct 6:50. https://doi.org/10.1186/1745-6150-6-50

Seviour RJ, Mino T, Onuki M (2003) The microbiology of biological phosphorus removal in activated sludge systems. FEMS Microbiol Rev 27:99–127. https://doi.org/10.1016/S0168-6445(03)00021-4

Shiba T, Tsutsumi K, Yano H, Ihara Y, Kameda A, Tanaka K, Takahashi H, Munekata M, Rao NN, Kornberg A (1997) Inorganic polyphosphate and the induction of rpoS expression. Proc Natl Acad Sci U S A 94:11210–11215. https://doi.org/10.1073/pnas.94.21.11210

Smith IW, Wilkinson JF, Duguid JP (1954) Volutin production in *Aerobacter aerogenes* due to nutrient imbalance. J Bacteriol 68:450–463

Steinbüchel A, Aerts K, Babel W, Follner C, Liebergesell M, Madkour MH, Mayer F, Pieper-Furst U, Pries A, Valentin HE (1995) Considerations on the structure and biochemistry of bacterial polyhydroxyalkanoic acid inclusions. Can J Microbiol 41(Suppl 1):94–105

Strohl WR, Larkin JM (1978) Enumeration, isolation, and characterization of beggiatoa from freshwater sediments. Appl Environ Microbiol 36:755–770

Tocheva EI, Dekas AE, McGlynn SE, Morris D, Orphan VJ, Jensen GJ (2013) Polyphosphate storage during sporulation in the gram-negative bacterium *Acetonema longum*. J Bacteriol 195:3940–3946. https://doi.org/10.1128/JB.00712-13

Tumlirsch T (2017) Identifizierung neuer Proteine des Polyphosphat Granulumkomplexes von *Ralstonia eutropha* H16. Thesis, University Stuttgart

Tumlirsch T, Jendrossek D (2017) Proteins with CHADs (conserved Histidine α-helical domains) are attached to polyphosphate granules in vivo and constitute a novel family of polyphosphate-associated proteins (Phosins). Appl Environ Microbiol 83:e03399–16–14. https://doi.org/10.1128/AEM.03399-16

Tumlirsch T, Sznajder A, Jendrossek D (2015) Formation of polyphosphate by polyphosphate kinases and its relationship to poly(3-hydroxybutyrate) accumulation in *Ralstonia eutropha* strain H16. Appl Environ Microbiol 81:8277–8293. https://doi.org/10.1128/AEM.02279-15

Tzeng CM, Kornberg A (2000) The multiple activities of polyphosphate kinase of *Escherichia coli* and their subunit structure determined by radiation target analysis. J Biol Chem 275(6):3977–3983

Varela C, Mauriaca C, Paradela A, Albar JP, Jerez CA, Chávez FP (2010) New structural and functional defects in polyphosphate deficient bacteria: a cellular and proteomic study. BMC Microbiol 10:7. https://doi.org/10.1186/1471-2180-10-7

Werten S, Rustmeier NH, Gemmer M, Virolle M-J, Hinrichs W (2019) Structural and biochemical analysis of a phosin from *Streptomyces chartreusis* reveals a combined polyphosphate- and metal-binding fold. FEBS Lett 593:2019–2029. https://doi.org/10.1002/1873-3468.13476

Wiame JM (1947) The metachromatic reaction of Hexametaphosphate. J Am Chem Soc 69:3146–3147. https://doi.org/10.1021/ja01204a508

Widra A (1959) Metachromatic granules of microorganisms. J Bacteriol 78:664–670

Yang Y, Shi X, Ballent W, Mayer BK (2017) Biological phosphorus recovery: review of current Progress and future needs. Water Environ Res 89:2122–2135. https://doi.org/10.2175/106143017X15054988926424

Yoo NG, Dogra S, Meinen BA, Tse E, Haefliger J, Southworth DR, Gray MJ, Dahl J-U, Jakob U (2018) Polyphosphate stabilizes protein unfolding intermediates as soluble amyloid-like oligomers. J Mol Biol 430:4195–4208. https://doi.org/10.1016/j.jmb.2018.08.016

Zhang HY, Ishige K, Kornberg A (2002) A polyphosphate kinase (PPK2) widely conserved in bacteria. Proc Natl Acad Sci USA 99(26):16678–16683. https://doi.org/10.1073/pnas.262655199

Zhu Y, Huang W, Lee SSK, Xu W (2005) Crystal structure of a polyphosphate kinase and its implications for polyphosphate synthesis. EMBO Rep 6(7):681–687. https://doi.org/10.1038/sj.embor.7400448

Bacterial Intracellular Sulphur Globules

Christiane Dahl

Contents

1 Introduction 20
2 History 20
3 Elemental Sulphur 21
4 Organisms Forming Intracellular Sulphur Globules 22
5 Subcellular Localization of Sulphur Globules 29
6 Properties and Function of Sulphur Globule Proteins 32
7 Speciation of Sulphur 33
8 Formation of Stored Sulphur 35
9 Degradation of Stored Sulphur 38
9.1 Oxidative Degradation 38
9.2 Reductive Degradation 39
10 Outlook 40
References 40

Abstract Reduced sulphur compounds such as sulphide, polysulphides, thiosulphate, and elemental sulphur are oxidized by a large and diverse group of prokaryotes. In many cases, intracellular globules of polymeric, water-insoluble sulphur are accumulated either as a transient product *en route* to sulphate or as the final product. Sulphur globule formation is especially widespread among sulphur-oxidizing Proteobacteria and occurs in purple sulphur bacteria of the family Chromatiaceae, in *Beggiatoa* species as well as in other "morphologically conspicuous" sulphur bacteria (e.g. *Thioploca, Achromatium, Thiovulum*). Sulphur globules are typically enclosed by a surface layer consisting of highly repetitive glycine-rich structural proteins (sulphur globule proteins, Sgps) and reside in the bacterial periplasm. Here, an overview of recent findings on the speciation of stored sulphur, the occurrence of Sgps and the enzymes involved in the formation and breakdown of bacterial sulphur globules is given.

C. Dahl (✉)
Institut für Mikrobiologie & Biotechnologie, Rheinische Friedrich-Wilhelms-Universität Bonn, Bonn, Germany
e-mail: ChDahl@uni-bonn.de

D. Jendrossek (ed.), *Bacterial Organelles and Organelle-like Inclusions*, Microbiology Monographs 34, https://doi.org/10.1007/978-3-030-60173-7_2

1 Introduction

Sulphur is the 16th element on the periodic table and the tenth most abundant element in the universe (Steudel and Chivers 2019). Sulphur serves essential functions in all living cells. In proteins it occurs not only in the form of cysteine and methionine, but also in iron-sulphur clusters, in several sulphur-containing cofactors like thiamine, biotin, coenzyme A and lipoic acid and is furthermore indispensable in tRNAs through a variety of modifications (Shigi 2014, 2018). Sulphur is a very versatile chemical element and undergoes permanent cycling in terrestrial as well as in marine environments. Dissimilatory sulphate reduction is the primary driver of the biogeochemical sulphur cycling. In this anaerobic respiratory process, sulphate is used as an electron acceptor instead of oxygen, nitrate or manganese [Mn(IV)] (Henkel et al 2019; Rabus et al 2015). In turn, hydrogen sulphide, polysulphides, thiosulphate, elemental sulphur and polythionates serve as electron donors for a huge array of chemo- and photolithotrophic bacteria and archaea such as *Acidithiobacillus* or *Acidianus* species (Dahl et al. 2008; Mangold et al. 2011; Kletzin et al. 2004; Frigaard and Dahl 2009; Dahl 2017). A large portion of these organisms forms sulphur globules both extracellularly and intracellularly (Dahl and Prange 2006; Dahl 2017; Maki 2013). Whether the sulphur accumulates as a transient or the final product varies depending on the species, the culture conditions and the reduced sulphur substrate.

Here, I attempt to give an update about the different sulphur-forming prokaryotes, the structure and chemical nature of bacterial sulphur inclusions and the metabolic pathways related to sulphur globule formation and degradation. An exclusive focus will be laid on sulphur globules deposited within the confines of the cell wall, i.e. sulphur present as a bacterial inclusion *senso strictu*. For further detailed information, the reader is referred to a number of reviews on oxidative sulphur metabolism (Frigaard and Dahl 2009; Dahl 2017; Wang et al. 2019; Friedrich et al. 2005; Dahl et al. 2008; Dahl and Prange 2006).

2 History

Internal sulphur globules are easily recognized even via light microscopy as they are highly light refractive (Fig. 1) and can reach diameters of several micrometres. Accordingly, the first mentioning of these conspicuous structures dates back to 1786, when Müller described intracellular spherical inclusions of unknown composition in 'colourless', egg-shaped algae (Müller 1786), later identified as *Thiovulum majus* (Rivière and Schmidt 2006). Over the following decades, several additional microorganisms were mentioned to contain similar inclusions (Ehrenberg 1838; Trevisan 1842; Perty 1852) and differentiated due to the presence of colour (the later *Thiospirillum*, *Chromatium*, *Lamprocystis*) and lack of colour (*Beggiatoa*, *Thiothrix* and *Thiovulum*). About a century after their discovery the cellular

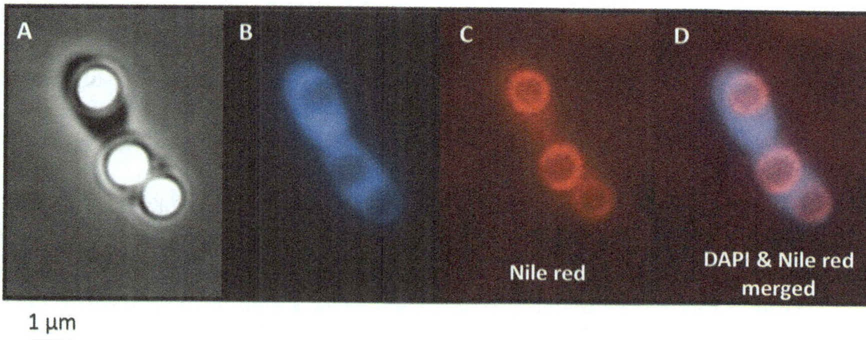

Fig. 1 *Allochromatium vinosum* DSM 180^T cells with sulphur globules visualized by light microscopy: **(a)** phase contrast **(b)** DNA stain with DAPI. In panel **(c)**, staining with Nile red highlights sulphur globules. **(d)**, DAPI and Nile red stain merged. Microscopy was carried out at room temperature using a Zeiss Axio Observer Z1 microscope (Zeiss, Jena, Germany) equipped with HXP 120 V light source and Axio Cam MR3 camera. Standard filter sets were used for DAPI (335–383 nm excitation and 420–470 nm emission) and Nile red (510–560 nm excitation and 590 nm long pass emission). Image acquisition and analysis were performed with Zen 2 software (Zeiss). Nile red is a useful probe of hydrophobic sites on proteins (Sackett and Wolff 1987) and probably interacts with the hydrophobic Sgps. The red halo in the Nile red image requires further investigation and currently remains unexplained. Photos courtesy of Fabian Grein

inclusions were proven to consist of elemental sulphur in *Beggiatoa* (C. Cramer in (Müller 1870); in *Allochromatium vinosum* (Cohn 1875); and in *Thiovulum muelleri* (Warming 1875)). A first systematic analysis of uncoloured and coloured 'sulphobacteria' and their sulphur globules was provided by Winogradsky (Winogradsky 1887), who also demonstrated the oxidation of hydrogen sulphide to stored sulphur under microaerophilic conditions in the chemotrophic *Beggiatoa* (Winogradsky 1889). Pioneering studies on the oxidation of sulphur in bacterial photosynthesis were done by van Niel whose classic studies about phototrophic sulphur bacteria and accumulation of elemental sulphur can be considered as milestones and provided the basis for further studies about sulphur compounds in photosynthesis (van Niel 1936; van Niel 1931). Those interested in the early research on sulphur bacteria are referred to discussions by (Waksman 1922; Waksman and Joffe 1922; Shively et al. 2006; Dahl and Prange 2006; Trüper 2008).

3 Elemental Sulphur

Naturally, sulphur occurs in a huge variety of environments (Cosmidis et al. 2019; Nims et al. 2019) such as volcanic areas including sulphidic springs (Macur et al. 2013; Kamyshny et al. 2014; Lau et al. 2017), deep-sea hydrothermal vents (Taylor et al. 1999), deep-sea hydrocarbon seeps (Eichinger et al. 2014) or marine sediments and salt marshes (Kamyshny and Ferdelman 2010; Zopfi et al. 2004; Jørgensen and Nelson 2004; Taylor and Wirsen 1997).

Elemental sulphur can be formed through abiogenic processes when sulphide is oxidized by molecular oxygen, possibly catalyzed by oxidized metals (Luther et al. 2011). Its presence in the environment is often associated with microbial oxidation of reduced sulphur compounds (Kleinjan et al. 2003).

Sulphur forms more than 30 solid allotropes, more than any other element. It exists in many forms, from cyclic octamers (S_8 rings) that crystallize in different structures to sulphur chains with varying numbers of S-S bonds, and polysulphides (Sn^{2-}) (Kamyshny and Ferdelman 2010; Meyer 1976; Trofimov et al. 2009). Only a few sulphur allotropes occur in biological systems. The thermodynamically most stable form at standard conditions is homocyclic, orthorhombic crystalline α-sulphur (α-S_8) (*cyclo*-octasulphur) (Roy and Trudinger 1970; Steudel 1996a, b). S_6, S_7 and S_{12} rings have also been detected in samples of biological origin, while bigger rings up to S_{20} were made accessible by chemical synthesis (Steudel 1987, 2000). Commercially available sulphur consists mainly of S_8 rings, traces of S_7 rings that are responsible for the bright yellow colour (Steudel and Holz 1988) and polymeric sulphur. Polymeric sulphur consists of very long helically wound chains of almost all sizes (Steudel 2000). Regardless of the molecular size, all sulphur allotropes are hydrophobic, are not wetted by water and have very low solubilities in water (Steudel and Eckert 2003).

4 Organisms Forming Intracellular Sulphur Globules

Sulphur can accumulate in the form of water-insoluble globules as a transient or the final product during the oxidation of reduced sulphur compounds (sulphide, polysulphides, thiosulphate, polythionates and elemental sulphur). Accordingly, sulphur-forming bacteria share environments characterized by elevated levels of hydrogen sulphide mainly produced by bacterial sulphate reduction in anoxic sediments rich in organic nutrients or originating from hydrothermal vents or cold seeps.

Concerning their physiology, two large groups of sulphur-storing bacteria can be differentiated: The first are phototrophic prokaryotes that use sulphur compounds as electron donors for CO_2 fixation in the light (Dahl 2017). Among these, purple sulphur bacteria of the family Chromatiaceae form intracellular sulphur deposits. On the other hand, chemotrophic (the classical "colourless") sulphur-oxidizing prokaryotes use the energy derived from the oxidation of sulphur compounds with either oxygen, nitrate or Mn(IV) oxide as electron acceptors to fix carbon dioxide (Henkel et al. 2019; Dahl et al. 2008; Friedrich 1998; Kletzin et al. 2004; Wang et al. 2019). Sulphur compounds are also an important energy source for symbiotic associations of chemoautotrophic sulphur bacteria with marine organisms from unicellular protists (Ott et al. 2004) to metazoans, such as meduzoans (Abouna et al. 2015), bivalves (Frenkiel et al. 1996) and nematodes (Himmel et al. 2009).

Although first discovered at hydrothermal vents, symbiosis with sulphur oxidizers is not limited to these highly specialized environments but has also been found in shallow subtidal sands, macrophyte debris, deep sea cold seeps, mangrove swamps, sea grass beds, anoxic marine basins, sewage outfalls and even rotting whale carcasses (Distel 1998; Kleiner et al. 2012; Petersen et al. 2016; Seah et al. 2019; Cavanaugh et al. 1981; Felbeck 1981; Nelson and Fisher 1995).

Concerning their systematic affiliation, the vast majority of organisms with reported capability for the formation of intracellular sulphur globules belong to the Proteobacteria (Table 1). Notable exceptions are the Gram-positives *Thermoanaerobacter sulfurigignens* and *Thermoanaerobacterium thermosulfurigienes* (Lee et al. 2007), which fall into the class Clostridia within the Firmicutes phylum. Sulphur globules within the confines of the cell have also been detected in *Thermus scotoductus* (Skirnisdottir et al. 2001) belonging to the class Deinococci within the phylum Thermus-Deinococcus.

Most reports about intracellular sulphur deposition are available for members of the α-, β-, γ- and ε-Proteobacteria (Table 1). The trait is widespread though not ubiquitous in alphaproteobacterial, microaerophilic, autotrophic magnetotactic bacteria which form the globules upon growth on sulphide and/or thiosulphate (Bazylinski et al. 2004, 2013; Bazylinski and Williams 2006; Keim et al. 2005; Williams et al. 2006; Lefevre et al. 2012; Spring and Bazylinski 2000). Another example among the Alphaproteobacteria is *Azospirillum thiophilum* for which intracellular sulphur globule formation has been described upon growth in the presence of sulphide (Lavrinenko et al. 2010). Within the β-Proteobacteria, we find the genus *Macromonas* (La Riviere and Schmidt 1999). The large cells of this genus are characterized by voluminous inclusions of calcium oxalate. In addition, sulphur globules may be present. *Macromonas bipunctata* can oxidize sulphide to sulphur by means of hydrogen peroxide; however, this process does not allow energy conservation (Willems 2014). Furthermore, the Betaproteobacterium *Thermothrix azorensis*, an aerobic, thermophilic, obligately chemolithoautotrophic sulphur oxidizer, appears to form inclusions of sulphur under certain growth conditions (incomplete thiosulphate oxidation, pH above 7.0) (Odintsova et al. 1996). *Thiovulum* is a spectacular genus belonging to the ε-branch of the Proteobacteria that has primarily been defined observationally by its large egg-shaped cells that can reach a length of 5–25μm. In the cells, sulphur globules are often concentrated at one cell pole (Marshall et al. 2012; La Riviere and Schmidt 1999). *Thiovulum* has so far evaded isolation in pure culture but appears to be a chemolithoautotrophic microaerophile. A single-cell genome is available (Marshall et al. 2012).

Among the Gammaproteobacteria, formation of intracellular sulphur globules is especially widespread (Table 1). Many of these bacteria belong to families within the order Chromatiales. Sulphur deposition is a characteristic trait of many purple sulphur bacteria of the family Chromatiaceae (Dahl 2017), while *Thiorhodospira sibirica* is the only phototrophic member of the family Ectothiorhodospiraceae that is capable of intracellular sulphur deposition (Bryantseva et al. 1999). In fact, this organism also forms extracellular sulphur deposits, the name-giving feature of the family. The Ectothiorhodospiraceae harbour additional species that store sulphur

Table 1 Systematic affiliation of selected proteobacteria with the capability of sulphur globule formation

Organism	Properties	(Predicted) *sgp* genes[a]	Molecular masses calculated without signal peptide [Da]	Enzyme (systems) for sulphur oxidation	References
Alphaproteobacteria					
Rhodospirillales					
Rhodospirillaceae					
Magnetovibrio blakemorei DSM 18854^{T})	Chemolithoautotrophic growth on sulphide and thiosulphate, internal sulphur-rich globules upon growth on sulphide	BEN30_14055, 07955, 10335	13.253, 12.979, 12.497	SqrA, SqrC, SqrF, FccAB, SoxXYZAB, Dsr, Soe, Apr, Sat	Bazylinski et al. (2013), Trubitsyn et al. (2016)
Azospirillum thiophilum DSM 21654^{T}	Simultaneous utilization of organic substrates and thiosulphate under microaerobic conditions, sulphur globules upon growth in the presence of sulphide	VY88_24200, 06955	9.610, 9.204	SqrA, SoxCDYZAXB	Kwak and Shin (2016), Lavrinenko et al. (2010), Frolov et al. (2013)
Betaproteobacteria					
Burkholderiales					
Comamonadaceae					
Macromonas sp. BK-30	Aerobic chemoorganotroph, oxidation of sulphide to sulphur does not provide useful energy	None identified		SoxY, partial SoxZ, SoxB	Willems (2014), Dubinina and Grabovich (1984) NZ_NWBQ01000022

Gammaproteobacteria					
Chromatiales					
Chromatiaceae					
Allochromatium vinosum DSM 180^T	Phototrophic purple sulphur bacterium	**Alvin_0358 (SgpB), 1325 (SgpC), _1905 (SgpA), _2515 (SgpD)**	10.650, 8.482, 10.497, 21.033	SqrD, SqrF, FccAB, SoxBXYZA, Dsr, Soe, Apr, Sat	Weissgerber et al. (2014), Weissgerber et al. (2011), Brune (1995a)
Thiocapsa roseopersicina DSM 217^T	Phototrophic purple sulphur bacterium	SAMN05421783_111111 (SgpA), 1099 (SgpA), 113134 (SgpA), 101470 (SgpC), 11156 (SgpD)	11.158, 9.914, 10.380, 8.936, 24.160	SqrC, SqrD, SqrF, FccAB, SoxBXYZA, Dsr, Soe, Apr, Sat	Pfennig and Trüper (1971), Brune (1995a)
Ectothiorhodospiraceae					
Thioalkalivibrio paradoxus ARh 1 (DSM 13531^T)	Obligately aerobic chemolithoautotrophs, sulphur globules as obligate intermediates on thiosulphate and thiocyanate	THITH_04870, 06175, 07650, 08045 (SgpA)	10.045, 10.508, 10.537, 11,546	SqrF, FccAB, SoxYZ, SoxB, Dsr, Soe, Apr, Sat, TcDH	Berben et al. (2015)
Thiorhodospira sibirica ATCC 70058	Photolithoautotroph, sulphur deposition in periplasm and extracellularly, thiosulphate not oxidized	ThisiDRAFT_0642 (SgpA), 2422	11.560, 12.810	SqrB, SqrD, FccB, SoxYZ, sHdr, Soe	Bryantseva et al. (1999)
Thiotrichales					
Thiotrichaceae					
Achromatium sp. (*Candidatus* Achromatium palustre)	Single-cell genome, sulphur inclusions, presumably microaerophilic chemolithotrophs, oxidizes reduced sulphur compounds to sulphate	TI05_14580 (SgpA), 15635 (SgpA), 11440 (SgpA)	11.641, 10.438, 10.617	SqrA, SoxBYZXA, Dsr, Apr, Sat	Mansor et al. (2015), Salman et al. (2016)

(continued)

Table 1 (continued)

Organism	Properties	(Predicted) *sgp* genes[a]	Molecular masses calculated without signal peptide [Da]	Enzyme (systems) for sulphur oxidation	References
Beggiatoa leptomitoformis D-401	Chemolithoautotrophic growth with sulphide and thiosulphate, intracellular sulphur globules	BLE401_16800 (SgpA)	12.649	SqrA, SqrF, FccB, SoxYZ, SoxB	Dubinina et al. (2017)
Thiothrix caldifontis DSM 21228[T]	Capacity for chemolithoautotrophic growth with thiosulphate and sulphide, intracellular elemental sulphur and sulphate formed	SAMN05660964_00436, 01322, 01876, 01943, 00727, 02054	12.348, 13.846, 24.730, 23.558, 17.120, 17.382	SqrA, SqrF, FccA, FccB, SoxBZYAZAX, Dsr, Soe, Apr, Sat	Chernousova et al. (2009)
Candidatus Thiomargarita nelsonii	Chemolithotroph, oxidizes sulphide to sulphate, intracellular sulphur globules as intermediates	PN36_24595 (SgpA), 24600	11.726, 14.521	SqrF, FccB, SoxYZAXB, Dsr, Soe, Apr, Sat	Flood et al. (2016)
Thiolinaceae					
Thiolinea disciformis DSM 14473[T]	Thiosulphate and sulphide oxidized during chemoheterotrophic growth, sulphur within invaginated inner membrane, sulphate formation weak	A3IE_RS0105985	23.978	SqrA, SqrF, FccB, SoxB, SoxXA, SoxYZ, Dsr, Soe	Boden and Scott (2018)
Thiofilaceae					
Thiofilum flexile DSM 14609[T]	Thiosulphate and sulphide oxidized during chemoheterotrophic growth, sulphur within invaginated inner membrane	A3IK_RS21110	24.794	SqrA, SoxYZAXB, Dsr, Soe	Boden and Scott (2018)

Epsilonproteobacteria					
Campylobacteriales					
Thiovulaceae					
Thiovulum	Aerobic chemoautotroph, oxidizes hydrogen sulphide to elemental sulphur	None identified		SqrD, SqrF	Marshall et al. (2012)
Unclassified Gammaproteobacteria					
Sedimenticola selenatireducens DSM 17993^T	Chemolithoautotrophic growth with thiosulphate, sulphide, tetrathionate, under hypoxic or anaerobic conditions, sulphur globules formed as intermediates	FHP88_07550 (SgpA), FHP88_10320 (SgpA)	8.737, 19.305,	SqrA, SqrF, FccAB, SoxBYZAX, Dsr, Soe, Apr, Sat	Flood et al. (2015)
Thiolapillus brandeum Hiromi 1	Facultative chemolioautotroph, growth on sulphur, thiosulphate and tetrathionate with oxygen or nitrate	TBH_C1660 (SgpA), C0689 (SgpA), (SgpA)	10.985, 10.494, 10.265	SqrA, SqrF, FccAB, SoxXAZYB, Dsr, Soe, Apr, Sat	Nunoura et al. (2014)
Endosymbiont of *Riftia pachyptila*	Trophosome tissue of *Riftia pachyptila* (Annelida, Siboglinidae)	Rifp1Sym_bd00100 (SgpA), at00150 (SgpA), fc00030 (SgpD)	9.039, 7.598, 23.147	SqrA, FccAB, SoxB, SoxAX, SoxYZ, dsr, Soe, Apr, Sat	Gardebrecht et al. (2012), Markert et al. (2007)

(continued)

Table 1 (continued)

Organism	Properties	(Predicted) *sgp* genes[a]	Molecular masses calculated without signal peptide [Da]	Enzyme (systems) for sulphur oxidation	References
Solemya velesiana gill symbiont Sveles-Q1	Gill tissue from *Solemya velesiana* (Mollusca, Solemyidae)	BOW51_07440 (SgpA), 08055 (SgpA/B), 05805 (SgpA), 05810 (SgpA), 07780 (SgpA)	9.094, 9.232, 17.092, 9.741, 20.658	SqrA, SqrF, FccAB, SoxXAYZB, Dsr, Soe, Apr, Sat	Russell et al. (2017)

[a]Genes predicted to encode sulphur globule proteins were identified by BLAST searches without compositional adjustments and without filtering for low complexity regions using the resources provided by Integrated Microbial Genomes (DOE Joint Genomes Institute, http://img.jgi.doe.gov) and GenBank (http://www.ncbi.nlm.nih.gov). The established sulphur globule proteins from *Allochromatium vinosum* (SgpA, Alvin_1905, SgpB, Alvin_0358, SgpC, Alvin_1325, SgpD, Alvin_2515) are printed in bold and were used as bait. For those predicted sulphur globule proteins, which clearly resemble one of the established *A. vinosum* Sgps more than the others, the most closely related *A.* vinosum protein is given in brackets. All listed predicted *sgp* genes encode Sec-dependent signal peptides. At this point, it cannot be excluded that the predicted *sgp* genes for the listed organisms are still incomplete. Baits for Blast searches for sulphur-oxidizing enzyme systems: FccAB from *Allochromatium vinosum* (AAA23316, AAB86576), SqrA from *Aquifex aeolicus* (NP_214500), SqrB from *Halorhodospira halophila* (WP_011814451), SqrC from *Chlorobaculum tepidum* (NP_661917), SqrD from *C. tepidum K(NP_661023), SqrE* from *C. tepidum* (NP_661769), SqrF from *A. aeolicus* (NP_213539), Dsr proteins (Alvin_1251 to Alvin_1262), AprBA (Alvin_1119–1120), Sat (Alvin_1118), SoeABC (Alvin_2492–2489) from *A. vinosum*, sHdr proteins from *Acidithiobacillus caldus* (Atc_2352–234

internally; these belong to the chemoautotrophic genus *Thioalkalivibrio* (Sorokin et al. 2003; Berben et al. 2015; Mu et al. 2016; Ahn et al. 2017). Recently, a member of the Thioalkalispiraceae, *Endothiovibriovibrio diazotrophicus*, was also described as containing intracellular sulphur globules (Bazylinski et al. 2017). The Thiotrichales are the second gammaproteobacterial order containing a variety of sulphur-storing chemotrophic sulphur oxidizers. Among these, the family Thiotrichaceae features some of the most conspicuous bacteria in nature. Species of the genera *Thiomargerita* and *Achromatium* as well as the filamentous sulphur-oxidizing bacteria of the genera *Beggiatoa*, *Thiothrix* and *Thioploca* are among the largest known prokaryotes (Mansor et al. 2015; Schulz and Jørgensen 2001; Schulz et al. 1999; Salman et al. 2016) and characterized by massive sulphur formation. The only representatives of the families Thiofilaceae and Thiolinaceae described so far also form intracellular sulphur globules (Boden and Scott 2018).

A vast majority of bacterial partners in thiotrophic symbioses with eukaryotes are taxonomically unclassified Gammaproteobacteria (Table 1). Regardless of whether the host is a protist or an invertebrate and whether the bacteria are associated as endo- or as ectosymbionts, formation of sulphur globules inside of their cells has often been noted (Grimonprez et al. 2018; Rinke et al. 2006, 2009; Bergin et al. 2018; Seah et al. 2019; Markert et al. 2011; Krieger et al. 2000; Frenkiel et al. 1996).

5 Subcellular Localization of Sulphur Globules

Internal sulphur globules are easily recognized even via light microscopy as they are highly light refractive (Fig. 1). Cultures and colonies of cells containing sulphur globules, therefore, exhibit a characteristic milky appearance (Fig. 2). Usually the diameter of sulphur globules is in the range of 1–3μm (Fig. 1), but sizes exceeding 15μm have also been reported (Williams et al. 1987; Head et al. 1996; Remsen 1978;

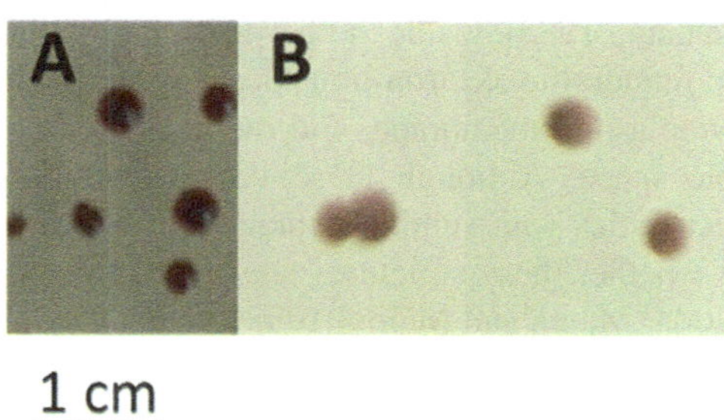

Fig. 2 Colonies *Allochromatium vinosum* DSM 180^T grown (**a**) on malate in the absence of reduced sulphur compounds (**b**) grown in the presence of sulphide and thiosulphate. Colonies appear milky-white due to massive accumulation of sulphur globules inside of the cells. *A. vinosum* was cultivated for 10 days on plates solidified with 1% (w/v) phytagel as described by (Pattaragulwanit and Dahl 1995)

Skirnisdottir et al. 2001). The sulphur can comprise 20–34% of the cell dry mass of *Beggiatoa* sp. and purple sulphur bacteria, respectively (Nelson and Castenholz 1981; Overmann 1997). While sulphur globules appear to be randomly localised in many bacterial species, specific cellular localizations have also been reported. In *Thiovulum* for example, the globules accumulate toward one cell pole (Marshall et al. 2012; La Riviere and Schmidt 1999).

A series of technical problems must be considered when it comes to elucidating the subcellular localization of sulphur globules using microscopic techniques. Sulphur dissolves during the preparation of biological samples for electron microscopy, and in addition, any remaining sulphur is subject to thermal degradation under the electron beam. Therefore, sulphur deposits appear as a conspicuous, empty, and electron-lucent space in electron micrographs (Strohl et al. 1981; Remsen and Trüper 1973; Vetter 1985; Pasteris et al. 2001). Nevertheless, microscopic evaluation has resolved the periplasm as the intracellular compartment harbouring sulphur globules in many cases. In studies with free-living filamentous sulphur bacteria, including *Thiothrix* (Bland and Staley 1978; Larkin and Shinabarger 1983; Williams et al. 1987), *Thioploca* (Maier and Murray 1965), *Thiofilum* and *Thiolinea* (Boden and Scott 2018) as well as *Beggiatoa* (de Albuquerque et al. 2010; Maier and Murray 1965; Larkin and Strohl 1983), sulphur inclusions were found to be located within invaginated pockets of the cytoplasmic membrane. In some cases, the sulphur globules appeared as a membrane-bound inclusion in the cytoplasm with no apparent connection to the cytoplasmic membrane (Strohl et al. 1981), which may be an effect of the specific sectioning plane (Shively et al. 1989). Other examples for which a periplasmic localization of sulphur globules has been settled are species of the genera *Thioalkalivibrio* (Sorokin et al. 2001) and *Thermus* (Skirnisdottir et al. 2001).

In some cases, it has been problematic to distinguish putative sulphur vesicles from other vesicle-like storage structures such as polyhydroxyalkanoate bodies. This applies especially to chemoautotrophic sulphur–oxidizing endosymbionts that reside in animal organs, e.g., in specialized gills of Vesicomyid clams (Goffredi and Barry 2002) or in so-called trophosomes in Vestimeniferan worms like *Riftia pachyptila* (Felbeck 1981; Cavanaugh 1983). Inside these organs, the symbiontic bacterial cells exhibit roundish to polymorphic electron-translucent vesicles whose membranes are infoldings of the cytoplasmic membrane, and the enclosed spaces are contiguous with the periplasmic space. Although these vesicles obviously share common ultrastructural characteristics with sulphur-containing globules of other organisms, it has been debated whether these structures are indeed related to sulphur storage (Bright and Sorgo 2003; Maina and Maloyi 1998; Vetter 1985). On the other hand, electron spectroscopic imaging pictures clearly identified sulphur in the globules of gutless oligochaete worm endosymbionts. A cytoplasmic localization was inferred for the globules without analysis via cryo-EM (Krieger et al. 2000).

Interpretation of electron micrographs of phototrophic bacteria containing sulphur globules is complicated by the dense packing of these cells with intracytoplasmic membranes harbouring the photosynthetic apparatus. These so-called chromatophores are associated with sulphur globules in a highly organized manner. Careful inspection of electron micrographs revealed that some chromatophores, the insides of which are extracytoplasmic or periplasmic (depending on whether the insides are continuous with the periplasm or not), open into the space enclosing the sulphur globules, thus implying an extracytoplasmic location for the globules themselves (Pattaragulwanit et al. 1998).

It is obvious from the last paragraphs that high-resolution microscopy has so far not provided assignment of the correct subcellular compartment for sulphur deposition in all cases. Fortunately, another very valuable information resource is available. As early as 1963, an envelope was reported for the sulphur globules of the purple sulphur bacterium *A. vinosum* (Kran et al. 1963) that was soon identified as a protein envelope (Nicolson and Schmidt 1971; Schmidt and Kamen 1970). In *A. vinosum*, the envelope consists of four different proteins of 8.5–20.8 kDa named SgpA, SgpB, SgpC and SgpD (Brune 1995a; Pattaragulwanit et al. 1998; Weissgerber et al. 2014). Relative transcript abundances for all four of the corresponding genes strongly increase upon the exposure of the cells to sulphide, thiosulphate and elemental sulphur compared to photoorganoheterotrophic growth on malate in the absence of reduced sulphur compounds (Weissgerber et al. 2013; Weissgerber et al. 2014). All four proteins are synthesized as precursors carrying amino-terminal signal peptides mediating Sec-dependent transport across the cytoplasmic membrane (Weissgerber et al. 2014; Pattaragulwanit et al. 1998). The proposed targeting process was experimentally confirmed with a *sgpA-phoA* fusion in *E. coli* (Pattaragulwanit et al. 1998) which finally resolved the subcellular localization of the globules in purple sulphur bacteria of the family Chromatiaceae. Single-layered electron-dense envelopes of 2–5 nm have also been observed for the sulphur globules of *Thioalkalivibrio paradoxus* (Berben et al. 2015) as well as for *Beggiatoa* and *Thiothrix* species (Strohl et al. 1981; Williams et al. 1987). In *Beggiatoa alba* BL15D, the envelope is pentalaminar, 12–14 nm thick and consists of three electron dense layers of 3.5, 2.1 and 3.5 nm thickness (Strohl et al. 1982). Sulphur inclusion envelopes have been described as being fragile in fixatives used for transmission electron microscopy (Strohl et al. 1981) which may explain why they are not always visible in electron micrographs of sulphur-depositing bacteria. In fact, recently performed BLAST searches revealed the presence of genes encoding putative sulphur globule proteins targeted to the periplasm in almost all genome-sequenced, globule-forming Proteobacteria analyzed (unpublished) with *Thiovulum* and *Macromonas* as the only notable exceptions. Table 1 provides an overview of selected species. The number of predicted *sgp* genes in a given organism can vary from only one, e.g., in *Beggiatoa alba* or in *Thiolinea disciformis* to six in *Thiothrix caldifontis* (Table 1) and even 15 in *Thiothrix lacustris* (not shown). Taken together, these observations provide strong indication that the general target compartment for sulphur storage is not the bacterial cytoplasm, but that deposited sulphur is separated from the cytoplasm by a unit membrane which may be continuous with the cytoplasmic membrane, depending on the organism.

6 Properties and Function of Sulphur Globule Proteins

Brune already noted in 1995 that SgpA, SgpB and SgpC from the purple sulphur bacteria *A. vinosum* and *Thiocapsa roseopersicina* exhibit sequence similarity with structural proteins containing repetitive amino acid sequences rich in regularly spaced glycine-like cytoskeletal keratins, insect and blood fluke egg-shell proteins and plant cell wall proteins (Brune 1995a). SgpC shows some sequence similarity to Gly and Trp-rich regions of prion proteins (Brune 1995a). SgpD from *A. vinosum* appears to be a coiled-coil protein (Weissgerber et al. 2014). The coiled coil is a protein motif characterized by superhelical twisting of two or more alpha helices around one another. They can form rod-like tertiary structures and include the intermediate filaments of the metazoan cytoskeleton as well as bacteria specific cytoskeletal proteins that typically assemble into stable macromolecular scaffolds (Lin and Thanbichler 2013; Rose and Meier 2004). In general, sulphur globule proteins appear to be rich in glycine, alanine and asparagine. Tyrosine and glutamine and proline can also be major constituents, depending on the protein (Brune 1995a).

In plant cell walls, glycine-rich proteins form an important group of structural protein components (Ringli et al. 2001). The primary sequences of these proteins contain more than 60% glycine, which is considerably higher than the glycine content in bacterial Sgps (Brune 1995a). Just as in glycine-rich proteins from plants, the sequences of Sgps often follow the motif $(Gly\text{-}X)_n$ in the glycine rich regions. In the proteins forming sulphur globule envelopes, Ala, Pro, Ser and Tyr are common at the X position. In some cases the motif varies, e.g. $(G\text{-}G\text{-}X)_n$, or is more complex. The structures proposed for the glycine-rich plant cell-wall proteins are antiparallel beta-pleated structures analogous to that of silk fibroin, in which the side chains of the X residues in the GXGX repeats all lie on one side of the sheet (Condit and Meagher 1986; Keller et al. 1988). If the sulphur globule proteins fold in a similar way, this may help to explain how they aid preserving the enclosed hydrophobic sulphur in a reactive state and at the same time impart hydrophilic properties to the globule surface (Steudel 1989). In the future, it will certainly be necessary to study secondary structure of Sgps in detail.

The presence of a protein envelope around the sulphur inclusions in sulphur-oxidizing bacteria suggests an important structure–function relationship. Indeed, mutants of *A. vinosum* lacking SgpB and SgpC are no longer able to oxidize sulphide and thiosulphate and to form sulphur inclusions from these sulphur compounds (Prange et al. 2004). In *A. vinosum* SgpA and SgpB can replace each other in the presence of SgpC. Still, SgpB and SgpA are not fully competent to replace each other as sulphur globule formation is not possible in mutants possessing soley SgpA or SgpB (Prange et al. 2004). A mutant containing SgpA and SgpB but lacking SgpC can grow on sulphide and thiosulphate. As this mutant forms significantly smaller sulphur globules, SgpC probably plays an important role in sulphur globule expansion. The construction of mutants lacking SgpA and SgpC or SgpA, SgpB and SgpC was not possible, leading to the conclusion that a basic level of Sgps is obligatory for cell survival even under conditions that do not allow sulphur globule formation

(Prange et al. 2004). SgpD appears to be the most abundant of the *A. vinosum* sulphur globule proteins (Weissgerber et al. 2014). Genetic information about its role is not available because mutants lacking the respective gene have not yet been analyzed. The analysis of the *A. vinsoum* sulphur globule proteome revealed SgpB as the second most abundant sulphur globule protein in this organism while peptides originating from SgpA and SgpC were less frequently detected (Weissgerber et al. 2014). In *A. vinosum*, all four *sgp* genes form separate transcriptional units (Pattaragulwanit et al. 1998; Prange et al. 2004; Weissgerber et al. 2013). All four genes are constitutively expressed and their expression is significantly enhanced in the presence of sulphide and thiosulphate (Weissgerber et al. 2013; Prange et al. 2004).

Interestingly, cells of *Beggiatoa alba* grown in the absence of sulphur compounds apparently contained small rudimentary sulphur inclusion envelopes. It was hypothesized that these envelopes were present in collapsed form until a reduced sulphur source became available (Strohl et al. 1982). A direct/covalent attachment of chains of stored sulphur to the proteins enclosing the globules does not appear to occur as a vast majority of studied or predicted Sgps do not contain any cysteine residues (Brune 1995a; Weissgerber et al. 2014). It has been speculated that protein envelope may provide binding sites for sulphur-metabolizing enzymes (Schmidt et al. 1971). To elucidate the possibility that enzymes taking part in sulphur globule formation and/or oxidation are bound to or interact with the envelope proteins, similar to the situation found for polyhydroxyalkanoate (PHA) granules (Jendrossek 2009), the sulphur globule proteome of *A. vinosum* was enriched and analyzed (Weissgerber et al. 2014). While this approach identified 78 proteins that occur exclusively in the sulphur globule proteome and were not detected in the soluble and membrane fractions, none of the established components of the periplasmic sulphide- and thiosulphate-metabolizing enzymes appeared to be enriched with the globules.

7 Speciation of Sulphur

The chemical nature of the sulphur in the globules has been the subject of intensive controversy (Pickering et al. 1998; Prange et al. 1999b, 2002; George et al. 2002, 2008; Berg et al. 2014; Pasteris et al. 2001). It has been recognized several decades ago that this sulphur has several properties that do not go along with those of elemental sulphur outside of biological systems. The first discrepancy relates to its low density of 1.2 (Guerrero et al. 1984), compared to 2.1 for the common α-sulphur (Meyer 1976). Moreover, globule sulphur has been described as 'liquid' and 'hydrophilic' (Steudel 1989; Hageage et al. 1970), while all allotropes of sulphur are solid and virtually water-insoluble at room temperature. In fact, analysis of sulphur in biological systems is generally hampered by the variety of possible reactions, the high reactivity and short lifetime of sulphur compounds with intermediate oxidation states that may be formed during these reactions, the allotropic enantiotropy of S_8 and its ability to catenate (Steudel 1982).

The first study focussing on the speciation of sulphur in sulphur globules dates back to 1970, applied polarizing microscopy and X-ray diffraction to a purple sulphur bacterium, A. *vinosum* (Hageage et al. 1970), and culminated in the conclusion that the sulphur is present in a 'liquid' or 'liquid-like' state. Later, this was questioned because Raman spectroscopy provided evidence for predominance of S_8 sulphur in the globules from *Beggiatoa* and *Thioploca* (Pasteris et al. 2001). Several other studies applied synchrotron-based X-ray absorption near-edge structure spectroscopy (XANES) at the sulphur K-edge to investigate the nature of intracellular sulphur globules (George et al. 2008; Pickering et al. 2001; Prange et al. 1999a, 2002; Lee et al. 2007), however, with contradicting results. Our own work indicated different speciation depending on metabolic properties of the organisms and the environmental conditions: long sulphur chains very probably terminated by organic residues (mono- or bisorganyl polysulphanes) in purple sulphur bacteria, *cyclo*-octasulphur in chemotrophic sulphur oxidizers like *Beggiatoa alba* and *Thiomargararita namibiensis* and long chain polythionates in the aerobically grown acidophilic sulphur oxidizer *Acidithiobacillus ferrooxidans* (Prange et al. 1999a, 2002). Others pointed out shortcomings of the detection method applied that may suffer from spectroscopic distortions dependent upon particle size and compositions. Experimental data was provided suggesting that the spectral differences observed for the sulphur globules from different organisms are not due to differences in sulphur speciation but are solely due to differences in the particle sizes of the sulphur globules (George et al. 2008). More recently, Raman spectroscopy has been applied to various sulphur globule-forming bacteria. This non-destructive analytical technique circumvents many of the problems associated with other characterization methods, as measurements can be collected on solid, liquid and live samples at room temperature and atmospheric pressure. Characteristic internal vibrational (molecular) spectra make elemental sulphur easy to detect and characterize (Eichinger et al. 2014; Pasteris et al. 2001; Berg et al. 2014; Oren et al. 2015; Maurin et al. 2010; Himmel et al. 2009). Additionally, Raman mapping produces high spatial (~1 μm) and spectral resolution. A first Raman study on the globules of *Thioploca* and *Beggiatoa* indicated the presence of S_8 in a nano-crystalline form (Pasteris et al. 2001). Another study also identified S_8 as the main form of elemental sulphur in sulphur globules produced in *Beggiatoa* filaments, but only in subpopulations located at the sulphide–oxygen interface of gradient tubes or in early growth stage cultures (Berg et al. 2014). *Beggiatoa* mats in a deeper sulphide-rich, anoxic zone, and freshwater gradient cultures gave rise to Raman signals suggesting a mixture of S_8 rings and linear polysulphides (S_n^{2-}) within the globules. In vivo Raman spectra for *Thiothrix* presented the characteristic S_8 structure previously described for the crystalline S_8 standards and molten sulphur in the internal modes (Nims et al. 2019). The significance of the speciation of sulphur lies in the bioavailability of the different forms, i.e. amorphous polymeric or (nano)crystalline elemental sulphur. It has been shown, for example, that the purple sulphur bacterium *Allochromatium vinosum*, uses only the polymeric component and not the cyclo-octasulphur component when commercially available sulphur is provided as the substrate, i.e. the organism has difficulties to attack the comparatively stable S_8 rings

(Franz et al. 2007). It is possible that this also applies to sulphur stored inside of cells such that it is deposited in a chemical form that is readily available for further degradation.

8 Formation of Stored Sulphur

In general, sulphur stored in sulphur globules is formed by the oxidation of more reduced sulphur species. The location of the sulphur inclusions in the periplasmic space implies that globule formation from sulphide, polysulphides, thiosulphate, elemental sulphur and also the much less widely used substrate thiocyanate must occur in this cellular compartment (Fig. 3).

The main characterized enzymes for sulphide oxidation, FAD-containing flavocytochrome *c* (FccAB) and sulphide:quinone oxidoreductases (SQR) indeed

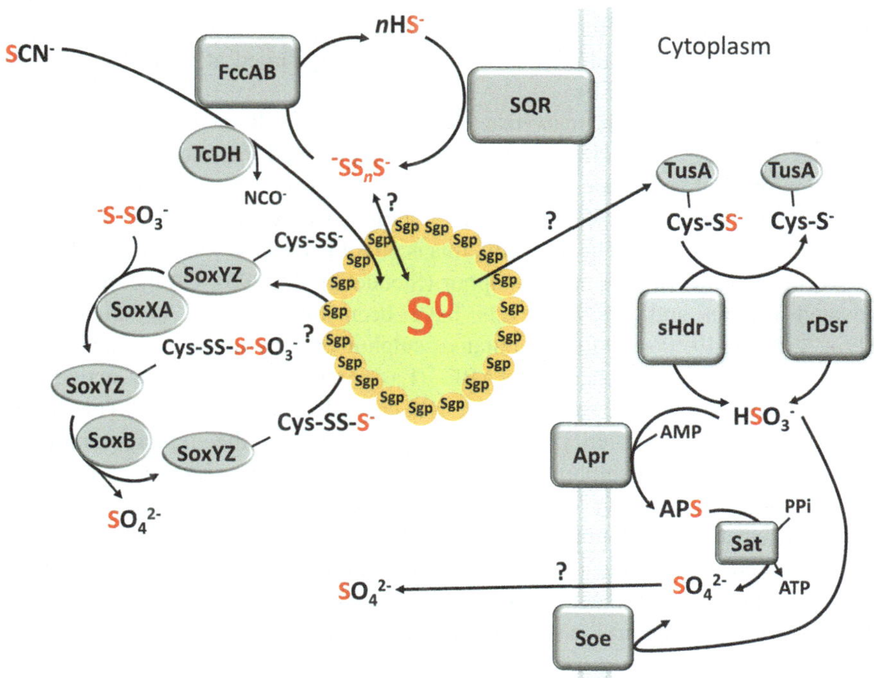

Fig. 3 Pathways of sulphur globule formation and degradation. A given organism may contain all or only a subset of the pathways depicted (cf. Table 1). For clarity, reactions are not given with exact stoichiometries. TcDh, thiocyanate dehydrogenase. SQR, sulphide:quinone oxidoreductase, FccAB, flavoctochrome *c* sulphide dehydrogenase, rDsr, sulphur oxidizing Dsr system, sHdr, sulphur-oxidizing heterodisulphide reductase-like system, Apr, APS reductase, Sat, ATP sulfurylase. Question marks indicate that neither the pathways of sulphane sulphur transport into the cytoplasm nor sulphate export from the cytoplasm have been experimentally clarified

reside in or are oriented towards the periplasm. All characterized SQRs are single-subunit flavoproteins asscociated with the cytoplasmic membrane. Based on protein structure, six distinct types of single subunit flavoprotein SQRs were identified (Marcia et al. 2009, 2010a, b; Shahak and Hauska 2008). Flavocytochrome is present as a soluble protein or as a membrane-bound enzyme and shows sulphide: cytochrome *c* oxidoreductase activity in vitro (Bosshard et al. 1986). Usually, the protein consists of a larger flavoprotein (FccB) and a smaller haemoprotein (FccA) (Castillo et al. 1994; Fukumori and Yamanaka 1979). All sulphur-globule-forming bacteria contain at least one and often several types of SQR. Flavocytochrome *c* can be present in addition (Table 1, Fig. 3). In *A. vinosum*, the gene for SqrF is followed in the same direction of transcription by two genes (Alvin_1196/97) each encoding a very short (32 amino acid) transmembrane protein. Deletion of the genes led to a ~ 60% reduced rate of sulphur formation from sulphide indicating a direct functional relation with SQR. Related genes occur in other purple sulphur bacteria, but nothing is known about their abundance and relevance in other sulphur-globule-forming bacteria (Weissgerber et al. 2013). In *A. vinosum*, mutants lacking flavocytochrome *c* sulphide oxidation proceeds with wild-type rates indicating that SQRs play a major role (Reinartz et al. 1998). Flavocytochrome *c* may represent a high affinity system for sulphide oxidation especially suited at very low sulphide concentration (Brune 1995b).

Polysulphides are the primary reaction products of SQR- and FccAB-catalyzed sulphide oxidation, and indeed, they are well-documented intermediates during the formation of sulphur globules from sulphide in *A. vinosum* (Prange et al. 2004). It is still unclear whether their conversion into sulphur stored in sulphur globules is a purely chemical process. Theoretically, this is possible because longer polysulphides are in equilibrium with elemental sulphur (Steudel et al. 1990). On the other hand, elevated protein and mRNA levels have been observed in *A. vinosum* for Alvin_1317–1319, constituting a putative sulphur or polysulphide reductase with highest similarity to archaeal SreABC (Laska et al. 2003). The active site molybdopterin-containing subunit PsrA is localized in the periplasm. This led to the proposal that this enzyme may be involved in the transformation of polysulphides to stored sulphur (Weissgerber et al. 2013; Weissgerber et al. 2014). However, only a minority of sulphur-globule-forming bacteria contains closely related genes shedding doubt on a general role of the encoded enzyme.

As apparent from Table 1, utilization of thiosulphate is very widespread among sulphur-globule-forming bacteria. When thiosulphate is oxidized, only its sulphane group is stored as sulphur; the sulphone sulphur is immediately excreted as sulphate. Thiosulphate oxidation is catalyzed by the periplasmic Sox system, consisting minimally of the proteins SoxXAK, SoxYZ and SoxB (Fig. 1). The heterodimeric SoxYZ protein acts as the central player and serves as a carrier of pathway intermediates (Sauvé et al. 2007). Recently, it has been shown that these intermediates are not simply bound to a cysteine residue located near the carboxy-terminus of the SoxY subunit as previously assumed but that the true carrier species is a SoxYZ-S-sulphane adduct (Grabarczyk and Berks 2017). The *c*-type cytochrome SoxXA (K) catalyzes the oxidative formation of a disulphide linkage between the sulphane sulphur of thiosulphate and the persulphurated active site cysteine residue of SoxY (Bamford et al. 2002; Ogawa et al. 2008; Grabarczyk and Berks 2017). Then, the sulphone group is hydrolytically released as sulphate. This reaction is catalyzed by SoxB (Grabarczyk et al. 2015; Sauvé et al. 2009) and leaves the original sulphane sulphur of thiosulphate bound to SoxY (Fig. 3). From here, the sulphur is transferred to the sulphur globules by an unknown mechanism, possibly involving the rhodoanese-like protein SoxL (Welte et al. 2009). It may be important to note in this regard that polysulphurated SoxY(S_{3-4})Z species occur as intermediates of thiosulphate oxidation catalyzed by a reconstituted Sox system in vitro (Grabarczyk and Berks 2017). Such polysulphurated species could serve as direct donors for sulphur globule formation. In organisms that do not form sulphur globules from thiosulphate, the Sox pathway involves one further crucial enzyme, SoxCD. This hemomolybdoprotein acts as a sulphane dehydrogenase and oxidizes the SoxY-bound sulphane sulphur stemming from thiosulphate to the level of a sulphone which is finally hydrolytically released as sulphate in a reaction catalyzed by SoxB. Among the sulphur-storing organisms tabulated in Table 1, *Azospirillum thiophilum* is the only one containing *soxCD*-homologous genes. Notably, this organism forms sulphur globules only in the presence of sulphide but not on thiosulphate (Kwak and Shin 2016; Lavrinenko et al. 2010).

Many phototrophic and also chemotrophic sulphur oxidizers use external elemental sulphur as a substrate and transform it into intracellular sulphur deposits before further oxidation (Franz et al. 2007). How external elemental sulphur is transformed into internally stored sulphur is currently completely unclear. *A. vinosum* needs direct cell–sulphur contact for the uptake of elemental sulphur (Franz et al. 2007). Further details remain to be investigated.

Some *Thioalkalivibrio* species are able to form sulphur globules from thiocyanate (SCN^-) (Berben et al. 2017; Sorokin et al. 2002). Two different pathways for thiocyanate degradation have been described. In the first, a periplasmic cobalt-dependent enzyme, thiocyanate dehydrogenase, catalyzes direct oxidation of the sulphane atom, forming cyanate and sulphur (Berben et al. 2017; Tsallagov et al. 2019) (Fig. 3). The second pathway occurs in *Thioalkalivibrio thiocyanodenitrificans* (Berben et al. 2017) and involves hydrolysis of the $C \equiv N$ bond by thiocyanate hydrolase to form carbonyl sulphide (COS) and ammonia. The carbonyl sulphide is further hydrolyzed to CO_2 and sulphide by carbonyl sulphide hydrolase. *T. thiocyanodenitrificans* has not been reported to form sulphur globules from thiocyanate (Sorokin et al. 2004). In addition, none of the three genes for the subunits of its thiocyanate hydrolase encode signal peptides mediating transport into the periplasm (Berben et al. 2017). The carbonyl sulphide hydrolase from *Thiobacillus thioparus* also resides in the cytoplasm (Ogawa et al. 2013). It is therefore highly unlikely that the pathway is relevant for sulphur globule formation and it is not integrated into Fig. 3.

9 Degradation of Stored Sulphur

9.1 Oxidative Degradation

The majority of sulphur-storing organisms has the capacity to completely oxidize sulphur to sulphate (Table 1, Fig. 3). The enzyme systems involved reside in the cytoplasm necessitating sulphur transfer from the periplasm as the storage compartment to the cytoplasm as the compartment for further oxidation. How this transfer is achieved has not been not clarified. Low-molecular-weight organic persulphide such as glutathione amide persulphide has been proposed as a carrier molecule; however although potential transporters for such molecules are encoded in the genome of *A. vinosum,* they have not been genetically or biochemically characterized from this or any other sulphur-oxidizing prokaryote (Weissgerber et al. 2014).

Sulphur is never processed in a free form in the cytoplasm but rather in a protein-bound persulphidic state (Dahl 2015; Tanabe et al. 2019). A cascade of sulphur transfer reactions usually involving a rhodanese-like protein, a protein of the DsrE family and a TusA homolog delivers the sulphur to an oxidizing enzyme machinery that generates sulphite (Venceslau et al. 2014; Liu et al. 2014; Tanabe et al. 2019; Dahl 2015; Stockdreher et al. 2014). The sulphur carrier protein TusA has been recognized as a central element in these reactions (Tanabe et al. 2019). For better clarity, Fig. 3 shows only this central sulphur carrier protein instead of each single sulphur transferase. The pathway employed for the oxidation of protein-bound sulphane sulphur to sulphite can vary (Fig. 3).

The best-characterized cytoplasmic sulphite-generating pathway involves reverse-acting dissimilatory sulphite reductase rDsrAB as a central player (Pott and Dahl 1998; Dahl et al. 2005; Stockdreher et al. 2012). This pathway occurs in a majority of sulphur-globule-forming organisms (Table 1). The protein DsrC serves as the substrate-binding entity (Cort et al. 2008; Stockdreher et al. 2012). Presumably, the membrane-bound DsrMKJOP electron-transporting complex oxidizes persulphurated DsrC, thus generating a DsrC trisulphide, in which a sulphur atom is bridging two strictly conserved cysteine residues. As DsrC trisulphide has been identified as the reaction product of DsrAB in a sulphate reducer (Santos et al. 2015) and very probably serves as the substrate for oxidation catalysed by rDsrAB which releases sulphite and the reduced DsrC protein as products. The two released electrons are used to generate NADH. This reaction is catalyzed by the iron-sulphur flavoprotein DsrL, an intimate interaction partner of rDsrAB (Löffler et al. 2020).

The second sulphite-generating pathway, the so-called sulphur-oxidizing heterodisulphide reductase-like (sHdr) pathway (Cao et al. 2018; Koch and Dahl 2018), is much less studied and occurs in only a few organisms forming intracellular sulphur globules like *Thiorhodospira sibirica* (Table 1) and several *Thioalkalivibrio* species (Berben et al. 2019). The central element of this pathway is an enzyme complex resembling heterodisulphide reductase HdrABC from methanogenic archaea (Kaster et al. 2011; Wagner et al. 2017). The other crucial component of the pathway is a novel lipoate-binding protein (Cao et al. 2018). Both the sHdr complex and the lipoate-binding protein have been identified as indispensable for sulphur compound oxidation in the Alphaproteobacterium *Hyphomicrobium denitrificans* (Cao et al. 2018; Koch and Dahl 2018). The reaction mechanism of the Hdr-LbpA-based sulphur oxidation system is currently unclear although an experimentally testable hypothesis has recently been put forward (Tanabe et al. 2019).

Neither the Dsr nor the sHdr pathway is confined to sulphur oxidizers with the capacity for depositing intracellular sulphur globules. Both pathways also occur in sulphur oxidizers that do not form sulphur deposits, e.g. species of the genera *Thiobacillus* or *Acidithiobacillus* (Quatrini et al. 2009; Beller et al. 2006a, b). The Dsr and sHdr pathways occur virtually exclusively. Only very few organisms bear the genetic potential for both oxidation routes (Berben et al. 2019; Koch and Dahl 2018).

Sulphite is usually oxidized to the final product sulphate. All the sulphur-storing organisms tabulated in Table 1 that contain sulphite-generating enzyme systems in the cytoplasm also have the ability to oxidize sulphite in this compartment. Again, two pathways exist (Fig. 3) that can either occur individually or in parallel. The first pathway involves direct oxidation of sulphite to sulphate via the cytoplasm-oriented membrane-bound iron-sulphur molybdoenzyme SoeABC (Dahl et al. 2013). The second pathway proceeds via formation of the intermediate adenosine-5-′-phosphosulphate (APS) and is catalyzed by APS reductase and ATP sulphurylase (Sat) (Dahl 1996; Parey et al. 2013). In *A. vinosum*, the periplasmic substrate-binding protein SoxYZ is needed in parallel to the cytoplasmic enzymes for effective sulphite oxidation (Dahl et al. 2013). Whether this also applies to other sulphur-oxidizing bacteria has not been elucidated.

9.2 *Reductive Degradation*

In purple sulphur bacteria, sulphur globules serve as an electron acceptor reserve that allow rudimentary anaerobic respiration under anoxic conditions leading to production of sulphide (van Gemerden 1968). *Beggiatoa* OH-75-2a used sulphur globules that were accumulated during aerobic thiosulphate oxidation to sustain anaerobic metabolism and several days of anoxia. Reduction of stored sulphur to sulphide with concomitant de novo synthesis of cell material was also found during anoxic incubation of *Beggiatoa alba* BL18LD. Furthermore, elemental sulphur stored as globules in thioautotrophic symbionts may serve as an electron sink, leading to production of sulphide during temporary anoxia (Gardebrecht et al. 2012; Arndt et al. 2001; Duplessis et al. 2004). Similar processes have been suggested for the sulphur globules in those organisms that lack enzymes to further oxidize stored sulphur, i.e. Dsr or sHdr systems, as is the case for *Thiovulum* for example (Table 1). *Thiovulum* may have to oscillate between an aerobic mode of energy conservation in which elemental sulphur accumulates in the cell and an anaerobic mode of energy conservation in which intracellular sulphur serves as an electron acceptor, perhaps with formate acting as an electron donor or via anaerobic sulphur disproportionation (Marshall et al. 2012). The mechanisms underlying reductive degradation of stored sulphur are unresolved.

10 Outlook

Much remains to be learned on bacterial sulphur globules. This especially applies to the abundance, function and structure of the proteins in the globule envelopes. As evident from Table 1, most—if not all—organisms depositing intracellular sulphur encode periplasmic sulphur globule proteins; however, the only organisms for which the proteins have been unambiguously identified are *A. vinosum* and *Thiocapsa roseopersicina* (Brune 1995a). None of the proteins have been structurally characterized nor have their interactions been analysed. Further research should also finally clarify the question whether any other proteins involved in formation or degradation of the globules may be specifically attached to the protein envelope.

Acknowledgements This work was supported by the Deutsche Forschungsgemeinschaft (grants Da 351/6-2 and Da 351/8-1).

References

Abouna S, Gonzalez-Rizzo S, Grimonprez A, Gros O (2015) First description of sulphur-oxidizing bacterial symbiosis in a cnidarian (Medusozoa) living in sulphidic shallow-water environments. PLoS One 10(5):e0127625. https://doi.org/10.1371/journal.pone.0127625

Ahn AC, Meier-Kolthoff JP, Overmars L, Richter M, Woyke T, Sorokin DY, Muyzer G (2017) Genomic diversity within the haloalkaliphilic genus *Thioalkalivibrio*. PLoS One 12(3): e0173517. https://doi.org/10.1371/journal.pone.0173517

Arndt C, Gaill F, Felbeck H (2001) Anaerobic sulfur metabolism in thiotrophic symbioses. J Exp Biol 204:741–750

Bamford VA, Bruno S, Rasmussen T, Appia-Ayme C, Cheesman MR, Berks BC, Hemmings AM (2002) Structural basis for the oxidation of thiosulfate by a sulfur cycle enzyme. EMBO J 21 (21):5599–5610

Bazylinski DA, Williams TJ (2006) Ecophysiology of magnetotactic bacteria. In: Schüler D (ed) Magnetoreception and magnetosomes in bacteria, Microbiology monographs, vol 3. Springer, Berlin, pp 37–75

Bazylinski DA, Dean AJ, Williams TJ, Long LK, Middleton SL, Dubbels BL (2004) Chemolithoautotrophy in the marine, magnetotactic bacterial strains MV-1 and MV-2. Arch Microbiol 182(5):373–387

Bazylinski DA, Williams TJ, Lefevre CT, Trubitsyn D, Fang J, Beveridge TJ, Moskowitz BM, Ward B, Schubbe S, Dubbels BL, Simpson B (2013) *Magnetovibrio blakemorei* gen. nov., sp. nov., a magnetotactic bacterium (Alphaproteobacteria: Rhodospirillaceae) isolated from a salt marsh. Int J Syst Evol Microbiol 63(Pt 5):1824–1833. https://doi.org/10.1099/ijs.0.044453-0

Bazylinski DA, Morillo V, Lefevre CT, Viloria N, Dubbels BL, Williams TJ (2017) *Endothiovibrio diazotrophicus* gen. nov., sp. nov., a novel nitrogen-fixing, sulfur-oxidizing gammaproteobacterium isolated from a salt marsh. Int J Syst Evol Microbiol 67(5):1491–1498. https://doi.org/10.1099/ijsem.0.001743

Beller HR, Chai PSG, Letain TE, Chakicherla A, Larimer FW, Richardson PM, Coleman MA, Wood AP, Kelly DP (2006a) The genome sequence of the obligately chemolithoautotrophic, facultatively anaerobic bacterium *Thiobacillus denitrificans*. J Bacteriol 188(4):1473–1488

Beller HR, Letain TE, Chakicherla A, Kane SR, Legler TC, Coleman MA (2006b) Whole-genome transcriptional analysis of chemolithoautotrophic thiosulfate oxidation by *Thiobacillus denitrificans* under aerobic versus denitrifying conditions. J Bacteriol 188(19):7005–7015

Berben T, Sorokin DY, Ivanova N, Pati A, Kyrpides N, Goodwin LA, Woyke T, Muyzer G (2015) Complete genome sequence of *Thioalkalivibrio paradoxus* type strain ARh 1^T, an obligately chemolithoautotrophic haloalkaliphilic sulfur-oxidizing bacterium isolated from a Kenyan soda lake. Stand Genomic Sci 10:105. https://doi.org/10.1186/s40793-015-0097-7

Berben T, Overmars L, Sorokin DY, Muyzer G (2017) Comparative genome analysis of three thiocyanate oxidizing *Thioalkalivibrio* species isolated from soda lakes. Front Microbiol 8:254. https://doi.org/10.3389/fmicb.2017.00254

Berben T, Overmars L, Sorokin DY, Muyzer G (2019) Diversity and distribution of sulfur oxidation-related genes in *Thioalkalivibrio*, a genus of chemolithoautotrophic and haloalkaliphilic sulfur-oxidizing bacteria. Front Microbiol 10:160. https://doi.org/10.3389/fmicb.2019.00160

Berg JS, Schwedt A, Kreutzmann AC, Kuypers MM, Milucka J (2014) Polysulfides as intermediates in the oxidation of sulfide to sulfate by *Beggiatoa* spp. Appl Environ Microbiol 80 (2):629–636. https://doi.org/10.1128/AEM.02852-13

Bergin C, Wentrup C, Brewig N, Blazejak A, Erseus C, Giere O, Schmid M, De Wit P, Dubilier N (2018) Acquisition of a novel sulfur-oxidizing symbiont in the gutless marine worm *Inanidrilus exumae*. Appl Environ Microbiol 84(7):e02267–e02217. https://doi.org/10.1128/AEM.02267-17

Bland JA, Staley JT (1978) Observations on the biology of *Thiothrix*. Arch Microbiol 117:79–87

Boden R, Scott KM (2018) Evaluation of the genus *Thiothrix* Winogradsky 1888 (approved lists 1980) emend. Aruga et al. 2002: reclassification of *Thiothrix disciformis* to *Thiolinea disciformis* gen. nov., comb. nov., and of *Thiothrix flexilis* to *Thiofilum flexile* gen. nov., comb nov., with emended description of *Thiothrix*. Int J Syst Evol Microbiol 68 (7):2226–2239. https://doi.org/10.1099/ijsem.0.002816

Bosshard HR, Davidson MW, Knaff DB, Millett F (1986) Complex formation and electron transfer between mitochondrial cytochrome *c* and flavocytochrome c_{552} from *Chromatium vinosum*. J Biol Chem 261:190–193

Bright M, Sorgo A (2003) Ultrastructural reinvestigation of the trophosome in adults of *Riftia pachyptila* (Annelida, Siboglinidae). Invertebr Biol 122(4):345–366

Brune DC (1995a) Isolation and characterization of sulfur globule proteins from *Chromatium vinosum* and *Thiocapsa roseopersicina*. Arch Microbiol 163:391–399

Brune DC (1995b) Sulfur compounds as photosynthetic electron donors. In: Blankenship RE, Madigan MT, Bauer CE (eds) Anoxygenic photosynthetic bacteria, Advances in photosynthesis, vol 2. Kluwer Academic Publishers, Dordrecht, pp 847–870

Bryantseva IA, Gorlenko VM, Kompantseva EI, Imhoff JF, Sling J, Mityushina L (1999) *Thiorhodospira sibirica* gen. nov., sp. nov., a new alkaliphilic purple sulfur bacterium from a Siberian soda lake. Int J Syst Bacteriol 49:697–703. https://doi.org/10.1099/00207713-49-2-697

Cao X, Koch T, Steffens L, Finkensieper J, Zigann R, Cronan JE, Dahl C (2018) Lipoate-binding proteins and specific lipoate-protein ligases in microbial sulfur oxidation reveal an atpyical role for an old cofactor. eLife 7:e37439. https://doi.org/10.7554/eLife.37439

Castillo MCG, Lou BS, Ondrias MR, Robertson DE, Knaff DB (1994) Characterization of flavocytochrome c_{552} from the thermophilic photosynthetic bacterium *Chromatium tepidum*. Arch Biochem Biophys 315:262–266

Cavanaugh CM (1983) Symbiontic chemoautotrophic bacteria in marine invertebrates from sulfide-rich habitats. Nature 302:58–61

Cavanaugh CM, Gardiner SL, Jones ML, Jannasch HW, Waterbury JB (1981) Prokaryotic cells in the hydrothermal vent tube worm *Riftia pachyptila* Jones: possible chemoautotrophic symbionts. Science 213:340–342

Chernousova E, Gridneva E, Grabovich M, Dubinina G, Akimov V, Rossetti S, Kuever J (2009) *Thiothrix caldifontis* sp. nov. and *Thiothrix lacustris* sp. nov., gammaproteobacteria isolated from sulfide springs. Int J Syst Evol Microbiol 59(12):3128–3135. https://doi.org/10.1099/ijs.0.009456-0

Cohn F (1875) Untersuchungen über Bakterien II. Beitr z Biol d Pflanzen 1:141–207

Condit CM, Meagher RB (1986) A gene encoding a novel glycine-rich structural protein of petunia. Nature 323:178–181. https://doi.org/10.1038/323178a0

Cort JR, Selan UM, Schulte A, Grimm F, Kennedy MA, Dahl C (2008) *Allochromatium vinosum* DsrC: solution-state NMR structure, redox properties and interaction with DsrEFH, a protein essential for purple sulfur bacterial sulfur oxidation. J Mol Biol 382:692–707

Cosmidis J, Nims CW, Diercks D, Templeton AS (2019) Formation and stabilization of elemental sulfur through organomineralization. Geochim Cosmochim Acta 247:59–82. https://doi.org/10.1016/j.gca.2018.12.025

Dahl C (1996) Insertional gene inactivation in a phototrophic sulphur bacterium: APS-reductase-deficient mutants of *Chromatium vinosum*. Microbiology 142:3363–3372. https://doi.org/10.1099/13500872-142-12-3363

Dahl C (2015) Cytoplasmic sulfur trafficking in sulfur-oxidizing prokaryotes. IUBMB Life 67 (4):268–274. https://doi.org/10.1002/iub.1371

Dahl C (2017) Sulfur metabolism in phototrophic bacteria. In: Hallenbeck PC (ed) Modern topics in the phototrophic prokaryotes: metabolism, bioenergetics and omics. Springer International Publishing, Cham, pp 27–66. https://doi.org/10.1007/978-3-319-51365-2_2

Dahl C, Prange A (2006) Bacterial sulfur globules: occurrence, structure and metabolism. In: Shively JM (ed) Inclusions in prokaryotes, Microbiology monographs, vol 1. Springer, Berlin, Heidelberg, pp 21–51

Dahl C, Engels S, Pott-Sperling AS, Schulte A, Sander J, Lübbe Y, Deuster O, Brune DC (2005) Novel genes of the *dsr* gene cluster and evidence for close interaction of Dsr proteins during sulfur oxidation in the phototrophic sulfur bacterium *Allochromatium vinosum*. J Bacteriol 187 (4):1392–1404

Dahl C, Friedrich CG, Kletzin A (2008) Sulfur oxidation in prokaryotes. In: Encyclopedia of life sciences (ELS). John Wiley & Sons, Chichester., http://www.els.net/. https://doi.org/10.1002/9780470015902.a9780470021155

Dahl C, Franz B, Hensen D, Kesselheim A, Zigann R (2013) Sulfite oxidation in the purple sulfur bacterium *Allochromatium vinosum*: identification of SoeABC as a major player and relevance of SoxYZ in the process. Microbiology 159:2626–2638

de Albuquerque JP, Keim CN, Lins U (2010) Comparative analysis of *Beggiatoa* from hypersaline and marine environments. Micron 41(5):507–517. https://doi.org/10.1016/j.micron.2010.01.009

Distel DL (1998) Evolution of chemoautotrophic endosymbioses in bivalves. Bioscience 48 (4):277–286

Dubinina GA, Grabovich MY (1984) Isolation, cultivation and characteristics of *Macromonas bipunctata*. Mikrobiologiya 53:748–755

Dubinina G, Savvichev A, Orlova M, Gavrish E, Verbarg S, Grabovich M (2017) *Beggiatoa leptomitoformis* sp. nov., the first freshwater member of the genus capable of chemolithoautotrophic growth. Int J Syst Evol Microbiol 67(2):197–204. https://doi.org/10.1099/ijsem.0.001584

Duplessis MR, Ziebis W, Gros O, Caro A, Robidart J, Felbeck H (2004) Respiration strategies utilized by the gill endosymbiont from the host lucinid *Codakia orbicularis* (Bivalvia: Lucinidae). Appl Environ Microbiol 70(7):4144–4150

Ehrenberg CG (1838) Die Infusionsthierchen als vollkommene Organismen. Ein Blick in das tiefere organische Leben der Natur. Leopold Voss-Verlag, Leipzig

Eichinger I, Schmitz-Esser S, Schmid M, Fisher CR, Bright M (2014) Symbiont-driven sulfur crystal formation in a thiotrophic symbiosis from deep-sea hydrocarbon seeps. Environ Microbiol Rep 6(4):364–372. https://doi.org/10.1111/1758-2229.12149

Felbeck H (1981) Chemoautotrophic potential of the hydrothermal vent tube worm, *Riftia pachyptila* Jones (Vestimenifera). Science 213:336–338

Flood BE, Jones DS, Bailey JV (2015) *Sedimenticola thiotaurini* sp. nov., a sulfur-oxidizing bacterium isolated from salt marsh sediments, and emended descriptions of the genus *Sedimenticola* and *Sedimenticola selenatireducens*. Int J Syst Evol Microbiol 65 (8):2522–2530. https://doi.org/10.1099/ijs.0.000295

Flood BE, Fliss P, Jones DS, Dick GJ, Jain S, Kaster AK, Winkel M, Mußmann M, Bailey JL (2016) Single-cell (meta-)genomics of a dimorphic *Candidatus* Thiomargarita nelsonii reveals genomic plasticity. Front Microbiol 3(7):602. https://doi.org/10.3389/fmicb.2016.00603

Franz B, Lichtenberg H, Hormes J, Modrow H, Dahl C, Prange A (2007) Utilization of solid "elemental" sulfur by the phototrophic purple sulfur bacterium *Allochromatium vinosum*: a sulfur K-edge XANES spectroscopy study. Microbiology 153:1268–1274

Frenkiel L, Gros O, Mouëza M (1996) Gill ultrastructure in *Lucina pectinata* (Bivalvia: Lucinidae) with reference to hemoglobin in bivalves with symbiotic sulphur-oxidizing bacteria. Mar Biol 125:511–524

Friedrich CG (1998) Physiology and genetics of sulfur-oxidizing bacteria. Adv Microb Physiol 39:235–289

Friedrich CG, Bardischewsky F, Rother D, Quentmeier A, Fischer J (2005) Prokaryotic sulfur oxidation. Curr Opin Microbiol 8(3):253–259

Frigaard NU, Dahl C (2009) Sulfur metabolism in phototrophic sulfur bacteria. Adv Microb Physiol 54:103–200

Frolov EN, Belousova EV, Lavrinenko K, Dubinina GA, Grabovich MY (2013) Capacity of *Azospirillum thiophilum* for lithotrophic growth coupled to oxidation of reduced sulfur compounds. Microbiology 82(3):271–279

Fukumori Y, Yamanaka T (1979) Flavocytochrome *c* of *Chromatium vinosum*. J Biochem 85:1405–1414

Gardebrecht A, Markert S, Sievert SM, Felbeck H, Thürmer A, Albrecht D, Wollherr A, Kabisch J, Le Bris N, Lehmann R, Daniel R, Liesegang H, Hecker M, Schweder T (2012) Physiological homogeneity among the endosymbionts of *Riftia pachyptila* and *Tevnia jerichonana* revealed by proteogenomics. ISME J 6(4):766–776. https://doi.org/10.1038/ismej.2011.137

George GN, Pickering IJ, Yu EY, Prince RC (2002) X-ray absorption spectroscopy of bacterial sulfur globules. Microbiology 148:2267–2268

George GN, Gnida M, Bazylinski DA, Prince RC, Pickering IJ (2008) X-ray absorption spectroscopy as a probe of microbial sulfur biochemistry: the nature of bacterial sulfur globules revisited. J Bacteriol 190(19):6376–6383. https://doi.org/10.1128/JB.00539-08

Goffredi SK, Barry JP (2002) Species-specific variation in sulfide physiology between closely related Vesicomyid clams. Mar Ecol Prog Ser 225:227–238

Grabarczyk DB, Berks BC (2017) Intermediates in the sox sulfur oxidation pathway are bound to a sulfane conjugate of the carrier protein SoxYZ. PLoS One 12(3):e0173395. https://doi.org/10.1371/journal.pone.0173395

Grabarczyk DB, Chappell PE, Johnson S, Stelzl LS, Lea SM, Berks BC (2015) Structural basis for specificity and promiscuity in a carrier protein/enzyme system from the sulfur cycle. Proc Natl Acad Sci U S A 112(52):E7166–E7175

Grimonprez A, Molza A, Laurent MCZ, Mansot JL, Gros O (2018) Thioautotrophic ectosymbiosis in *Pseudovorticella* sp., a peritrich ciliate species colonizing wood falls in marine mangrove. Eur J Protistol 62:43–55. https://doi.org/10.1016/j.ejop.2017.11.002

Guerrero R, Mas J, Pedros-Alio C (1984) Boyant density changes due to intracellular content of sulfur in *Chromatium warmingii* and *Chromatium vinosum*. Arch Microbiol 137:350–356

Hageage GJ Jr, Eanes ED, Gherna RL (1970) X-ray diffraction studies of the sulfur globules accumulated by *Chromatium* species. J Bacteriol 101:464–469

Head IM, Gray ND, Clarke KJ, Pickup RW, Jones JG (1996) The phylogenetic position and ultrastructure of the uncultured bacterium *Achromatium oxaliferum*. Microbiology 142:2341–2354

Henkel JV, Dellwig O, Pollehne F, Herlemann DPR, Leipe T, Schulz-Vogt HN (2019) A bacterial isolate from the Black Sea oxidizes sulfide with manganese(IV) oxide. Proc Natl Acad Sci U S A 116(25):12153–12155. https://doi.org/10.1073/pnas.1906000116

Himmel D, Maurin LC, Gros O, Mansot JL (2009) Raman microspectrometry sulfur detection and characterization in the marine ectosymbiotic nematode *Eubostrichus dianae* (Desmodoridae, Stilbonematidae). Biol Cell 101(1):43–54. https://doi.org/10.1042/BC20080051

Jendrossek D (2009) Polyhydroxyalkanoate granules are complex subcellular organelles (carbonosomes). J Bacteriol 191(10):3195–3202

Jørgensen BB, Nelson DC (2004) Sulfide oxidation in marine sediments: geochemistry meets microbiology. In: Sulfur biochemistry – past and present. Geological Society of America, Boulder, Colorado, pp 63–81

Kamyshny A, Ferdelman TG (2010) Dynamics of zero-valent sulfur species including polysulfides at seep sites on intertidal sand flats (Wadden Sea, North Sea). Mar Chem 121(1–4):17–26. https://doi.org/10.1016/j.marchem.2010.03.001

Kamyshny A, Druschel G, Mansaray ZF, Farquhar J (2014) Multiple sulfur isotopes fractionations associated with abiotic sulfur transformations in Yellowstone National Park geothermal springs. Geochem T 15(1):7. https://doi.org/10.1186/1467-4866-15-7

Kaster AK, Moll J, Parey K, Thauer RK (2011) Coupling of ferredoxin and heterodisulfide reduction via electron bifurcation in hydrogenotrophic methanogenic archaea. Proc Natl Acad Sci U S A 108(7):2981–2986

Keim CN, Solorzano G, Farina M, Lins U (2005) Intracellular inclusions of uncultured magnetotactic bacteria. Int Microbiol 8(2):111–117

Keller B, Sauer N, Lamb CJ (1988) Glycine-rich cell wall proteins in bean: gene structure and association of the protein with the vascular system. EMBO J 7:3625–3633. https://doi.org/10.1002/j.1460-2075.1988.tb03243.x

Kleiner M, Petersen JM, Dubilier N (2012) Convergent and divergent evolution of metabolism in sulfur-oxidizing symbionts and the role of horizontal gene transfer. Curr Opin Microbiol 15 (5):621–631. https://doi.org/10.1016/j.mib.2012.09.003

Kleinjan WE, de Keizer A, Janssen AJH (2003) Biologically produced sulfur. In: Steudel R (ed) Elemental sulfur and sulfur-rich compounds I. Springer, Berlin, pp 167–187

Kletzin A, Urich T, Müller F, Bandeiras TM, Gomes CM (2004) Dissimilatory oxidation and reduction of elemental sulfur in thermophilic archaea. J Bioenerg Biomembr 36:77–91

Koch T, Dahl C (2018) A novel bacterial sulfur oxidation pathway provides a new link between the cycles of organic and inorganic sulfur compounds. ISME J 12(10):2479–2491. https://doi.org/10.1038/s41396-018-0209-7

Kran G, Schlote FW, Schlegel HG (1963) Cytologische Untersuchungen an *Chromatium okenii* Perty. Naturwissenschaften 50:728–730

Krieger J, Giere O, Dubilier N (2000) Localization of RubisCO and sulfur in endosymbiontic bacteria of the gutless marine oligochaete *Inanidrilus leukodermatus* (Annelida). Mar Biol 137:239–244

Kwak Y, Shin JH (2016) First *Azospirillum* genome from aquatic environments: whole-genome sequence of *Azospirillum thiophilum* BV-S^T, a novel diazotroph harboring a capacity of sulfur-chemolithotrophy from a sulfide spring. Mar Genomics 25:21–24. https://doi.org/10.1016/j.margen.2015.11.001

La Riviere JWM, Schmidt K (1999) Morphologically conspicuous sulfur-oxidizing eubacteria. In: Dworkin M (ed) The prokaryotes: an evolving electronic resource for the microbiological community, vol 3. Springer, New York. http://link.springer-ny.com/link/service/books/10125/

Larkin JM, Shinabarger DL (1983) Characterization of *Thiothrix nivea*. Int J Syst Bacteriol 33:841–846
Larkin JM, Strohl WR (1983) *Beggiatoa*, *Thiothrix*, and *Thioploca*. Annu Rev Microbiol 37:341–367
Laska S, Lottspeich F, Kletzin A (2003) Membrane-bound hydrogenase and sulfur reductase of the hyperthermophilic and acidophilic archaeon *Acidianus ambivalens*. Microbiology 149:2357–2371
Lau GE, Cosmidis J, Grasby SE, Trivedi CB, Spear JR, Templeton AS (2017) Low-temperature formation and stabilization of rare allotropes of cyclooctasulfur (β-S8 and γ-S8) in the presence of organic carbon at a sulfur-rich glacial site in the Canadian high Arctic. Geochim Cosmochim Acta 200:218–231. https://doi.org/10.1016/j.gca.2016.11.036
Lavrinenko K, Chernousova E, Gridneva E, Dubinina G, Akimov V, Kuever J, Lysenko A, Grabovich M (2010) *Azospirillum thiophilum* sp. nov., a diazotrophic bacterium isolated from a sulfide spring. Int J Syst Evol Microbiol 60(Pt 12):2832–2837. https://doi.org/10.1099/ijs.0.018853-0
Lee YJ, Prange A, Lichtenberg H, Rohde M, Dashti M, Wiegel J (2007) In situ analysis of sulfur species in sulfur globules produced from thiosulfate by *Thermoanaerobacter sulfurigignens* and *Thermoanaerobacterium thermosulfurigenes*. J Bacteriol 189(20):7525–7529
Lefevre CT, Viloria N, Schmidt ML, Posfai M, Frankel RB, Bazylinski DA (2012) Novel magnetite-producing magnetotactic bacteria belonging to the Gammaproteobacteria. ISME J 6 (2):440–450. https://doi.org/10.1038/ismej.2011.97
Lin L, Thanbichler M (2013) Nucleotide-independent cytoskeletal scaffolds in bacteria. Cytoskeleton 70(8):409–423
Liu LJ, Stockdreher Y, Koch T, Sun ST, Fan Z, Josten M, Sahl HG, Wang Q, Luo YM, Liu SJ, Dahl C, Jiang CY (2014) Thiosulfate transfer mediated by DsrE/TusA homologs from acidothermophilic sulfur-oxidizing archaeon *Metallosphaera cuprina*. J Biol Chem 289 (39):26949–26959. https://doi.org/10.1074/jbc.M114.591669
Löffler M, Feldhues J, Venceslau SS, Kammler L, Grein F, IAC P, Dahl C (2020) DsrL mediates electron transfer between NADH and rDsrAB in *Allochromatium vinosum*. Environ Microbiol 22(2):783–795. https://doi.org/10.1111/1462-2920.14899
Luther GW 3rd, Findlay AJ, Macdonald DJ, Owings SM, Hanson TE, Beinart RA, Girguis PR (2011) Thermodynamics and kinetics of sulfide oxidation by oxygen: a look at inorganically controlled reactions and biologically mediated processes in the environment. Front Microbiol 2:62. https://doi.org/10.3389/fmicb.2011.00062
Macur RE, Jay ZJ, Taylor WP, Kozubal MA, Kocar BD, Inskeep WP (2013) Microbial community structure and sulfur biogeochemistry in mildly-acidic sulfidic geothermal springs in Yellowstone National Park. Geobiology 11(1):86–99. https://doi.org/10.1111/gbi.12015
Maier S, Murray RG (1965) The fine structure of *Thioploca ingrica* and a comparison with *Beggiatoa*. Can J Microbiol 11:645–655
Maina JN, Maloyi GMO (1998) Adaptations of a tropical swamp worm, *Alma emini*, for subsistence in a H_2S-rich habitat: evolution of endosymbiontic bacteria, sulfide-metabolizing bodies, and a novel process of elimination of neutralized sulfide complexes. J Struct Biol 122(3):257–266
Maki JS (2013) Bacterial intracellular sulfur globules: structure and function. J Mol Microbiol Biotechnol 23(4–5):270–280
Mangold S, Valdés J, Holmes DS, Dopson M (2011) Sulfur metabolism in the extreme acidophile *Acidithiobacillus caldus*. Front Microbiol 2:17. https://doi.org/10.3389/fmcib.2011.00017
Mansor M, Hamilton TL, Fantle MS, Macalady JL (2015) Metabolic diversity and ecological niches of *Achromatium* populations revealed with single-cell genomic sequencing. Front Microbiol 6:822. https://doi.org/10.3389/fmicb.2015.00822
Marcia M, Ermler U, Peng GH, Michel H (2009) The structure of *Aquifex aeolicus* sulfide:quinone oxidoreductase, a basis to understand sulfide detoxification and respiration. Proc Natl Acad Sci U S A 106(24):9625–9630

Marcia M, Ermler U, Peng GH, Michel H (2010a) A new structure-based classification of sulfide: quinone oxidoreductases. Proteins: Struct Funct Bioinf 78(5):1073–1083

Marcia M, Langer JD, Parcej D, Vogel V, Peng GH, Michel H (2010b) Characterizing a monotopic membrane enzyme. Biochemical, enzymatic and crystallization studies on *Aquifex aeolicus* sulfide:quinone oxidoreductase. BBA-Biomembranes 1798(11):2114–2123

Markert S, Arndt C, Felbeck H, Becher D, Sievert SM, Hugler M, Albrecht D, Robidart J, Bench S, Feldman RA, Hecker M, Schweder T (2007) Physiological proteomics of the uncultured endosymbiont of *Riftia pachyptila*. Science 315(5809):247–250

Markert S, Gardebrecht A, Felbeck H, Sievert SM, Klose J, Becher D, Albrecht D, Thurmer A, Daniel R, Kleiner M, Hecker M, Schweder T (2011) Status quo in physiological proteomics of the uncultured *Riftia pachyptila* endosymbiont. Proteomics 11(15):3106–3117

Marshall IP, Blainey PC, Spormann AM, Quake SR (2012) A single-cell genome for *Thiovulum* sp. Appl Environ Microbiol 78(24):8555–8563. https://doi.org/10.1128/AEM.02314-12

Maurin LC, Himmel D, Mansot JL, Gros O (2010) Raman microspectrometry as a powerful tool for a quick screening of thiotrophy: an application on mangrove swamp meiofauna of Guadeloupe (F.W.I.). Mar Environ Res 69(5):382–389. https://doi.org/10.1016/j.marenvres.2010.02.001

Meyer B (1976) Elemental sulfur. Chem Rev 76(3):367–388. https://doi.org/10.1021/cr60301a003

Mu T, Zhou J, Yang M, Xing J (2016) Complete genome sequence of *Thioalkalivibrio versutus* D301 isolated from soda Lake in northern China, a typical strain with great ability to oxidize sulfide. J Biotechnol 227:21–22. https://doi.org/10.1016/j.jbiotec.2016.04.019

Müller C (1870) Chemisch-physikalische Beschreibung der Thermen von Baden in der Schweiz (Canton Aargau). Zehnder, Baden

Müller OF (1786) Animalcula infusoria fluviatilia et marina, quae detexit, systematice descripsit et ad vivum delineari curavit. Mülleri, Hauniae

Nelson DC, Castenholz RW (1981) Use of reduced sulfur compounds by *Beggiatoa* sp. J Bacteriol 147:140–154

Nelson DC, Fisher CR (1995) Chemoautotrophic and methanoautotrophic endosymbiontic bacteria at deep-sea vents and seeps. In: Karl DM (ed) Deep-sea hydrothermal vents. CRC Press, Boca Raton, FL, pp 125–167

Nicolson GL, Schmidt GL (1971) Structure of the *Chromatium* sulfur particle and its protein membrane. J Bacteriol 105:1142–1148

Nims C, Cron B, Wetherington M, Macalady J, Cosmidis J (2019) Low frequency Raman spectroscopy for micron-scale and in vivo characterization of elemental sulfur in microbial samples. Sci Rep 9(1):7971. https://doi.org/10.1038/s41598-019-44353-6

Nunoura T, Takaki Y, Kazama H, Kakuta J, Shimamura S, Makita H, Hirai M, Miyazaki M, Takai K (2014) Physiological and genomic features of a novel sulfur-oxidizing Gammaproteobacterium belonging to a previously uncultivated symbiotic lineage isolated from a hydrothermal vent. PLoS One 9(8):e104959. https://doi.org/10.1371/journal.pone.0104959

Odintsova EV, Jannasch H, Mamone JA, Langworthy TA (1996) *Thermothrix azorensis* sp. nov., an obligately chemolithoautotrophic, sulfur-oxidizing, thermophilic bacterium. Int J Syst Bacteriol 46(2):422–428

Ogawa T, Furusawa T, Nomura R, Seo D, Hosoya-Matsuda N, Sakurai H, Inoue K (2008) SoxAX binding protein, a novel component of the thiosulfate-oxidizing multienzyme system in the green sulfur bacterium Chlorobium tepidum. J Bacteriol 190(18):6097–6110

Ogawa T, Noguchi K, Saito M, Nagahata Y, Kato H, Ohtaki A, Nakayama H, Dohmae N, Matsushita Y, Odaka M, Yohda M, Nyunoya H, Katayama Y (2013) Carbonyl sulfide hydrolase from T*hiobacillus thioparus* strain THI115 is one of the beta-carbonic anhydrase family enzymes. J Am Chem Soc 135(10):3818–3825. https://doi.org/10.1021/ja307735e

Oren A, Mana L, Jehlicka J (2015) Probing single cells of purple sulfur bacteria with Raman spectroscopy: carotenoids and elemental sulfur. FEMS Microbiol Lett 362(6):fnv021. https://doi.org/10.1093/femsle/fnv021

Ott J, Bright M, Bulgheresi S (2004) Marine microbial thiotrophic ectosymbioses. In: Gibson RN, RJA A, JDM G (eds) Oceanography and marine biology – an annual review, vol 42. CRC Press, Boca Raton, pp 95–118

Overmann J (1997) Mahoney Lake: a case study of the ecological significance of phototrophic sulfur bacteria. Adv Microb Ecol 15:251–288

Parey K, Demmer U, Warkentin E, Wynen A, Ermler U, Dahl C (2013) Structural, biochemical and genetic characterization of ATP sulfurylase from *Allochromatium vinosum*. PLoS One 8(9): e74707

Pasteris JD, Freeman JJ, Goffredi SK, Buck KR (2001) Raman spectroscopic and laser confocal microscopic analysis of sulfur in living sulfur-precipitating marine bacteria. Chem Geol 180 (1–4):3–18

Pattaragulwanit K, Dahl C (1995) Development of a genetic system for a purple sulfur bacterium: conjugative plasmid transfer in *Chromatium vinosum*. Arch Microbiol 164:217–222

Pattaragulwanit K, Brune DC, Trüper HG, Dahl C (1998) Molecular genetic evidence for extracytoplasmic localization of sulfur globules in *Chromatium vinosum*. Arch Microbiol 169:434–444

Perty M (1852) Zur Kenntnis kleinster Lebensformen nach Bau, Funktionen, Systematik, mit Spezialverzeichnis der in der Schweiz beobachteten. Jent und Reinert, Bern

Petersen JM, Kemper A, Gruber-Vodicka H, Cardini U, van der Geest M, Kleiner M, Bulgheresi S, Mussmann M, Herbold C, Seah BK, Antony CP, Liu D, Belitz A, Weber M (2016) Chemosynthetic symbionts of marine invertebrate animals are capable of nitrogen fixation. Nat Microbiol 2:16195. https://doi.org/10.1038/nmicrobiol.2016.195

Pfennig N, Trüper HG (1971) Type and neotype strains of the species of phototrophic bacteria maintained in pure culture. Int J Syst Bacteriol 21:19–24

Pickering TJ, Prince RC, Divers T, George GN (1998) Sulfur K-edge X-ray absorption spectroscopy for determining the chemical speciation of sulfur in biological systems. FEBS Lett 441:11–14

Pickering IJ, George GN, Yu EY, Brune DC, Tuschak C, Overmann J, Beatty JT, Prince RC (2001) Analysis of sulfur biochemistry of sulfur bacteria using x-ray absorption spectroscopy. Biochemistry 40:8138–8145

Pott AS, Dahl C (1998) Sirohaem-sulfite reductase and other proteins encoded in the *dsr* locus of *Chromatium vinosum* are involved in the oxidation of intracellular sulfur. Microbiology 144:1881–1894

Prange A, Arzberger I, Engemann C, Modrow H, Schumann O, Trüper HG, Steudel R, Dahl C, Hormes J (1999a) In situ analysis of sulfur in the sulfur globules of phototrophic sulfur bacteria by X-ray absorption near edge spectroscopy. Biochim Biophys Acta 1428:446–454

Prange A, Modrow H, Dahl C, Steudel R, Trüper HG, Hormes J (1999b) Structural analysis of sulfur in the sulfur globules of sulfur bacteria by X-Ray Near Edge Absorption Spectroscopy (XANES). Biospektrum Sonderausgabe zur Fr hjahrstagung der VAAM, Göttingen, 85

Prange A, Chauvistr R, Modrow H, Hormes J, Trüper HG, Dahl C (2002) Quantitative speciation of sulfur in bacterial sulfur globules: X-ray absorption spectroscopy reveals at least three different speciations of sulfur. Microbiology 148:267–276

Prange A, Engelhardt H, Trüper HG, Dahl C (2004) The role of the sulfur globule proteins of *Allochromatium vinosum*: mutagenesis of the sulfur globule protein genes and expression studies by real-time RT PCR. Arch Microbiol 182:165–174

Quatrini R, Appia-Ayme C, Denis Y, Jedlicki E, Holmes DS, Bonnefoy V (2009) Extending the models for iron and sulfur oxidation in the extreme acidophile *Acidithiobacillus ferrooxidans*. BMC Genomics 10:394. https://doi.org/10.1186/1471-2164-10-394

Rabus R, Venceslau SS, Wohlbrand L, Voordouw G, Wall JD, Pereira IA (2015) A post-genomic view of the ecophysiology, catabolism and biotechnological relevance of sulphate-reducing prokaryotes. Adv Microb Physiol 66:55–321

Reinartz M, Tschäpe J, Brüser T, Trüper HG, Dahl C (1998) Sulfide oxidation in the phototrophic sulfur bacterium *Chromatium vinosum*. Arch Microbiol 170:59–68

Remsen CC (1978) Comparative subcellular architecture of photosynthetic bacteria. In: Clayton RK, Sistrom WR (eds) The photosynthetic bacteria. Plenum Press, New York, pp 31–62

Remsen CC, Trüper HG (1973) The fine structure of *Chromatium buderi*. Arch Mikrobiol 90:269–280

Ringli C, Keller B, Ryser U (2001) Glycine-rich proteins as structural components of plant cell walls. Cell Mol Life Sci 58(10):1430–1441. https://doi.org/10.1007/PL00000786

Rinke C, Schmitz-Esser S, Stoecker K, Nussbaumer AD, Molnar DA, Vanura K, Wagner M, Horn M, Ott JA, Bright M (2006) "*Candidatus* Thiobios zoothamnicoli," an ectosymbiotic bacterium covering the giant marine ciliate *Zoothamnium niveum*. Appl Environ Microbiol 72 (3):2014–2021. https://doi.org/10.1128/AEM.72.3.2014-2021.2006

Rinke C, Schmitz-Esser S, Loy A, Horn M, Wagner M, Bright M (2009) High genetic similarity between two geographically distinct strains of the sulfur-oxidizing symbiont '*Candidatus* Thiobios zoothamnicoli'. FEMS Microbiol Ecol 67:229–241

Rivière JWM, Schmidt K (2006) Morphologically consicuous sulfur-oxidizing eubacteria. In: Dworkin M, Falkow S, Rosenberg E, Schleifer KH, Stackebrandt E (eds) The prokaryotes, 3rd edn. Springer, New York, NY

Rose A, Meier I (2004) Scaffolds, levers, rods and springs: diverse cellular functions of long coiled-coil proteins. Cell Mol Life Sci 61(16):1996–2009

Roy AB, Trudinger PA (1970) The biochemistry of inorganic compounds of sulfur. Cambridge University Press, London

Russell SL, Corbett-Detig RB, Cavanaugh CM (2017) Mixed transmission modes and dynamic genome evolution in an obligate animal-bacterial symbiosis. ISME J 11(6):1359–1371. https://doi.org/10.1038/ismej.2017.10

Sackett DL, Wolff J (1987) Nile red as a polarity-sensitive fluorescent probe of hydrophobic protein surfaces. Anal Biochem 167:228–234

Salman V, Berben T, Bowers RM, Woyke T, Teske A, Angert ER (2016) Insights into the single cell draft genome of "*Candidatus* Achromatium palustre". Stand Genomic Sci 11:28. https://doi.org/10.1186/s40793-016-0146-x

Santos AA, Venceslau SS, Grein F, Leavitt WD, Dahl C, Johnston DT, Pereira IA (2015) A protein trisulfide couples dissimilatory sulfate reduction to energy conservation. Science 350 (6267):1541–1545

Sauvé V, Bruno S, Berks BC, Hemmings AM (2007) The SoxYZ complex carries sulfur cycle intermediates on a peptide swinging arm. J Biol Chem 282(32):23194–23204. https://doi.org/10.1074/jbc.M701602200

Sauvé V, Roversi P, Leath KJ, Garman EF, Antrobus R, Lea SM, Berks BC (2009) Mechanism for the hydrolysis of a sulfur-sulfur bond based on the crystal structure of the thiosulfohydrolase SoxB. J Biol Chem 284(32):21707–21718

Schmidt GL, Kamen MD (1970) Variable cellular composition of *Chromatium* in growing cultures. Arch Mikrobiol 73:1–18

Schmidt GL, Nicolson GL, Kamen MD (1971) Composition of the sulfur particle of *Chromatium vinosum*. J Bacteriol 105:1137–1141

Schulz HN, Jørgensen BB (2001) Big bacteria. Annu Rev Microbiol 55:105–137

Schulz HN, Brinkhoff T, Ferdelman TG, Hernndez Marin M, Teske A, Jørgensen BB (1999) Dense populations of a giant sulfur bacterium in namibian shelf sediments. Science 284:493–495

Seah BKB, Antony CP, Huettel B, Zarzycki J, Schada von Borzyskowski L, Erb TJ, Kouris A, Kleiner M, Liebeke M, Dubilier N, Gruber-Vodicka HR, Giovannoni SJ (2019) Sulfur-oxidizing symbionts without canonical genes for autotrophic CO_2 fixation. mBio 10(3). https://doi.org/10.1128/mBio.01112-19

Shahak Y, Hauska G (2008) Sulfide oxidation from cyanobacteria to humans: sulfide-quinone oxidoreductase (SQR). In: Hell R, Dahl C, Knaff DB, Leustek T (eds) Sulfur metabolism in phototrophic organisms, Advances in photosynthesis and respiration, vol 27. Springer, Dordrecht, pp 319–335

Shigi N (2014) Biosynthesis and functions of sulfur modifications in tRNA. Front Genet 5:67. https://doi.org/10.3389/fgene.2014.00067

Shigi N (2018) Recent advances in our understanding of the biosynthesis of sulfur modifications in tRNAs. Front Microbiol 9:2679. https://doi.org/10.3389/fmicb.2018.02679

Shively JM, Bryant DA, Fuller RC, Konopka AE, Stevens JSE, Strohl WR (1989) Functional inclusions in prokaryotic cells. Int Rev Cytol 113:35–100

Shively JM, Cannon GC, Heinhorst S, Bryant DA, DasSarma S, Bazylinski D, Preiss J, Steinbüchel A, Docampo R, Dahl C (2006) Bacterial inclusions. In: Encyclopedia of life sciences. John Wiley & Sons, Chichester., http://www.els.net/. https://doi.org/10.1038/npg.els.0004268

Skirnisdottir S, Hreggvidsson GO, Holst O, Kristjansson JK (2001) Isolation and characterization of a mixotrophic sulfur-oxidizing *Thermus scotoductus*. Extremophiles 5(1):45–51. https://doi.org/10.1007/s007920000172

Sorokin DY, Lysenko AM, Mityushina LL, Tourova TP, Jones BE, Rainey FA, Robertson LA, Kuenen GJ (2001) *Thioalkalimicrobium aerophilum* gen. Nov., sp. nov. and *Thioalkalimicrobium sibericum* sp. nov., and *Thioalkalivibrio versutus* gen. nov., sp. nov., *Thioalkalivibrio nitratis* sp. nov. and *Thioalkalivibrio denitrificans* sp. nov., novel obligately alkaliphilic and obligately chemolithoautotrophic sulfur-oxidizing bacteria from soda lakes. Int J Syst Evol Microbiol 51:565–580

Sorokin DY, Tourova TP, Lysenko AM, Mityushina LL, Kuenen JG (2002) *Thioalkalivibrio thiocyanoxidans* sp. nov. and *Thioalkalivibrio paradoxus* sp. nov., novel alkaliphilic, obligately autotrophic, sulfur-oxidizing bacteria capable of growth on thiocyanate, from soda lakes. Int J Syst Evol Microbiol 52(Pt 2):657–664. https://doi.org/10.1099/00207713-52-2-657

Sorokin DY, Tourova TP, Sjollema KA, Kuenen JG (2003) *Thialkalivibrio nitratireducens* sp. nov., a nitrate-reducing member of an autotrophic denitrifying consortium from a soda lake. Int J Syst Evol Microbiol 53(Pt 6):1779–1783. https://doi.org/10.1099/ijs.0.02615-0

Sorokin DY, Tourova TP, Antipov AN, Muyzer G, Kuenen JG (2004) Anaerobic growth of the haloalkaliphilic denitrifying sulfur-oxidizing bacterium *Thialkalivibrio thiocyanodenitrificans* sp. nov. with thiocyanate. Microbiology 150(Pt 7):2435–2442. https://doi.org/10.1099/mic.0.27015-0

Spring S, Bazylinski DA (2000) Magnetotactic bacteria. In: Dworkin M (ed) The prokaryotes: an evolving electronic resource for the microbiological community, vol 3. Springer, New York. http://link.springer-ny.com/link/service/books/10125/

Steudel R (1982) Homocyclic sulfur molecules. Top Curr Chem 102:149–176

Steudel R (1987) Sulfur homocycles. In: Haiduc I, Sowerby DB (eds) The chemistry of inorganic homo- and heterocycles. Academic Press, London, pp 737–768

Steudel R (1989) On the nature of the "elemental sulfur" (S^0) produced by sulfur-oxidizing bacteria- a model for S^0 globules. In: Schlegel HG, Bowien B (eds) Autotrophic bacteria. Sciene Tech Publishers, Madison, WI, pp 289–303

Steudel R (1996a) Das gelbe Element und seine erstaunliche Vielseitigkeit. Chemie Unserer Zeit 30:226–234

Steudel R (1996b) Mechanism for the formation of elemental sulfur from aqueous sulfide in chemical and microbiological desulfurization processes. Ind Eng Chem Res 35:1417–1423

Steudel R (2000) The chemical sulfur cycle. In: Lens P, Hulshoff Pol W (eds) Environmental technologies to treat sulfur pollution. IWA Publishing, London, pp 1–31

Steudel R, Chivers T (2019) The role of polysulfide dianions and radical anions in the chemical, physical and biological sciences, including sulfur-based batteries. Chem Soc Rev 48 (12):3279–3319. https://doi.org/10.1039/c8cs00826d

Steudel R, Eckert B (2003) Solid sulfur allotropes. In: Steudel R (ed) Elemental sulfur and sulfur-rich compounds. Springer, Berlin, pp 1–79

Steudel R, Holz B (1988) Detection of reactive sulfur molecules (S_6, S_7, S_9, S_∞) in commercial sulfur, in sulfur minerals, and in sulfur metals slowly cooled to 20°C. Z Naturforsch B43:581–589

Steudel R, Holdt G, Visscher PT, van Gemerden H (1990) Search for polythionates in cultures of *Chromatium vinosum* after sulfide incubation. Arch Microbiol 155:432–437

Stockdreher Y, Venceslau SS, Josten M, Sahl HG, Pereira IAC, Dahl C (2012) Cytoplasmic sulfurtransferases in the purple sulfur bacterium *Allochromatium vinosum*: evidence for sulfur transfer from DsrEFH to DsrC. PLoS One 7(7):e40785

Stockdreher Y, Sturm M, Josten M, Sahl HG, Dobler N, Zigann R, Dahl C (2014) New proteins involved in sulfur trafficking in the cytoplasm of *Allochromatium vinosum*. J Biol Chem 289 (18):12390–12403. https://doi.org/10.1074/jbc.M113.536425

Strohl WR, Geffers I, Larkin JM (1981) Structure of the sulfur inclusion envelopes from four Beggiatoas. Curr Microbiol 6:75–79

Strohl WR, Howard KS, Larkin JM (1982) Ultrastructure of *Beggiatoa alba* strain B15LD. J Gen Microbiol 128:73–84

Tanabe TS, Leimkühler S, Dahl C (2019) The functional diversity of the prokaryotic sulfur carrier protein TusA. Adv Microb Physiol 75:233–277. https://doi.org/10.1016/bs.ampbs.2019.07.004

Taylor CD, Wirsen CO (1997) Microbiology and ecology of filamentous sulfur formation. Science 277(5331):1483–1485. https://doi.org/10.1126/science.277.5331.1483

Taylor CD, Wirsen CO, Gaill F (1999) Rapid microbial production of filamentous sulfur mats at hydrothermal vents. Appl Environ Microbiol 65(5):2235–2255

Trevisan V (1842) Prospetto della Flora Euganea. Coi Tipi del Seminario, Padua, pp 1–68

Trofimov BA, Sinegovskaya LM, Gusarova NK (2009) Vibrations of the S–S bond in elemental sulfur and organic polysulfides: a structural guide. J Sulfur Chem 30(5):518–554. https://doi.org/10.1080/17415990902998579

Trubitsyn D, Abreu F, Ward FB, Taylor T, Hattori M, Kondo S, Trivedi U, Staniland S, Lins U, Bazylinski DA (2016) Draft genome sequence of *Magnetovibrio blakemorei* strain MV-1, a marine vibrioid magnetotactic bacterium. Genome Announc 4(6). https://doi.org/10.1128/genomeA.01330-16

Trüper HG (2008) Sulfur and light? History and "thiology" of the phototrophic sulfur bacteria. In: Dahl C, Friedrich CG (eds) Microbial sulfur metabolism. Springer, Berlin, Heidelberg, pp 87–100

Tsallagov SI, Sorokin DY, Tikhonova TV, Popov VO, Muyzer G (2019) Comparative genomics of *Thiohalobacter thiocyanaticus* HRh1T and *Guyparkeria* sp. SCN-R1, halophilic chemolithoautotrophic sulfur-oxidizing Gammaproteobacteria capable of using thiocyanate as energy source. Front Microbiol 10:898. https://doi.org/10.3389/fmicb.2019.00898

van Gemerden H (1968) On the ATP generation by *Chromatium* in the dark. Arch Mikrobiol 64:118–124

van Niel CB (1931) On the morphology and physiology of the purple and green sulfur bacteria. Arch Mikrobiol 3:1–112

van Niel BC (1936) On the metabolism of the *Thiorhodaceae*. Arch Mikrobiol 7:323–358

Venceslau SS, Stockdreher Y, Dahl C, Pereira IAC (2014) The "bacterial heterodisulfide" DsrC is a key protein in dissimilatory sulfur metabolism. Biochim Biophys Acta 1837(7):1148–1164. https://doi.org/10.1016/j.bbabio.2014.03.007

Vetter RD (1985) Elemental sulfur in the gills of three species of clams containing chemoautotrophic symbiontic bacteria: a possible inorganic energy storage compound. Mar Biol 88:33–42

Wagner T, Koch J, Ermler U, Shima S (2017) Methanogenic heterodisulfide reductase (HdrABC-MvhAGD) uses two noncubane [4Fe-4S] clusters for reduction. Science 357(6352):699–703. https://doi.org/10.1126/science.aan0425

Waksman SA (1922) Microorganisms concerned in the oxidation of sulfur in the soil: I. Introductory. J Bacteriol 7(2):231–238

Waksman SA, Joffe JS (1922) Microorganisms concerned in the oxidation of sulfur in the soil: II. *Thiobacillus thiooxidans*, a new sulfur-oxidizing organism isolated from the soil. J Bacteriol 7(2):239–256

Wang R, Lin J-Q, Liu X-M, Pang X, Zhang C-J, Yang C-L, Gao X-Y, Lin C-M, Li Y-Q, Li Y, Lin J-Q, Chen L-X (2019) Sulfur oxidation in the acidophilic autotrophic *Acidithiobacillus* spp. Front Microbiol 9:3290. https://doi.org/10.3389/fmicb.2018.03290

Warming E (1875) Om nogle ved Danmarks kyster levede bakterier. Vidensk MeddDanNaturhistForen Khobenhavn 20–28:3–116

Weissgerber T, Zigann R, Bruce D, Chang YJ, Detter JC, Han C, Hauser L, Jeffries CD, Land M, Munk AC, Tapia R, Dahl C (2011) Complete genome sequence of *Allochromatium vinosum* DSM 180^T. Stand Genomic Sci 5(3):311–330. https://doi.org/10.4056/sigs.2335270

Weissgerber T, Dobler N, Polen T, Latus J, Stockdreher Y, Dahl C (2013) Genome-wide transcriptional profiling of the purple sulfur bacterium *Allochromatium vinosum* DSM 180^T during growth on different reduced sulfur compounds. J Bacteriol 195:4231–4245

Weissgerber T, Sylvester M, Kröninger L, Dahl C (2014) A comparative quantitative proteome study identifies new proteins relevant for sulfur oxidation in the purple sulfur bacterium *Allochromatium vinosum*. Appl Environ Microbiol 80(7):2279–2292. https://doi.org/10.1128/AEM.04182-13

Welte C, Hafner S, Krätzer C, Quentmeier AT, Friedrich CG, Dahl C (2009) Interaction between sox proteins of two physiologically distinct bacteria and a new protein involved in thiosulfate oxidation. FEBS Lett 583:1281–1286

Willems A (2014) The family *Comamonadaceae*. In: Rosenberg E, DeLong EF, Lory S, Stackebrandt E, Thompson F (eds) The prokaryotes – Alphaproteobacteria and Betaproteobacteria. Springer-Verlag, Berlin, Heidelberg, pp 777–851. https://doi.org/10.1007/978-3-642-30197-1_238

Williams TM, Unz RF, Doman T (1987) Ultrastructure of *Thiothrix* and "type 012N" bacteria. Appl Environ Microbiol 53(7):1560–1570

Williams TJ, Zhang CL, Scott JH, Bazylinski DA (2006) Evidence for autotrophy via the reverse tricarboxylic acid cycle in the marine magnetotactic coccus strain MC-1. Appl Environ Microbiol 72(2):1322–1329

Winogradsky SN (1887) Über Schwefelbakterien. Botanische Zeitung 45:489–508

Winogradsky SN (1889) Recherches physiologiques sur le sulfobactéries. Ann Inst Pasteur 3:49–60

Zopfi J, Ferdelman TG, Fossing H (2004) Distribution and fate of sulfur intermediates – sulfite, terathionate, thiosuflate and elemental sulfur – marine sediments. In: Sulfur biogeochemistry – past and present. Geological Society of America, Boulder, Colorado, pp 97–116

Biosynthesis and Intracellular Organization of Magnetosomes in Magnetotactic Bacteria

Dirk Schüler and Frank D. Müller

Contents

1 Introduction 54
2 Architecture and Biosynthesis of Magnetosomes 55
3 Formation of Magnetosome Membrane Vesicles 57
4 Biomineralization of Magnetite Crystals 58
5 Structure, Assembly, and Positioning of Magnetosome Chains 59
6 Three-dimensional Chain Positioning, Motility Axis, and Magnetotaxis 62
References 66

Abstract Magnetotactic bacteria are ubiquitous aquatic microorganisms capable of orientation within the earth's magnetic field. They receive their stunning proficiency from magnetosomes, which are unique organelles used to synthesize nanometer-sized crystals of magnetic iron minerals. Most of these microorganisms cannot be cultivated under laboratory conditions, much less genetically engineered with only few exceptions. However, two *Magnetospirillum* species have emerged as model organisms to study magnetosome formation and magnetotaxis on genetic level, and within the past decade, much has been revealed about the process of magnetosome biogenesis. In this chapter, we summarize these new insights and place the molecular mechanisms of magnetosome formation in the context of the complex cell biology of *Magnetospirillum* spp. After giving an overview of magnetosome vesicle synthesis and biomineralization, we focus on recent findings in positioning and dynamics of the organelles and the biological implications of it, which emphasize that magnetotactic spirilla have evolved sophisticated mechanisms to construct, incorporate, and inherit a navigational device perfectly.

D. Schüler (✉) · F. D. Müller
Department of Microbiology, University of Bayreuth, Bayreuth, Germany
e-mail: dirk.schueler@uni-bayreuth.de; frank.mueller@uni-bayreuth.de

D. Jendrossek (ed.), *Bacterial Organelles and Organelle-like Inclusions*,
Microbiology Monographs 34, https://doi.org/10.1007/978-3-030-60173-7_3

1 Introduction

Magnetotactic bacteria (MTB) are defined by the biomineralization of dedicated intracellular structures, the magnetosomes, which consist of membrane-enclosed crystals of a magnetic iron mineral. Magnetosomes function as sensors for the earth's magnetic field, which is assumed to direct the swimming of bacterial cells along vertical redox gradients within sediments of natural waters (Blakemore et al. 1980; Bazylinski and Frankel 2004; Simmons et al. 2006). The ability to form magnetosomes with diverse shapes and alignments has been found in many phylogenetic groups of prokaryotes (Jogler et al. 2011; Lefèvre et al. 2014; Lin et al. 2017), and magnetotactic microorganisms are abundant and widespread in almost any aquatic habitat.

Magnetosomes are the best-studied examples of lipid-bounded bacterial organelles, and recent research has revealed that they represent one of the most complex structures found in prokaryotic cells (Uebe and Schüler 2016). Besides their function in magnetotactic navigation and motility, magnetosomes have emerged as an effective model for studying cell biology and formation of prokaryotic organelles. In addition, bacterial magnetosomes that can be isolated from disrupted cells represent magnetic nanoparticles with exceptionally well-defined characteristics such as high crystallinity, strong magnetization, and a uniform size distribution, which is owing to the precise control that is exerted during all stages of biomineralization (Faivre and Schüler 2008; Staniland and Rawlings 2016). In addition, the defined composition of the enveloping magnetosome membrane provides an excellent target for functionalization by chemical and genetic coupling of diverse functional moieties such as fluorophores, enzymes or ligands, and antibody fragments (Mickoleit and Schüler 2018; Vargas et al. 2018). Therefore, much of the interest in their biosynthesis has been motivated by their potential use in several biotechnical and biomedical settings. For example, the use of bacterial magnetosomes has been successfully tested in pilot applications such as magnetic imaging techniques like magnet resonance imaging (MRI) and magnetic particle imaging (MPI) (Kraupner et al. 2017) or magnetic hyperthermia (Hergt et al. 2008; Le Fèvre et al. 2017), in which they outperformed abiogenic magnetic nanoparticles generated by chemical synthesis.

Most of our knowledge about magnetosome structure and biosynthesis comes from studies of the two closely related Alphaproteobacteria *Magnetospirillum magneticum* (*Mmag*) and *Magnetospirillum gryphiswaldense* (*M. gryphiswaldense*), which in contrast to most other MTB can be cultivated and genetically manipulated reasonably well. Both species are microaerophiles that can also grow anaerobically by denitrification, they share many similarities with respect to their magnetosome biosynthesis and cell morphology, yet they also exhibit several notable differences in magnetosome structure and intracellular organization. While increasing knowledge is accumulating also from other cultured and uncultured MTB, in this chapter we will mostly focus on magnetosome structure, biosynthesis, and biological function in magnetospirilla.

2 Architecture and Biosynthesis of Magnetosomes

In both *Mmag* and *M. gryphiswaldense*, magnetosomes consist of single cubo-octahedral crystals of magnetite (Fe_3O_4) which in their mature state are about 45 nm in size (Fig. 1). Early studies in the related *M. magnetotacticum* indicated that each magnetite particle is enveloped by a membrane containing phospholipids

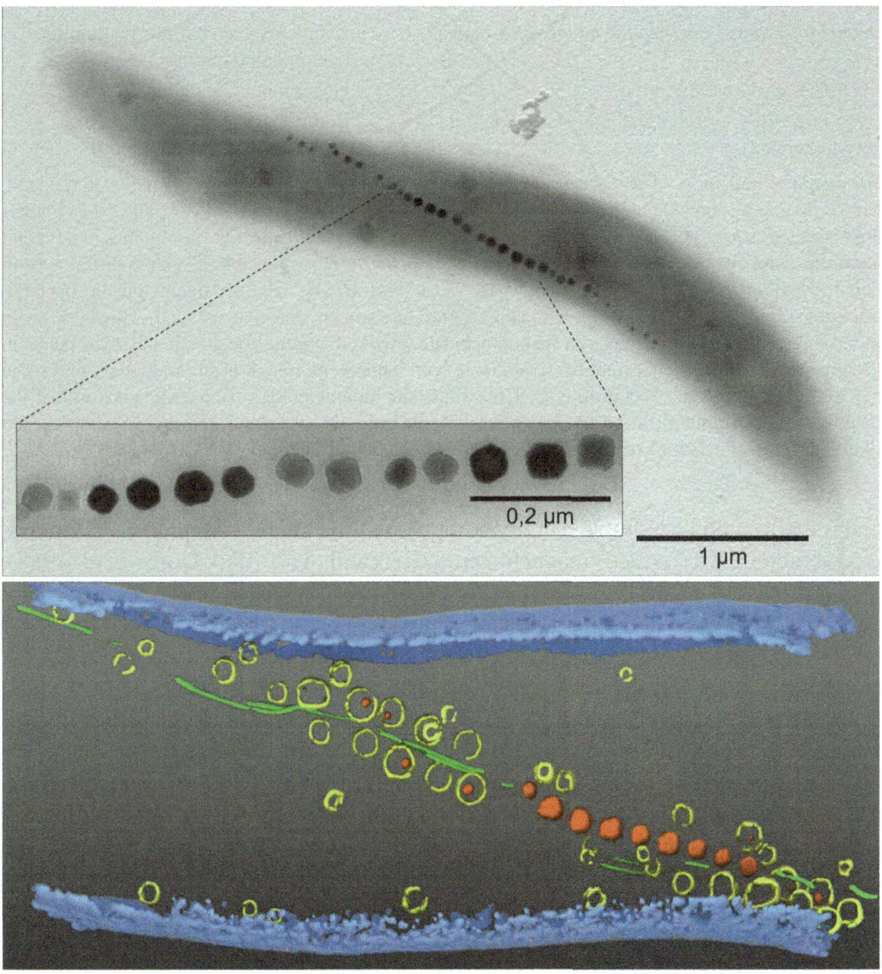

Fig. 1 Upper panel: Transmission electron micrograph of an *M. gryphiswaldense* wild-type cell. The magnetosomes are arranged in a straight chain within the curved bacterium. Inset: Magnification of the chain center highlighting the crystalline magnetite core of the magnetosomes. Lower panel: Rendered electron cryo-tomogaphy image of a section from an *M. gryphiswaldense* wt cell containing the magnetosome chain (mod. after (Scheffel et al. 2006)). The magnetite core of the magnetosomes is colored in red. The magnetosome vesicles (yellow) are cut and appear as open circles. The MamK filament is shown in green and the cellular membrane in blue

Table 1 Synopsis of genes known to control magnetosome biosynthesis in magnetotactic spirilla

Gene	Affected function upon deletion
mamA ("mms24"), mamB, mamQ/mamQ-like	Vesicle formation
mamC ("mms13"), mamD ("mms7")/mamD-like, mamE/limE/mamE-like, mamF/mamF-like, mamG, mamH, mamI, mamL/mamL-like, mamM, mamN, mamO/limO, mamP, mamR, mamS, mamT, mamX, mamZ, mms5, mms6, mms36, mms48, mmsF, mmxF, feoB1, feoB2, ftsZm, feR5, feR6, Amb411	Iron transport/magnetite biomineralization and crystal size control
mamJ/limJ/mamJ-like, mamK/mamK-like, mamY	Magnetosome chain formation, localization and dynamics
mamU, mamV, mamW, mms5	Uncertain
nap, nirS, nirN, norC, norB, fnr, cbb3, fur (not MAI-encoded)	Redox balance or iron homeostasis

All essential and most of the accessory genes cluster in a genomic island (magnetosome island, MAI). Few genes have been identified and named twice such as *mamA/mms24*. The "*-like*" genes are restricted to *Mmag* and have been identified as paralogs of the respective *mam*-genes in a second genomic "islet" of this organism. This redundancy is partially responsible for some differences in deletion mutant phenotypes of *M. gryphiswaldense* and *Mmag*. In other magnetotactic bacteria, homologs for most of the *mam* and *mms* genes have been detected as well. However, with increasing phylogenetic distance to magnetospirilla, distinctive genes occur such as *mad*-genes in greigite–mineralizing deltaproteobacteria. Moreover, some gene functions are redundant and some deletion phenotypes are pleiotropic so that unique functions cannot yet be assigned to all of the listed genes. Biomineralization phenotypes owing to *nap, nir, nor, fnr, cbb3* and *fur* deletion are likely an indirect effect of perturbed cellular redox balance or iron homeostasis

and proteins, and the complete entity comprising the mineral core plus the surrounding membrane was termed 'magnetosome' (Balkwill et al. 1980; Gorby et al. 1988). The presence of a similar magnetosome membrane was later confirmed in *Mmag*, *M. gryphiswaldense* and apparently all other MTB.

The biosynthesis of magnetosomes was subsequently revealed to be a complex, step-wise process which can be genetically dissected (Raschdorf et al. 2016; Cornejo et al. 2016; Uebe and Schüler 2016): First, the magnetosome membrane is invaginated from the cytoplasmic membrane. Second, a set of specific magnetosome proteins is sorted to the magnetosome membrane. Third, iron is transported into the magnetosome membrane vesicle and mineralized as magnetite crystals. Fourth, a magnetosome chain is assembled, positioned, and partitioned during cell division. All steps are highly controlled by a distinct set of about 35 genes, which by proteomic and comparative genomic studies were identified to be clustered within a conserved magnetosome gene island (MAI, (Ullrich et al. 2005; Schübbe et al. 2006)). Genetic studies revealed that these so-called *mam* and *mms* genes in fact do control nearly all features of magnetosome biosynthesis (Murat et al. 2010; Lohsse et al. 2011; Lohsse et al. 2014) (Table 1). Transfer of these genes into the photosynthetic Alphaprotebacterium *Rhodospirillum rubrum* resulted in its "magnetization"; that is, it caused the biosynthesis of magnetosomes resembling those of the donor *M. gryphiswaldense*, including the formation of well-ordered linear magnetosome chains, and the accumulation of cells near the pole of a magnet.

Thus, this demonstrated that the MAI genes are necessary and sufficient to confer magnetosome biosynthesis to this hitherto non-magnetic microbe (Kolinko et al. 2014).

3 Formation of Magnetosome Membrane Vesicles

Magnetosome vesicles are formed by the invagination of the cytoplasmic membrane and seemingly remain connected to the cytoplasmic membrane in *Mmag*, but apparently become pinched off at later stages of biogenesis in *M. gryphiswaldense* (Fig. 2).

Formation of the magnetosome membrane is independent of magnetite biomineralization, as shown by the presence of empty vesicles in iron-starved cells or biomineralization-defective mutants. Likewise, vesicles are formed when cells are cultivated under aerobic conditions which suppresses magnetite biomineralization likely due to inappropriate redox conditions (Heyen and Schüler 2003; Li et al. 2012, 2013). In both strains, these invaginations originated simultaneously from several nonspecific cellular locations, as was revealed by tracking de novo magnetosome biogenesis by time-lapse fluorescence microscopy and cryo-electron tomography (Raschdorf et al. 2016; Cornejo et al. 2016).

Under optimal growth conditions, cells contain up to 100 magnetosome particles per cell. However, the molecular mechanisms which ensure an optimal number and size of magnetosomes are not well understood, but seem to depend on the expression levels of biosynthetic genes, as simultaneous overexpression of almost all *mam* and *mms* gene clusters substantially increased the number (and size) of magnetosomes in *M. gryphiswaldense* (Lohsse et al. 2016).

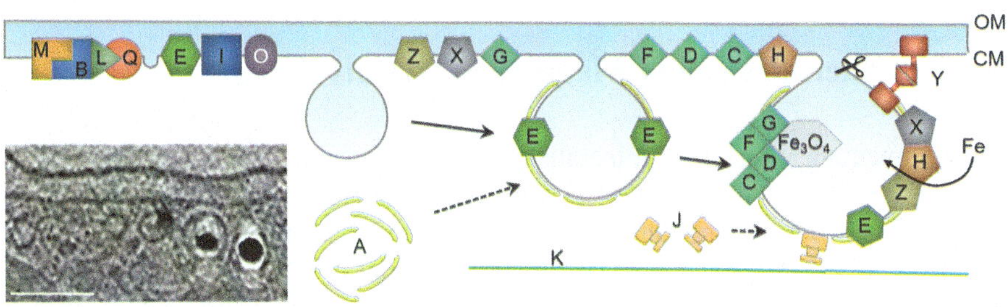

Fig. 2 Key proteins and steps of magnetosome vesicle biogenesis. Essential membrane-bound Mam-proteins (labeled with respective letters, see also Table 1) are thought to tightly interact at the cytoplasmic membrane and eventually facilitate formation and growth of vesicles. Later, further proteins that function in iron transport, redox control, and magnetite precipitation are recruited. Soluble proteins are associated with the periphery of the vesicles and play either a role in magnetosome membrane assembly such as MamA or they are involved in positioning and mobility of the vesicles, e.g., MamJ and MamK. In *M. gryphiswaldense*, vesicles eventually become pinched off the cytoplasmic membrane by an unknown mechanism

The magnetosome membrane has a similar lipid composition as the cytoplasmic membrane, but comprises a distinct set of proteins, with different functions in magnetite biomineralization which are encoded by *mam* and *mms* genes (Gorby et al. 1988; Grünberg et al. 2004). Magnetosome proteins are present in different quantities, from one or few up to 120 copies per particle (Grünberg et al. 2004; Raschdorf et al. 2016). The high protein content of the magnetosome membrane suggests a crowded composition and a tight packing with transmembrane domains of integral proteins. It has been proposed that a lipid raft-like association of magnetosome-membrane proteins takes place prior to the magnetosome invagination (Raschdorf et al. 2016). How proteins are specifically targeted to the magnetosome membrane is not known, and no conserved motifs that encode sorting signals to the magnetosome membrane have been found.

The four proteins MamB, MamI, MamL, and MamQ were identified as key factors in the early biogenesis of the magnetosome membrane (Murat et al. 2010; Lohsse et al. 2014; Raschdorf et al. 2016). Among them, elimination of only MamB completely abolished the formation of regular magnetosome membranes, while mutants of MamI, MamL, or MamQ still contained fewer immature vesicles in *M. gryphiswaldense* (Raschdorf et al. 2016). The crucial role of the cation diffusion facilitator (CDF) protein MamB in membrane biogenesis is independent from its function in iron transport, but involves interactions with other magnetosome proteins, including the paralogous CDF transporter MamM, and the protease MamE, which acts in control of protein sorting to magnetosomes (Uebe et al. 2011, 2018; Hershey et al. 2016a, b). From available genetic and biochemical data, a model for magnetosome membrane formation has been proposed, in which MamB may serve as landmark protein witch interacts with a subset of proteins at the inner cell membrane (Fig. 2). This initial protein complex then recruits further interaction partners that by protein crowding eventually induce lateral pressure to generate membrane curvature (Raschdorf et al. 2016).

4 Biomineralization of Magnetite Crystals

Compartmentalization of biomineralization by the magnetosome membrane provides a specialized "nanoreactor" in which the iron, redox, and pH environments of biomineralization can be strictly regulated for the formation of single crystals of magnetite. Most available evidence supports a model in which extracellular iron is first imported into the cytoplasm by generic transporters. Subsequently, iron is transported from the cytoplasm into magnetosome membrane vesicles by magnetosome-specific transporters MamB and MamM (for ferrous iron), which are members of the Fe/Zn-transporting subfamily of divalent metal CDF proteins (Uebe et al. 2018), MamH and MamZ (for ferric iron), which are members of the major facilitator superfamily (MFS) (Raschdorf et al. 2013). The proper Fe^{2+}/Fe^{3+} ratio for production of the mixed-valence iron mineral magnetite inside magnetosome membrane vesicles is thought to be regulated by MamE, MamP,

MamT, and MamX which are constituents of the magnetosome membrane (Fig. 2) (Grünberg et al. 2004). These proteins each contain two or three conserved MTB-specific CXXCH *c*-type cytochrome haem-binding motifs denoted the "magnetochrome" domain that may oxidize magnetosomal Fe^{2+} for biomineralization (Siponen et al. 2013; Jones et al. 2015). In addition, redox balance for magnetite biomineralization is also poised by the activity of cellular electron transport chains. For instance, magnetite formation is linked to dissimilatory nitrate reduction, and cells of *M. gryphiswaldense* are impaired in magnetosome biomineralization in the absence of nitrate, or upon deletion of genes encoding the periplasmic nitrate and nitrite reductases Nap and NirS, as well in cells lacking the fumarate and nitrate reduction regulator protein Fnr. In a similar manner, inactivation of the terminal oxidase Cbb3 involved in aerobic respiration caused pleiotropic effects on magnetosomes under microaerobic conditions, probably by disturbing the redox balance required for proper magnetite biomineralization (Li et al. 2012, 2013; Müller et al. 2014).

After nucleation of the magnetite crystal, several magnetosome proteins regulate their maturation into particles of defined size and shape in a positive (MamG, MamF, MamD, MamC, MamS, MamR, MamN, Mms6, and MmsF) or negative (Mms36 and Mms48) manner (Fig. 2). However, the understanding how they exactly interact with the surface of the nascent magnetite crystals is just emerging (Scheffel et al. 2008; Lohsse et al. 2011; Rong et al. 2012; Nudelman et al. 2018).

5 Structure, Assembly, and Positioning of Magnetosome Chains

During maturation, magnetosomes become organized and assembled into chains. Essential active parts of this assembly and positioning machinery consist of dedicated cytoskeletal proteins which have been identified some years ago. In *M. gryphiswaldense* and related magnetotactic bacteria, the magnetosome chains become concatenated by the joint action of the actin-like MamK, which polymerizes into cell-spanning dynamic filaments (Fig. 1) and MamJ, an adaptor protein with less-understood function (Komeili et al. 2006; Scheffel et al. 2006; Katzmann et al. 2010; Draper et al. 2011). In *Mmag*, both proteins exist as two paralogs with overlapping but also slightly different functions causing less pronounced phenotypes upon deletion compared to *M. gryphiswaldense* (Rioux et al. 2010; Abreu et al. 2014). *mamK* mutants fail to assemble wt-like continuous magnetosome chains but contain either disordered (in *Mmag*, (Komeili et al. 2006)) or short and fragmented chains in *M. gryphiswaldense* (Katzmann et al. 2010) (Fig. 3). Magnetosome chains connected to MamK filaments are only formed in the presence of MamJ, a poorly characterized cytoplasmic protein. It consists of a central acidic repetitive (CAR) domain, which is largely dispensable (Scheffel and Schüler 2007), and seems to be weakly structured (Nudelman and Zarivach 2014). The *mamJ* deletion mutant

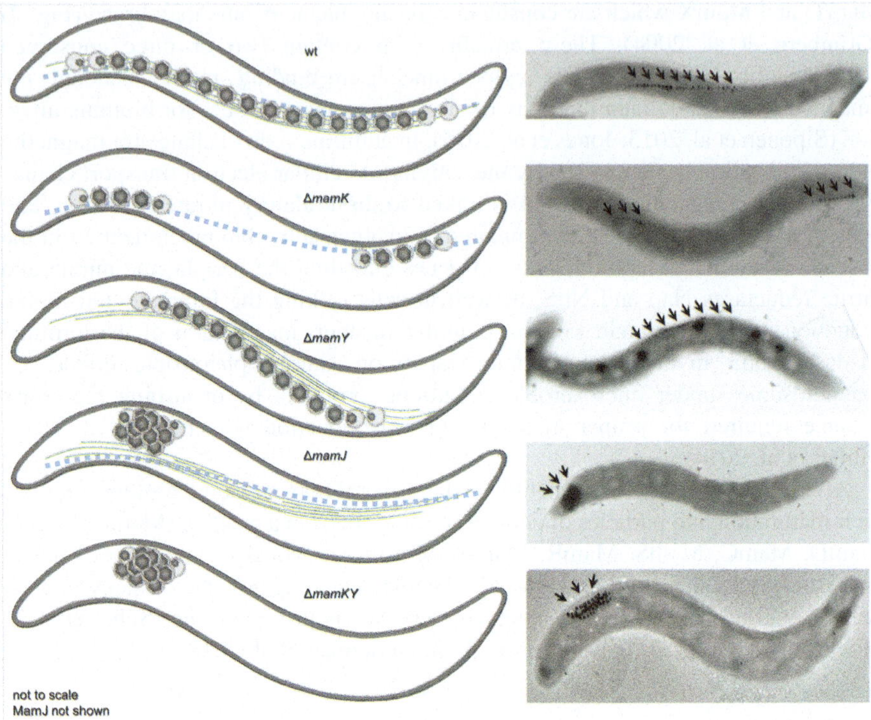

Fig. 3 Phenotypes of *mamK, Y, J,* and *KY* deletion mutants in *M. gryphiswaldense*. In the WT, magnetosomes are attached to cytoplasmic MamK filaments and to membrane-bound MamY structures via MamJ, thereby forming a straight and continuous chain. In the *mamK* mutant, chains are fragmented and off-center, but still attached to MamY and therefore at the geodetic cell axis. In the *mamJ* mutant, the magnetosomes are not attached to any structure and agglomerate by their own magnetic attraction. The *mamY* mutant shows continuous chains that are detached from the geodetic cell axis and in the *mamKY* double mutant; no chain stabilizing structure is present, resulting in agglomerated magnetosomes as in the *mamJ* mutant

phenotype in *M. gryphiswaldense* is striking as magnetosomes are completely disorganized and cluster irregularly due to their unconstrained magnetic interaction or form magnetic flux-closed rings, even in the presence of MamK (Scheffel and Schüler 2007; Bennet et al. 2015; Kiani et al. 2018) (Fig. 3). In *Mmag*, which contains two MamJ paralogs, the phenotype is less pronounced but magnetosome chain formation is also perturbed (Draper et al. 2011). These phenotypes and suggested MamK-MamJ interactions (Scheffel and Schüler 2007) led to the early model of an only two-part cytoskeletal structure, consisting of a presumably rather static backbone made of MamK filaments and MamJ, which attaches magnetosomes to that scaffold. However, this model recently proved incomplete in at least two aspects: First, the magnetosome chain was revealed to be highly dynamic; and second, localization of the chain in spirilla is much more controlled than initially assumed.

MamK has been analyzed extensively in vitro and the pure protein was found to polymerize into filaments when adenosine triphosphate (ATP) or guanosine triphosphate (GTP) are present (Sonkaria et al. 2012; Ozyamak et al. 2013; Bergeron et al. 2017). Recently, the crystal structure of the protein and molecular details of its polymeric form have been solved (Löwe et al. 2016) and underpinned its structural relatedness to eukaryotic actin proteins. In vivo studies suggest that MamK filaments not only concatenate magnetosome chains but also expedite splitting of the chain during cell division by distinct localization at the division plane: In magnetospirilla, cytokinesis occurs by unidirectional indentation and asymmetric septum formation at the division plane, resulting in a buckling-like deformation of the dividing cell (Staniland et al. 2010). Localization of the magnetosome chain opposite to the unilateral growing septum by MamK results in a leverage-like mechanism, which has been interpreted as specific adaptation required to overcome the magnetostatic interactions between separating daughter chains (Katzmann et al. 2011).

In vivo studies analyzed the dynamics of polymeric MamK filaments in its cellular context. By fluorescence recovery after photo bleaching (FRAP) experiments, it was found that MamK treadmilling speed in *M. gryphiswaldense* is about 300 nm/min (Toro-Nahuelpan et al. 2016) and depends on its ATP hydrolysis capacity. Similar observations were made in *Mmag* (Pradel et al. 2006; Draper et al. 2011; Abreu et al. 2014). Interestingly, dynamics of the MamK filaments was found to also depend on MamJ, which therefore seems not only necessary to tether magnetosomes to MamK filaments but also for MamK turnover rates.

The dynamics of MamK filaments in *M. gryphiswaldense* already suggested that a magnetosome chain is no static structure and that whole magnetosomes are moved along the track. In fact, active movement of the organelles seems a crucial prerequisite to assemble a chain: Magnetosome vesicles emerge more or less randomly at the cytoplasmic membrane (Raschdorf et al. 2016). Mature magnetosomes, however, are regularly found in chains suggesting an active collecting and positioning mechanism, although spontaneous incorporation of maturing magnetosomes based on their increasing magnetism may also play a role (Klumpp and Faivre 2012). The function of MamK in *Mmag* seems somewhat different, but still, MamK is needed to restrict movement and to position magnetosomes (Grant et al. 2018). There is, however, other evidence for active translocation of magnetosomes: When *M. gryphiswaldense* cells divide, each daughter cell receives exactly half the number of magnetosomes from the mother cell (Toro-Nahuelpan et al. 2016), indicating that cells distribute their magnetosomes with highest possible precision to their offspring. This is achieved by strict positioning of the chain center at the cell division site. After cytokinesis, the daughter chains undergo repositioning from the new cell poles to midcell, and they are maintained at this position during growth by an unknown mechanism. In mutants with severely impaired dynamics of MamK, chain dynamics is also perturbed (i.e., chains are not efficiently re-localized to midcell), leading to unequal distribution of the organelles (a phenotype that is also seen in *mamK* and *mamJ* deletion mutants) (Katzmann et al. 2011; Toro-Nahuelpan et al. 2016). Yet, the molecular mechanism by which magnetosomes move along MamK filaments is still unknown. Photobleaching experiments suggest that MamK filaments in

M. gryphiswaldense nucleate close to the cell poles and grow toward midcell, i.e., in the direction where magnetosomes migrate. Interestingly, the treadmilling speed of MamK filaments seems much higher than the speed of magnetosomes (Toro-Nahuelpan et al. 2016), suggesting that magnetosomes are not tightly coupled to MamK oligomers. On the other hand, cargo-carrying motor proteins reminiscent to myosin, which walks along actin filaments in eukaryotic cells, have not been identified in MTB. Potential factors that control nucleation of MamK filaments at the cell poles and their polarity are also unknown. It is also not clear what happens if the filaments of opposite polarity meet at midcell and if MamK filaments stop growing at all. For example, it has been observed that filaments that reach the opposite cell pole bend and turn, sometimes even under physiological expression levels (Komeili et al. 2006; Katzmann et al. 2011; Toro-Nahuelpan et al. 2016).

Furthermore, MamK and other cytoskeletal elements of the magnetosome chain were recently found to be linked to the generic cytoskeleton by the coiled-coil protein CcfM, suggesting an intricate network of magnetosome chain assembly determinants and cell shape control in magnetotactic spirilla (Pfeiffer et al. 2020).

6 Three-dimensional Chain Positioning, Motility Axis, and Magnetotaxis

Magnetotaxis differs from conventional chemotaxis paradigms known from, e.g., *E. coli*. Here, chemotactic swimming resembles a three-dimensional trial-and-error walk where periods of straight movement are interrupted by tumbling pauses where cells turn randomly before they resume straight swimming. These so called run-and-tumble sequences are biased by chemosensory signal cascades which control the frequency of runs and tumbling by interaction with flagellar motor proteins in response to detected gradients of nutrients, repellents, or electron acceptors (reviewed, for example, in (Bi and Sourjik 2018)). Magnetospirilla and all other characterized MTB so far are also motile by flagella, and aero- or even phototaxis has been described (Chen et al. 2010; Popp et al. 2014). So, what is the benefit of bearing a magnetosome chain in addition to extensive chemosensory networks and how does magnetotaxis feed into motility of MTB?

Magnetotaxis could, for example, be beneficial for a bacterium if it combined the positional information provided by the intracellular compass with decisions on locomotion, i.e., on run-and-tumble frequencies (conveyed by MamK, as suggested by (Philippe and Wu 2010)). However, up to now, neither deviant chemotaxis patterns in a *mamK* mutant nor direct evidence for a biochemical signal transduction between magnetosome chain and flagellar motor has been provided. This and the lack of any canonical signal transduction motif (as, for example, in kinases or methylases) in the magnetosome gene clusters suggests that magnetotaxis functions in a different way and exploits magnetic forces directly.

It has been shown that the force generated by a single magnetosome in the geomagnetic field is too weak to align a cell effectively (Frankel and Blakemore 1980). On the other hand, as described above, multiple disorganized magnetosomes within a cell agglomerate by magnetic attraction, likewise resulting in a net cellular magnetic moment close to zero (Kirschvink 1982; Kobayashi et al. 2006). Hence, to serve as an efficient magnetic field receptor, single magnetosomes become concatenated into a chain which in the model organism *M. gryphiswaldense* consists of ~45 particles (Zahn et al. 2017), adding the single magnetic moment of each unit to a functional magnetic dipole. This suggests that the number of magnetosomes per cell is controlled so that the chain forms a compass strong enough to passively align whole cells to the geomagnetic field, akin to a compass needle (Frankel and Blakemore 1980; Rosenblatt et al. 1982a, 1982b; Moskowitz et al. 1988). Correspondingly, bigger cells that belong to other phyla such as members of the nitrospirae or multicellular MTBs tend to possess a multitude of magnetosomes and chains (Jogler et al. 2011; Leão et al. 2017), which increases the magnetic force for their alignment. Therefore, it is believed that the force which aligns a magnetosome chain to the magnetic field is mechanically transmitted to the cell body and used to align whole cells passively (Kobayashi et al. 2006). As a result, there is limited need to stop the movement for tumbling in magnetospirilla. Their motility axis is preset by the position of flagella and arrangement of the magnetosome chain. The alignment to the geomagnetic field then reduces three-dimensional swimming to a linear movement along the vertical inclines of the earth's magnetic field; and by the integration of chemotactic responses such as aerotaxis, magnetotactic bacteria in stratified environments are efficiently guided to their preferred oxygen concentration. However, recently it became clear that in particular spirilla have evolved additional sophisticated means to optimize their magnetic navigation.

For efficient magnetotaxis, the magnetosome chain must adopt and maintain a very distinct position in the cell. First, a fixed position is important because a flexible, "floating" magnetosome chain could move within the cell rather than aligning it. A physical connection of magnetosomes to cytoplasmic content such as DNA (as described for carboxysomes, carbonosomes, and polyphosphate inclusions) seems therefore not sufficient. Tethering the chain to rigid and more static structures such as the cell envelope would meet this requirement much better. Second, for efficient magnetotaxis, the magnetic moment of the magnetosome chain must perfectly match the swimming direction of the cell, which is predefined by the position of flagella on the cell surface (Frankel and Blakemore 1980). Spirilla are propelled by polar flagella and hence move along the longitudinal cell axis. This propulsion is accompanied by fast rotations of the cell body. If swimming direction and magnetosome chain were misaligned, cells would tumble when they swim because two forces pulled in slightly different directions. Third, positioning of the linear magnetosome chain must safeguard that it is maintained straight. Straightness is necessary to maximize the net magnetic moment of the chain similar to a corresponding bar magnet. Strikingly, the chains of MTB seem to regularly meet all these criteria, even in helical cells of magnetospirilla. This is intriguing since unlike rod-shaped bacteria, spirilla lack any straight cell surface to support a rod-like magnetoreceptor. However, this question has remained unaddressed for a long time, and mechanisms for magnetosome chain positioning in curved cells have been unknown until recently. The first glance on the intriguing proficiency of magnetotactic spirilla to accommodate a straight magnetoreceptor in helical cells was possible through an in-depth analysis of the magnetosome protein MamY and the *mamY* mutant in *M. gryphiswaldense.*

Initially, three-dimensional analysis of magnetosome chain positioning in *M. gryphiswaldense* wt cells revealed that the chain tightly follows a path along the cytoplasmic membrane with highest convex curvature. This path coincides with the shortest connection between the cell poles and hence represents the geodetic axis of the helix. Importantly, this path also coincides with the cellular motility axis. Upon the deletion of *mamY*, magnetosome chains lose their straight appearance and detach from the inner (convex) curvature of the helical cell shifting from the geodetic path to the outer (concave) curvature (Fig. 3). This indicates that MamY is involved in tethering the magnetosome chain to a trail that is perfectly congruent with the motility axis. Consequently, when *mamY* is deleted and magnetosome chains are displaced, the ability of the cells to align to the magnetic field drops similar to the *mamK* mutant (Pfeiffer and Schüler 2019). Another indication for the scaffolding function of MamY is the phenotype of a *mamYK* double mutant. This strain lost its ability to form magnetosome chains completely and phenocopies a *mamJ* mutant where all magnetosomes agglomerate (Fig. 3).

MamY, initially proposed to be involved in vesicle formation in *Mmag* (Tanaka et al. 2010), is a protein of the inner and the magnetosome membranes in *M. gryphiswaldense*, The protein self-interacts and forms higher ordered structures at the membrane. Upon a certain size, these polymers are supposed to become curvature sensitive and further enrich along the membrane with highest positive curvature, eventually forming an extended assembly reaching from pole to pole and following the geodetic cell axis. With its cytoplasmic domain, MamY recruits the magnetosome chain made by MamK and J and forces it to the geodetic cell axis (Toro-Nahuelpan et al. 2019).

The function of MamY could also explain why in the mamK mutant there are still short magnetosome chains: The magnetosomes are likely attached to MamY structures but are not concatenated into coherent chains because the cytomotive filaments of MamK are missing. Correspondingly, the short chains of the *mamK* mutant are still observed at sites of inner positive cell curvature. The formation of short chains instead of clusters could be explained by magnetic attraction of the particles rather than by active assembly (Klumpp and Faivre 2012).

Taken together, MamY seems the key to (1) connect the magnetosome chain to the cell envelope, ensuring efficient force transmission and cell alignment, (2) keep the magnetosome chain straight to maximize its magnetic moment, and (3) fit it to the cellular motility axis. Finally, it reconciles the different phenotypes of the *mamK* and *mamJ* mutants.

Altogether, a sophisticated "magnetoskeleton" can now be defined which to date consists of three cytoskeletal factors dedicated for magnetotaxis (Fig. 4). The actin-

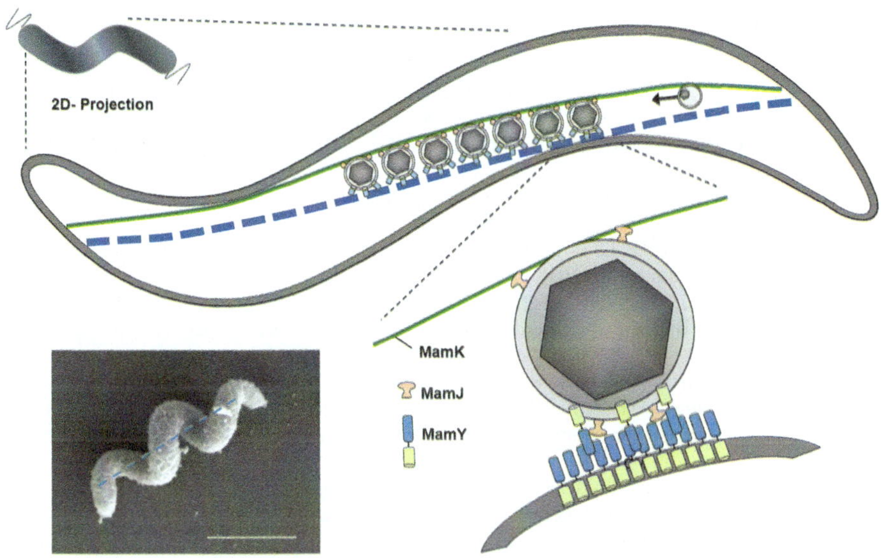

Fig. 4 Scheme of the current view on the tripartite magnetoskeleton in *M. gryphiswaldense*. The actin-like MamK (green) polymerizes in cell-spanning dynamic filaments. Magnetosomes are attached to this structure with the help of MamJ (beige). MamY, however, concentrates along the inner positive membrane curvature and in the magnetosome membrane and recruits the magnetosome chain to the geodetic cell axis. Note that the scheme represents a 2-dimensional projection of a helical cell meaning that in reality, the magnetosome chain stays close to the cytoplasmic membrane. The inset shows a scanning electron microscopy image of an *M. gryphiswaldense* cell to highlight the corkscrew-like cell morphology. The geodetic path is indicated as dashed line. Bar: 1μm

like MamK and its adaptor MamJ represent the dynamic part of the magnetoskeleton as they actively concatenate magnetosomes and position the chain at midcell, thereby also ensuring precise splitting and equipartitioning of the magnetoreceptor upon cell division. The static MamY structure by contrast is responsible to identify and to tether the dynamic assembly to the geodetic cell axis, i.e., to fix it along the shortest path connecting the cell poles. This straightens the “compass needle” within the helix and aligns it with the motility axis, thereby perfectly superimposing the earth’s magnetic field vector on swimming movements.

These recent results highlight that magnetosomes—in contrast to other bacterial organelles—do not function as single entity but require dedicated cytoskeletal elements to form a dynamic and distinctly localized higher ordered structure suitable for perfect magnetic navigation.

Acknowledgements We are grateful to the Deutsche Forschungsgemeinschaft grant Schu1080/9-2 and the European Research Council Advanced Grant 692637 (Syntomagx).

References

Abreu N, Mannoubi S, Ozyamak E, Pignol D, Ginet N, Komeili A (2014) Interplay between two bacterial actin homologs, MamK and MamK-like, is required for the alignment of magnetosome organelles in Magnetospirillum magneticum AMB-1. J Bacteriol 196:3111–3121. https://doi.org/10.1128/JB.01674-14

Balkwill DL, Maratea D, Blakemore RP (1980) Ultrastructure of a magnetotactic spirillum. J Bacteriol 141:1399–1408

Bazylinski DA, Frankel RB (2004) Magnetosome formation in prokaryotes. Nat Rev Microbiol 2:217–230. https://doi.org/10.1038/nrmicro842

Bennet M, Bertinetti L, Neely RK, Schertel A, Körnig A, Flors C, Müller FD, Schüler D, Klumpp S, Faivre D (2015) Biologically controlled synthesis and assembly of magnetite nanoparticles. Faraday Discuss 181:71–83. https://doi.org/10.1039/C4FD00240G

Bergeron JRC, Hutto R, Ozyamak E, Hom N, Hansen J, Draper O, Byrne ME, Keyhani S, Komeili A, Kollman JM (2017) Structure of the magnetosome-associated actin-like MamK filament at subnanometer resolution. Protein Sci Publ Protein Soc 26:93–102. https://doi.org/10.1002/pro.2979

Bi S, Sourjik V (2018) Stimulus sensing and signal processing in bacterial chemotaxis. Curr Opin Microbiol 45:22–29. https://doi.org/10.1016/j.mib.2018.02.002

Blakemore RP, Frankel RB, Kalmijn AJ (1980) South-seeking magnetotactic bacteria in the southern hemisphere. Nature 286:384–385. https://doi.org/10.1038/286384a0

Chen C, Ma Q, Jiang W, Song T (2010) Phototaxis in the magnetotactic bacterium Magnetospirillum magneticum strain AMB-1 is independent of magnetic fields. Appl Microbiol Biotechnol 90:269–275. https://doi.org/10.1007/s00253-010-3017-1

Cornejo E, Subramanian P, Li Z, Jensen GJ, Komeili A (2016) Dynamic remodeling of the Magnetosome membrane is triggered by the initiation of biomineralization. mBio 7:e01898-01815. https://doi.org/10.1128/mBio.01898-15

Draper O, Byrne ME, Li Z, Keyhani S, Barrozo JC, Jensen G, Komeili A (2011) MamK, a bacterial actin, forms dynamic filaments in vivo that are regulated by the acidic proteins MamJ and LimJ. Mol Microbiol 82:342–354. https://doi.org/10.1111/j.1365-2958.2011.07815.x

Faivre D, Schüler D (2008) Magnetotactic Bacteria and Magnetosomes. Chem Rev 108:4875–4898. https://doi.org/10.1021/cr078258w

Frankel RB, Blakemore RP (1980) Navigational compass in magnetic bacteria. J Magn Magn Mater 15–18:1562–1564. https://doi.org/10.1016/0304-8853(80)90409-6

Gorby YA, Beveridge TJ, Blakemore RP (1988) Characterization of the bacterial magnetosome membrane. J Bacteriol 170:834–841. https://doi.org/10.1128/jb.170.2.834-841.1988

Grant CR, Wan J, Komeili A (2018) Organelle formation in Bacteria and Archaea. Annu Rev Cell Dev Biol 34:217–238. https://doi.org/10.1146/annurev-cellbio-100616-060908

Grünberg K, Müller E-C, Otto A, Reszka R, Linder D, Kube M, Reinhardt R, Schüler D (2004) Biochemical and proteomic analysis of the magnetosome membrane in Magnetospirillum gryphiswaldense. Appl Environ Microbiol 70:1040–1050

Hergt R, Dutz S, Röder M (2008) Effects of size distribution on hysteresis losses of magnetic nanoparticles for hyperthermia. J Phys Condens Matter Inst Phys J 20:385214. https://doi.org/10.1088/0953-8984/20/38/385214

Hershey DM, Browne PJ, Iavarone AT, Teyra J, Lee EH, Sidhu SS, Komeili A (2016a) Magnetite biomineralization in Magnetospirillum magneticum is regulated by a switch-like behavior in the HtrA protease MamE. J Biol Chem 291:17941–17952. https://doi.org/10.1074/jbc.M116.731000

Hershey DM, Ren X, Melnyk RA, Browne PJ, Ozyamak E, Jones SR, Chang MCY, Hurley JH, Komeili A (2016b) MamO is a repurposed serine protease that promotes magnetite biomineralization through direct transition metal binding in Magnetotactic Bacteria. PLoS Biol 14:e1002402. https://doi.org/10.1371/journal.pbio.1002402

Heyen U, Schüler D (2003) Growth and magnetosome formation by microaerophilic Magnetospirillum strains in an oxygen-controlled fermentor. Appl Microbiol Biotechnol 61:536–544. https://doi.org/10.1007/s00253-002-1219-x

Jogler C, Wanner G, Kolinko S, Niebler M, Amann R, Petersen N, Kube M, Reinhardt R, Schüler D (2011) Conservation of proteobacterial magnetosome genes and structures in an uncultivated member of the deep-branching Nitrospira phylum. Proc Natl Acad Sci 108:1134–1139. https://doi.org/10.1073/pnas.1012694108

Jones SR, Wilson TD, Brown ME, Rahn-Lee L, Yu Y, Fredriksen LL, Ozyamak E, Komeili A, Chang MCY (2015) Genetic and biochemical investigations of the role of MamP in redox control of iron biomineralization in Magnetospirillum magneticum. Proc Natl Acad Sci U S A 112:3904–3909. https://doi.org/10.1073/pnas.1417614112

Katzmann E, Scheffel A, Gruska M, Plitzko JM, Schüler D (2010) Loss of the actin-like protein MamK has pleiotropic effects on magnetosome formation and chain assembly in Magnetospirillum gryphiswaldense. Mol Microbiol 77:208–224. https://doi.org/10.1111/j.1365-2958.2010.07202.x

Katzmann E, Müller FD, Lang C, Messerer M, Winklhofer M, Plitzko JM, Schüler D (2011) Magnetosome chains are recruited to cellular division sites and split by asymmetric septation. Mol Microbiol 82:1316–1329. https://doi.org/10.1111/j.1365-2958.2011.07874.x

Kiani B, Faivre D, Klumpp S (2018) Self-organization and stability of magnetosome chains—a simulation study. PLoS One 13:e0190265. https://doi.org/10.1371/journal.pone.0190265

Kirschvink JL (1982) Paleomagnetic evidence for fossil biogenic magnetite in western Crete. Earth Planet Sci Lett 59:388–392. https://doi.org/10.1016/0012-821X(82)90140-6

Klumpp S, Faivre D (2012) Interplay of magnetic interactions and active movements in the formation of magnetosome chains. PLoS One 7:e33562. https://doi.org/10.1371/journal.pone.0033562

Kobayashi A, Kirschvink JL, Nash CZ, Kopp RE, Sauer DA, Bertani LE, Voorhout WF, Taguchi T (2006) Experimental observation of magnetosome chain collapse in magnetotactic bacteria: Sedimentological, paleomagnetic, and evolutionary implications. Earth Planet Sci Lett 245:538–550. https://doi.org/10.1016/j.epsl.2006.03.041

Kolinko I, Lohße A, Borg S, Raschdorf O, Jogler C, Tu Q, Pósfai M, Tompa E, Plitzko JM, Brachmann A, Wanner G, Müller R, Zhang Y, Schüler D (2014) Biosynthesis of magnetic nanostructures in a foreign organism by transfer of bacterial magnetosome gene clusters. Nat Nanotechnol 9:193–197. https://doi.org/10.1038/nnano.2014.13

Komeili A, Li Z, Newman DK, Jensen GJ (2006) Magnetosomes are cell membrane invaginations organized by the actin-like protein MamK. Science 311:242–245. https://doi.org/10.1126/science.1123231

Kraupner A, Eberbeck D, Heinke D, Uebe R, Schüler D, Briel A (2017) Bacterial magnetosomes – nature's powerful contribution to MPI tracer research. Nanoscale 9:5788–5793. https://doi.org/10.1039/c7nr01530e

Le Fèvre R, Durand-Dubief M, Chebbi I, Mandawala C, Lagroix F, Valet J-P, Idbaih A, Adam C, Delattre J-Y, Schmitt C, Maake C, Guyot F, Alphandéry E (2017) Enhanced antitumor efficacy of biocompatible magnetosomes for the magnetic hyperthermia treatment of glioblastoma. Theranostics 7:4618–4631. https://doi.org/10.7150/thno.18927

Leão P, Chen Y-R, Abreu F, Wang M, Zhang W-J, Zhou K, Xiao T, Wu L-F, Lins U (2017) Ultrastructure of ellipsoidal magnetotactic multicellular prokaryotes depicts their complex assemblage and cellular polarity in the context of magnetotaxis. Environ Microbiol 19:2151–2163. https://doi.org/10.1111/1462-2920.13677

Lefèvre CT, Bennet M, Landau L, Vach P, Pignol D, Bazylinski DA, Frankel RB, Klumpp S, Faivre D (2014) Diversity of magneto-aerotactic behaviors and oxygen sensing mechanisms in cultured magnetotactic bacteria. Biophys J 107:527–538. https://doi.org/10.1016/j.bpj.2014.05.043

Li Y, Katzmann E, Borg S, Schüler D (2012) The Periplasmic nitrate Reductase nap is required for anaerobic growth and involved in redox control of magnetite biomineralization in

Magnetospirillum gryphiswaldense. J Bacteriol 194:4847–4856. https://doi.org/10.1128/JB.00903-12

Li Y, Bali S, Borg S, Katzmann E, Ferguson SJ, Schüler D (2013) Cytochrome cd1 nitrite reductase NirS is involved in anaerobic magnetite biomineralization in Magnetospirillum gryphiswaldense and requires NirN for proper d1 heme assembly. J Bacteriol 195:4297–4309. https://doi.org/10.1128/JB.00686-13

Lin W, Pan Y, Bazylinski DA (2017) Diversity and ecology of and biomineralization by magnetotactic bacteria. Environ Microbiol Rep 9:345–356. https://doi.org/10.1111/1758-2229.12550

Lohsse A, Ullrich S, Katzmann E, Borg S, Wanner G, Richter M, Voigt B, Schweder T, Schüler D (2011) Functional analysis of the magnetosome island in Magnetospirillum gryphiswaldense: the mamAB operon is sufficient for magnetite biomineralization. PLoS One 6:e25561. https://doi.org/10.1371/journal.pone.0025561

Lohsse A, Borg S, Raschdorf O, Kolinko I, Tompa E, Pósfai M, Faivre D, Baumgartner J, Schüler D (2014) Genetic dissection of the mamAB and mms6 operons reveals a gene set essential for magnetosome biogenesis in Magnetospirillum gryphiswaldense. J Bacteriol 196:2658–2669. https://doi.org/10.1128/JB.01716-14

Lohsse A, Kolinko I, Raschdorf O, Uebe R, Borg S, Brachmann A, Plitzko JM, Müller R, Zhang Y, Schüler D (2016) Overproduction of Magnetosomes by genomic amplification of biosynthesis-related gene clusters in a Magnetotactic bacterium. Appl Environ Microbiol 82:3032–3041. https://doi.org/10.1128/AEM.03860-15

Löwe J, He S, Scheres SHW, Savva CG (2016) X-ray and cryo-EM structures of monomeric and filamentous actin-like protein MamK reveal changes associated with polymerization. Proc Natl Acad Sci U S A 113:13396–13401. https://doi.org/10.1073/pnas.1612034113

Mickoleit F, Schüler D (2018) Generation of nanomagnetic biocomposites by genetic engineering of bacterial magnetosomes. Bioinspir Biomim Nan 8:86–98. https://doi.org/10.1680/jbibn.18.00005

Moskowitz BM, Frankel RB, Flanders PJ, Blakemore RP, Schwartz BB (1988) Magnetic properties of magnetotactic bacteria. J Magn Magn Mater 73:273–288. https://doi.org/10.1016/0304-8853(88)90093-5

Müller FD, Raschdorf O, Nudelman H, Messerer M, Katzmann E, Plitzko JM, Zarivach R, Schüler D (2014) The FtsZ-like protein FtsZm of Magnetospirillum gryphiswaldense likely interacts with its generic homolog and is required for biomineralization under nitrate deprivation. J Bacteriol 196:650–659. https://doi.org/10.1128/JB.00804-13

Murat D, Quinlan A, Vali H, Komeili A (2010) Comprehensive genetic dissection of the magnetosome gene island reveals the step-wise assembly of a prokaryotic organelle. Proc Natl Acad Sci 107:5593–5598. https://doi.org/10.1073/pnas.0914439107

Nudelman H, Zarivach R (2014) Structure prediction of magnetosome-associated proteins. Front Microbiol 5:9. https://doi.org/10.3389/fmicb.2014.00009

Nudelman H, Lee Y-Z, Hung Y-L, Kolusheva S, Upcher A, Chen Y-C, Chen J-Y, Sue S-C, Zarivach R (2018) Understanding the biomineralization role of magnetite-interacting components (MICs) from Magnetotactic Bacteria. Front Microbiol 9:2480. https://doi.org/10.3389/fmicb.2018.02480

Ozyamak E, Kollman J, Agard DA, Komeili A (2013) The bacterial actin MamK: in vitro assembly behavior and filament architecture. J Biol Chem 288:4265–4277. https://doi.org/10.1074/jbc.M112.417030

Pfeiffer D, Schüler D (2019) Quantifying the benefit of a dedicated "magnetoskeleton" in bacterial magnetotaxis by live-cell motility tracking and soft agar swimming assay. Appl Environ Microbiol 86:e01976-19. https://doi.org/10.1128/AEM.01976-19

Pfeiffer D, Toro-Nahuelpan M, Awal RP, Müller F-D, Bramkamp M, Plitzko JM, Schüler D (2020) A bacterial cytolinker couples positioning of magnetic organelles to cell shape control. Proc Natl Acad Sci U S A. https://doi.org/10.1073/pnas.2014659117

Philippe N, Wu L-F (2010) An MCP-like protein interacts with the MamK cytoskeleton and is involved in Magnetotaxis in Magnetospirillum magneticum AMB-1. J Mol Biol 400:309–322. https://doi.org/10.1016/j.jmb.2010.05.011

Popp F, Armitage JP, Schüler D (2014) Polarity of bacterial magnetotaxis is controlled by aerotaxis through a common sensory pathway. Nat Commun 5:5398. https://doi.org/10.1038/ncomms6398

Pradel N, Santini C-L, Bernadac A, Fukumori Y, Wu L-F (2006) Biogenesis of actin-like bacterial cytoskeletal filaments destined for positioning prokaryotic magnetic organelles. Proc Natl Acad Sci U S A 103:17485–17489. https://doi.org/10.1073/pnas.0603760103

Raschdorf O, Müller FD, Pósfai M, Plitzko JM, Schüler D (2013) The magnetosome proteins MamX, MamZ and MamH are involved in redox control of magnetite biomineralization in Magnetospirillum gryphiswaldense. Mol Microbiol 89:872–886. https://doi.org/10.1111/mmi.12317

Raschdorf O, Forstner Y, Kolinko I, Uebe R, Plitzko JM, Schüler D (2016) Genetic and Ultrastructural analysis reveals the key players and initial steps of bacterial Magnetosome membrane biogenesis. PLoS Genet 12:e1006101. https://doi.org/10.1371/journal.pgen.1006101

Rioux J-B, Philippe N, Pereira S, Pignol D, Wu L-F, Ginet N (2010) A second actin-like MamK protein in Magnetospirillum magneticum AMB-1 encoded outside the genomic Magnetosome Island. PLoS One 5:e9151. https://doi.org/10.1371/journal.pone.0009151

Rong C, Zhang C, Zhang Y, Qi L, Yang J, Guan G, Li Y, Li J (2012) FeoB2 functions in Magnetosome formation and oxidative stress protection in Magnetospirillum gryphiswaldense strain MSR-1. J Bacteriol 194:3972–3976. https://doi.org/10.1128/JB.00382-12

Rosenblatt C, de Araujo FFT, Frankel RB (1982a) Light scattering determination of magnetic moments of magnetotactic bacteria (invited). J Appl Phys 53:2727–2729. https://doi.org/10.1063/1.330948

Rosenblatt C, de Araujo FFT, Frankel RB (1982b) Birefringence determination of magnetic moments of Magnetotactic Bacteria. Biophys J 40:83–85. https://doi.org/10.1016/S0006-3495(82)84461-5

Scheffel A, Schüler D (2007) The acidic repetitive domain of the Magnetospirillum gryphiswaldense MamJ protein displays Hypervariability but is not required for Magnetosome chain assembly. J Bacteriol 189:6437–6446. https://doi.org/10.1128/JB.00421-07

Scheffel A, Gruska M, Faivre D, Linaroudis A, Plitzko JM, Schüler D (2006) An acidic protein aligns magnetosomes along a filamentous structure in magnetotactic bacteria. Nature 440:110–114. https://doi.org/10.1038/nature04382

Scheffel A, Gärdes A, Grünberg K, Wanner G, Schüler D (2008) The major magnetosome proteins MamGFDC are not essential for magnetite biomineralization in Magnetospirillum gryphiswaldense but regulate the size of magnetosome crystals. J Bacteriol 190:377–386. https://doi.org/10.1128/JB.01371-07

Schübbe S, Würdemann C, Peplies J, Heyen U, Wawer C, Glöckner FO, Schüler D (2006) Transcriptional organization and regulation of magnetosome operons in Magnetospirillum gryphiswaldense. Appl Environ Microbiol 72:5757–5765. https://doi.org/10.1128/AEM.00201-06

Simmons SL, Bazylinski DA, Edwards KJ (2006) South-seeking Magnetotactic Bacteria in the northern hemisphere. Science 311:371–374. https://doi.org/10.1126/science.1122843

Siponen MI, Legrand P, Widdrat M, Jones SR, Zhang W-J, Chang MCY, Faivre D, Arnoux P, Pignol D (2013) Structural insight into magnetochrome-mediated magnetite biomineralization. Nature 502:681–684. https://doi.org/10.1038/nature12573

Sonkaria S, Fuentes G, Verma C, Narang R, Khare V, Fischer A, Faivre D (2012) Insight into the assembly properties and functional organisation of the magnetotactic bacterial actin-like homolog, MamK. PLoS One 7:e34189. https://doi.org/10.1371/journal.pone.0034189

Staniland SS, Rawlings AE (2016) Crystallizing the function of the magnetosome membrane mineralization protein Mms6. Biochem Soc Trans 44:883–890. https://doi.org/10.1042/BST20160057

Staniland SS, Moisescu C, Benning LG (2010) Cell division in magnetotactic bacteria splits magnetosome chain in half. J Basic Microbiol 50:392–396. https://doi.org/10.1002/jobm.200900408

Tanaka M, Arakaki A, Matsunaga T (2010) Identification and functional characterization of liposome tubulation protein from magnetotactic bacteria. Mol Microbiol 76:480–488. https://doi.org/10.1111/j.1365-2958.2010.07117.x

Toro-Nahuelpan M, Müller FD, Klumpp S, Plitzko JM, Bramkamp M, Schüler D (2016) Segregation of prokaryotic magnetosomes organelles is driven by treadmilling of a dynamic actin-like MamK filament. BMC Biol 14:88. https://doi.org/10.1186/s12915-016-0290-1

Toro-Nahuelpan M, Giacomelli G, Raschdorf O, Borg S, Plitzko JM, Bramkamp M, Schüler D, Müller F-D (2019) MamY is a membrane-bound protein that aligns magnetosomes and the motility axis of helical magnetotactic bacteria. Nat Microbiol 4(11):1978–1989. https://doi.org/10.1038/s41564-019-0512-8

Uebe R, Schüler D (2016) Magnetosome biogenesis in magnetotactic bacteria. Nat Rev Microbiol 14:621–637. https://doi.org/10.1038/nrmicro.2016.99

Uebe R, Junge K, Henn V, Poxleitner G, Katzmann E, Plitzko JM, Zarivach R, Kasama T, Wanner G, Pósfai M, Böttger L, Matzanke B, Schüler D (2011) The cation diffusion facilitator proteins MamB and MamM of Magnetospirillum gryphiswaldense have distinct and complex functions, and are involved in magnetite biomineralization and magnetosome membrane assembly. Mol Microbiol 82:818–835. https://doi.org/10.1111/j.1365-2958.2011.07863.x

Uebe R, Keren-Khadmy N, Zeytuni N, Katzmann E, Navon Y, Davidov G, Bitton R, Plitzko JM, Schüler D, Zarivach R (2018) The dual role of MamB in magnetosome membrane assembly and magnetite biomineralization. Mol Microbiol 107:542–557. https://doi.org/10.1111/mmi.13899

Ullrich S, Kube M, Schübbe S, Reinhardt R, Schüler D (2005) A hypervariable 130-kilobase genomic region of Magnetospirillum gryphiswaldense comprises a magnetosome island which undergoes frequent rearrangements during stationary growth. J Bacteriol 187:7176–7184. https://doi.org/10.1128/JB.187.21.7176-7184.2005

Vargas G, Cypriano J, Correa T, Leão P, Bazylinski DA, Abreu F (2018) Applications of Magnetotactic Bacteria, Magnetosomes and Magnetosome crystals in biotechnology and nanotechnology: mini-review. Mol J Synth Chem Nat Prod Chem 23(10):2438. https://doi.org/10.3390/molecules23102438

Zahn C, Keller S, Toro-Nahuelpan M, Dorscht P, Gross W, Laumann M, Gekle S, Zimmermann W, Schüler D, Kress H (2017) Measurement of the magnetic moment of single Magnetospirillum gryphiswaldense cells by magnetic tweezers. Sci Rep 7:3558. https://doi.org/10.1038/s41598-017-03756-z

Gas Vesicles of Archaea and Bacteria

Felicitas Pfeifer

Contents

1 Introduction 74
2 Ecology of Gas Vesicle–Producing Archaea and Bacteria 75
  2.1 Halophilic Archaea 76
  2.2 Phototrophs and Other Aquatic Bacteria 78
  2.3 The Enterobacterium *Serratia* sp. ATCC39006 79
3 Gas Vesicle Wall and Proposed Structure of GvpA 80
  3.1 Gas Permeability and Critical Collapse Pressure 80
  3.2 GvpA Protein and the Gas Vesicle Wall 81
  3.3 Effect of Point Mutations in GvpA 84
  3.4 The Stabilizing Protein GvpC 85
4 Gas Vesicle Genes and Proteins of Haloarchaea 87
  4.1 Gas Vesicle Gene Clusters 88
  4.2 Transcription of *gvp* Genes and Putative Functions of Their Gene Products 90
  4.3 Regulation of Gas Vesicle Formation by GvpDE and Other Factors 92
5 Gas Vesicle Genes and Proteins of Bacteria 94
  5.1 Gas Vesicle Genes of Cyanobacteria 94
  5.2 The Gas Vesicle Gene Cluster of *Serratia* Sp. ATCC39006 95
  5.3 The *gvp* Gene Clusters of Soil Bacteria 96
6 Applications of Gas Vesicles 97
  6.1 Antigen Display and Antibody Formation 97
  6.2 Novel Enhancement Tool in Acoustic Resonance Imaging 98
  6.3 Imaging of Microbes in the Mammalian Microbiome 100
7 Conclusion 100
References 101

Abstract Gas vesicles are hollow nanostructures of spindle or cylinder shape produced by cyanobacteria, some heterotrophic bacteria as well as archaea. The possession of gas vesicles enables aquatic microbes to avoid sinking and to maintain

F. Pfeifer (✉)
Microbiology and Archaea, Department Biology, Technische Universität Darmstadt, Darmstadt, Germany
e-mail: pfeifer@bio.tu-darmstadt.de

D. Jendrossek (ed.), *Bacterial Organelles and Organelle-like Inclusions*, Microbiology Monographs 34, https://doi.org/10.1007/978-3-030-60173-7_4

a certain position in the water column; some of the gas-vesiculate strains even float at the surface. The rigid gas vesicle wall is formed solely of protein and freely permeable to gas molecules. The inner surface of the wall is hydrophobic, and surface tension presumably prevents the formation of water droplets inside. Major component is the small and hydrophobic GvpA that forms 4.6-nm-wide ribs running perpendicular to the long axis of the gas vesicle. The wall is stabilized by the second structural protein, GvpC, attached to the exterior surface. An *in silico* structure of GvpA is available and has been tested by amino acid substitutions to deduce sequence positions essential to form a gas-filled vesicle. Additional proteins are required for gas vesicle formation, and the *gvp* gene cluster involved has been investigated in haloarchaea, cyanobacteria, *Serratia*, *Bacillus* and *Streptomyces.* Special applications of gas vesicles in biomedical research and clinical diagnostics are their usage as effective antigen-presentation systems, or as novel reporters in acoustic resonance imaging.

List of Non-common Abbreviations

Aa	amino acid
A_{mut}	mutant *gvpA* gene expressed under P_{fdx}-control in an expression vector
ΔA	vector construct containing except for *gvpA* all *gvp* genes of p-vac
ΔC	vector construct containing except for *gvpC* all *gvp* genes of p-vac
ΔD	transformant harbouring two vector plasmids with sequences of p-vac; one contains *gvpACNO* and the other one *gvpEFGHIJKLM*. Or: single construct harbouring mc-vac that incurred a deletion in *gvpD*
ΔE	transformant harbouring two compatible vectors with sequences of p-vac; one contains *gvpACNO-gvpD* and the other one *gvpFGHIJKLM*
ΔF	transformant harbouring two compatible vectors with sequences of p-vac; one contains *gvpACNO-gvpDE* and the other one *gvpGHIJKLM*
ΔG	transformant harbouring two vector plasmids with sequences of p-vac; one contains *gvpACNO-gvpDEF* and the other one *gvpHIJKLM*
ΔH	transformant harbouring two vector plasmids with sequences of p-vac; one contains *gvpACNO-gvpDEFG* and the other one *gvpIJKLM*
ΔI	transformant harbouring two vector plasmids with sequences of p-vac; one contains *gvpACNO-gvpDEFGH* and the other one *gvpJKLM*
ΔJ	transformant harbouring two vector plasmids with sequences of p-vac; one contains *gvpACNO-gvpDEFGHI* and the other one *gvpKLM*

ΔK	transformant harbouring two vector plasmids with sequences of p-vac; one contains *gvpACNO-gvpDEFGHIJ* and the other one *gvpLM*
ΔL	transformant harbouring two vector plasmids with sequences of p-vac; one construct contains *gvpACNO-gvpDEFGHIJK* and the other one *gvpM*
ΔM	vector plasmid containing p-vac that incurred a deletion of *gvpM*
ΔN	transformant harbouring two vector plasmids with sequences of p-vac; one construct contains *gvpACO* and the other one *gvpDEFGHIJKLM*
ΔO	transformant harbouring two vector plasmids with sequences of p-vac; one construct contains *gvpACN* and the other one *gvpDEFGHIJKLM*
D^{ex}	*gvpD* expressed under P_{fdx} promoter control in vector pJAS35
$GHIJKLM^{ex}$	*gvpGHIJKLM* genes expressed under P_{fdx}-control in vector pJAS35
Glu-C	Endopeptidase from *Staphylococcus aureus* (cleaves D-X and E-X)
Gvp	gas vesicle protein
gvp	gas vesicle protein gene
M^{ex}	*gvpM* expressed under P_{fdx} promoter control in vector pJAS35
M_{mut}	mutant *gvpM* gene expressed under P_{fdx}-control in an expression vector
ISH	insertion sequence in haloarchaea
P_A	promoter of *gvpA*
P_D	promoter of *gvpD*
P_F	promoter of *gvpF*
P_{fdx}	promoter of the halobacterial ferredoxin gene
RsmA	Small RNA-binding protein found in bacteria
Vac	gas vacuole phenotype; gas vacuoles consists of several gas vesicles
vac region	genomic region containing all *gvp* genes required for gas vesicle formation
p-vac	plasmid-encoded vac-region of *Hbt. salinarum* containing 14 *gvp* genes as oppositely oriented p-*gvpACNO* and p-*gvpDEFGHIJKLM* gene clusters
c-vac	(mini)chromosomally encoded vac region of *Hbt. salinarum* consisting of the oppositely oriented c-*gvpACNO* and c-*gvpDEFGHIJKLM* differing in sequence from p-vac and mc-vac (Englert et al. 1992a)
mc-vac	chromosomally encoded vac-region of *Haloferax mediterranei* consisting of the oppositely oriented mc-*gvpACNO* and mc-*gvpDEFGHIJKLM* differing in sequence from p-vac and c-vac (Englert et al. 1992a)
nv-vac	chromosomally encoded vac region of *Halorubrum (Natronobacterium) vacuolatum* consisting of the single-gene cluster nv-*gvpACNO-FGHIJKLM*

1 Introduction

Gas vesicles are intracellular, gas-filled protein structures found in some bacteria and archaea. Especially aquatic microbes form gas vesicles, such as the photosynthetic cyanobacteria and some extremely halophilic archaea (haloarchaea). Even a methanogen, an enterobacterium and a few soil bacteria produce these hollow nanoparticles. Gas vesicles provide buoyancy, allowing the cells to counteract sinking and maintain a position in the water body where the conditions are optimal for growth (Walsby 1994). When present in sufficient amounts, they facilitate floating and colonization at the air–liquid interface, where biofilm-like communities are formed. Groups of gas vesicles constitute a gas vacuole (Vac), easily recognized as white bodies inside the cells by phase contrast light microscopy or transmission electron microscopy imaging (Fig. 1).

Gas vesicles have been first described in cyanobacteria by Hans Klebahn, who demonstrated that the light-scattering, intracellular bodies indeed contain gas and

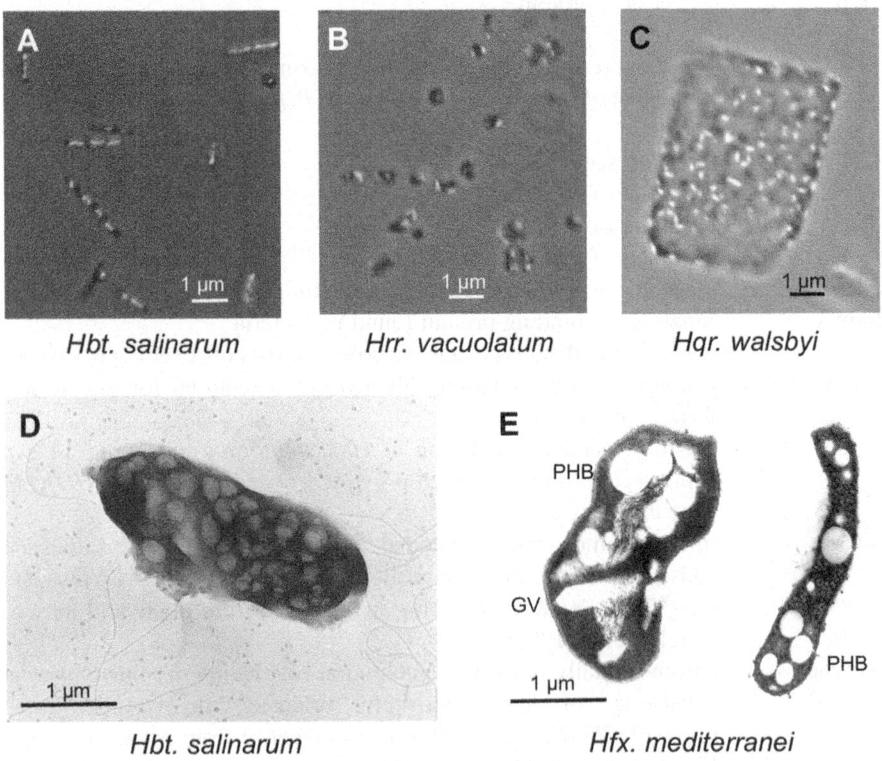

Fig. 1 Micrographs of haloarchaea. (**a–c**) Phase contrast (light) micrographs of *Hbt. salinarum*, *Hrr. vacuolatum* and *Hqu. walsbyi*. Gas vesicles are visible as white bodies inside the cells. (**d**, **e**) Electron micrographs of *Hbt. salinarum* (**d**) and thin sections of *Hfx. mediterranei* showing a top view (left) and a side view (right) of the sliced cells (**e**). Gas vesicles (GV) and spherical PHB granules (PHB) are marked

provide buoyancy, and similar results have been obtained by Helena Petter with the extremely halophilic archaeon *Halobacterium salinarum* (Klebahn 1895; Petter 1931). *Halobacterium* forms pink-opaque colonies on solid medium, and the cells contain up to 70 gas vesicles allowing them to float to the surface (Houwink 1956). The morphology of the gas-filled structures varies from spindle- to cylinder-shaped with sizes of 75–250 nm in diameter and 200–2000 nm in length. Treatment of cells with hydrostatic pressure results in the collapse of gas vesicles and paired "membranes" are found in their place (Bowen and Jensen 1965; Stoeckenius and Kunau 1968). The rigid wall is mainly constituted by the amphiphilic protein GvpA forming a single protein layer. Neither lipids nor carbohydrates are present. GvpA aggregates into a helix of low pitch seen as ribs running perpendicular to the long axis of the gas vesicle. The second structural protein, GvpC, strengthens the wall like a mesh by adhering to the exterior surface and stabilizing the wall formed by GvpA (Walsby and Hayes 1988). The formation of gas vesicles also requires several accessory Gvp proteins in minor amounts, especially at initial stages of their assembly.

This chapter will highlight features of these nanostructures, but also the diversity of gas-vesicle producers. Genetic analyses of gas vacuolate bacteria and haloarchaea uncovered up to 19 genes, and up to 11 of these gas vesicle protein (*gvp*) genes are required for gas vesicle formation as found for haloarchaea and the enterobacterium *Serratia* sp. ATCC39006 (Pfeifer 2012; Tashiro et al. 2016). Gas vesicle formation is regulated by environmental factors, and the regulatory systems differ in these microbes. Progress has been made with the 3D-structure of the haloarchaeal GvpA and its molecular arrangement in the gas vesicle wall, and also with the role of the accessory Gvp proteins acting as potential nucleation complex or as chaperones. Gas vesicles have also potential for special applications in biomedicine. They are used as an alternative antigen delivery system, but are also developed as novel reporters in acoustic resonance imaging or hyperpolarized xenon magnetic resonance imaging.

2 Ecology of Gas Vesicle–Producing Archaea and Bacteria

Gas vesicles supply aquatic microorganisms with a passive mode of movement. When present in sufficient amounts, cells float towards the air–water interface and form communities at or near the surface. Calculations on cyanobacteria and *Serratia* reveal that gas vesicles comprise 10–17% of the cell volume. The photosynthetic cyanobacteria use these vesicles for vertical movements in order to reach areas of higher light irradiance. Their cells are non-flagellated, whereas the haloarchaeon *Halobacterium salinarum* and the enterobacterium *Serratia* sp. ATCC39006 are motile due to archaella (archaeal flagella) or flagella, respectively. The ability to swim and to float requires coordination to avoid conflicting reactions in response to environmental factors.

2.1 Halophilic Archaea

Haloarchaea thrive in salt lakes containing sodium chloride up to saturation (30%). Such salty conditions are found in solar evaporation ponds, salt lakes like the Great Salt Lake in Utah, the Dead Sea, or shallow brine pools of arid regions. These water bodies are often coloured red due to the high cell density of the halophiles containing carotenoids to protect against UV-irradiation. Haloarchaea use the C_{50} carotenoid bacterioruberin. Oxygen is less soluble in brine than in water, resulting in steep diffusion gradients. Many haloarchaea are able to live under anoxic conditions and utilize nitrate or dimethyl sulfoxide (DMSO) as external electron acceptor, or ferment amino acids such as arginine. Only a small number of haloarchaea (8 out of 137 species; Oren 2013) produce gas vesicles that offer an efficient way for these aerobic microbes to reach the surface waters.

Gas vesicles of *Halobacterium salinarum* (basonym: *Hbt. halobium* or *Hbt. salinarium*) were first described by H. Petter who isolated red colonies from salted herring and dried cot (Petter 1931; Oren 2013). The rod-shaped cells are up to 5 μm in length and contain spindle-shaped gas vesicles throughout growth (Fig. 1a, d) (Stoeckenius and Kunau 1968). The gas vesicles of haloarchaea exhibit diameters up to 250 nm and are up to 2 μm in length (Larsen et al. 1967; Englert et al. 1990), and thus significantly larger in width compared to their bacterial counterparts (75–100 nm in diameter and up to 1 μm in length). A reason for this might be that the turgor pressure in haloarchaea is negligible due to a similarly high potassium chloride concentration inside as sodium chloride is found on the outside. The salt-in strategy adopts the cells to the high-salt environment. Colonies of *Hbt. salinarum* containing gas vesicles (Vac^+) are pink and opaque, whereas colonies of gas vesicle negative cells (Vac^-) are dark red and translucent (Fig. 2a). The mutation frequency

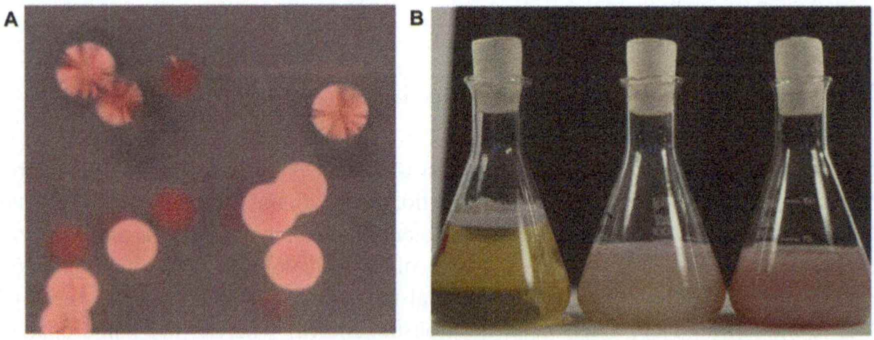

Fig. 2 *Hbt. salinarum* PHH1 colonies on solid medium and flotation of PHH1 in liquid cultures. (**a**) The colonies of Vac^+ cells are pink, whereas Vac^- cells form red and translucent colonies. Sectors are due to mutations in the *gvp* gene cluster caused by insertion elements. (**b**) Gas vesiculated cells remain swimming or floating in the culture and start to form a pellicle at the surface after 2 days standing on the lab bench. The culture on the left was standing without shaking for 3 weeks, whereas the culture on the right was moved from the shaker to the bench on the same day

for the Vac phenotype is with 1% very high and caused by insertion elements (ISH-elements) integrating in the *gvp* genes (Horne et al. 1991). Many copies of different, highly active ISH-elements are present in the *Hbt. salinarum* laboratory strains PHH1, NRC-1, and R1, whereas natural isolates such as GN101, or GRB, are genetically more stable (Pfeifer 2015).

Besides being red and gas-vacuolated, *Hbt. salinarum* produces the retinal protein bacteriorhodopsin (BR) acting as a light-driven proton pump under low oxygen tension and light (Oesterhelt 1998). The cells prefer regions of long-wavelength light (the absorption maximum of BR is 560 nm), whereas UV light is avoided. Two retinal-based photosensor proteins enable to monitor the quality of light, and the motility based on polar archaella is regulated by a signal transduction pathway resembling the bacterial chemotaxis system (Yao and Spudich 1992; Zhang et al. 1996). Gas-vacuolated *Hbt. salinarum* produce archaella and are able to swim (see Fig. 1d). However, to ensure that the cells flee from UV light and move deeper in the brine, gas vesicle formation is suppressed by UV light. Consequently, most haloarchaea position themselves in the salt water column rather than floating at the surface. In this way they also avoid exposure to rainfall that could lead to a sudden reduction in the outside salt concentration and cause lysis of the cells. In laboratory cultures, the gas-vesiculate *Hbt. salinarum* strains PHH1, NRC-1 and DSM3457 start to form a pellicle at the surface when liquid cultures are left standing at the bench for a few days (see Fig. 2b; the culture on the left is the oldest one).

Another gas-vesiculate haloarchaeon is the relatively flat *Haloferax mediterranei* forming cylinder-shaped gas vesicles in the stationary phase and only in media containing more than 17% salt (Fig. 1e) (Englert et al. 1990). Cylinder-shaped gas vesicles are also present in *Halorubrum vacuolatum* isolated from the Kenyan alkaline soda lake Magadi, but only when the salt concentration exceeds 15% (Fig. 1b) (Mwatha and Grant 1993; Mayr and Pfeifer 1997). The dependency of the Vac^{+} phenotype on salt suggests that gas vesicles are beneficial under high salt concentrations and support the migration of the cells to more oxygenated water near the surface. Other gas-vesiculate haloarchaea are *Halogeometricum borinquense* from a salt pond in Puerto Rico (Montalvo-Rodriguez et al. 1998), and *Haloplanus* species isolated from solar salterns (Elvi Bardavid et al. 2007; Cui et al. 2010, 2011).

The most unusual haloarchaeon containing gas vesicles is the large and square, but extremely thin *Haloquadratum walsbyi* (Fig. 1c), first isolated from a brine pool on the Sinai-peninsula (Parkes and Walsby 1981). *Hqr. walsbyi* is the dominant species in brine pools worldwide and is found in crystallizer ponds in Australia, Spain and Israel (Burns et al. 2004; Bolhuis et al. 2004; Walsby 2005). *Hqr. walsbyi* survives at extremely low water activities in media saturated with NaCl and magnesium chloride concentrations of up to 1 M. The thickness of the cells equals the diameter of gas vesicles that appear to keep the cellular membranes apart. Cells of *Hqr. walsbyi* do not float at the surface but remain distributed in the medium.

2.2 Phototrophs and Other Aquatic Bacteria

Cyanobacteria are found in seawater as well as in fresh water lakes. They gain energy by photosynthesis and move to positions where the light intensity is sufficient for phototrophic growth. Light decreases exponentially with depth due to the scattering or the absorption of light by the lake water (Walsby 1994). The filamentous species *Oscillatoria, Planktothrix, Anabaena* and *Calothrix,* but also the unicellular *Microcystis aeruginosa* and *Dactylococcopsis salina* produce gas vesicles. The cyanobacteria regulate their buoyancy in response to light, producing large amounts of gas vesicles at low irradiances and less gas vesicles at high irradiances. This leads to an upward movement of the cells from deeper regions of the lake, but avoids floating near the surface. Flotation at the lake surface exposes the cells to UV light, and sinking might be more favourable. *Anabaena flos-aquae* regulates buoyancy by the counteracting effects of gas vesicles and the accumulation of carbohydrate ballast formed during photosynthesis. Carbohydrates are considerably denser than water and an accumulation results in sinking, whereas the larger amounts of gas vesicles produced at low light intensity enable flotation (Walsby 1994). Both features allow the filamentous *Anabaena* to perform vertical migrations during the day. Calculations on the extent of these movements yield that single cells move up and down only a few centimetres per day, whereas large colonies may move tens of meters (Walsby and McAllister 1987). Spherical colonies of *Microcystis* with diameters between 2 and 4 mm exhibit the fastest floating velocities seen for cyanobacteria, with up to 1 mm per second. The average sinking velocity was calculated 52 m per day, allowing a migration to a depth of 8.6 m in 4 h, and a similar flotation rate allows return to the lakes surface in about the same time (Walsby and McAllister 1987). Thus, light is an important factor for gas vesicle and buoyancy regulation. About 3–10% of the cell volume must be occupied by gas vesicles to provide buoyancy (Oliver and Walsby 1984; Thomas and Walsby 1985). The abundance of the gas vesicles in *Microcystis* also increases in response to an elevated extracellular pH value, and the regulation occurs at the level of *gvp* gene expression (Gao et al. 2016).

A large variety of gas-vesiculate bacteria occur abundantly in the anaerobic hypolimnia of stratified lakes, such as purple sulphur bacteria (*Thiodictyon, Thiopedia, Ectothiorhodospira*), and green sulphur bacteria (*Pelodictyon, Ancalochloris, Chlorochromatium, Chloronema*) (Clarke and Walsby 1978; Overmann et al. 1991; Walsby 1994). These anoxigenic photosynthetic bacteria harvest the longer-wavelength far-red light and are often found in discrete layers in the anaerobic hypolimnia. Gradients of light, iron and sulphide may influence their vertical distribution. The green sulphur bacterium *Pelodictyon phaeoclathratiforme* regulates the cell density by the accumulation of carbohydrate ballast and a decrease in gas-vesicle production at high light intensity, similar to cyanobacteria (Overmann et al. 1991).

Gas-vesiculate bacteria are also found abundantly in cold marine and cold freshwater habitats (Staley et al. 1989; Gosink et al. 1997; Jung et al. 2004).

Filamentous gas-vesiculate cells constitute up to 83–91% of the total viable bacteria in sea ice of Antarctica. A remarkable diversity of phototrophic purple bacteria is present in the permanently ice-covered Lake Fryxell, Antarctica, where *Rhodoferax antarcticus* appears in high abundance at certain locations in the near-freezing water column (Karr et al. 2003; Jung et al. 2004). Gas vesicles are also observed in *Ancylobacter (Microcyclus) aquaticus, Aquabacter spiritensis* and in the appendaged bacteria *Stella* and *Prosthecomicrobium*, or the sheathed bacterium *Leptothrix* (Staley 1968; van Ert and Staley 1971; Konopka et al. 1977; Staley et al. 1987; Irgens et al. 1989). They are always used to position cells at a favourable depth in the water body.

2.3 *The Enterobacterium* Serratia *sp. ATCC39006*

Serratia sp. 39,006 was isolated from a salt marsh in New Jersey, USA. Besides containing gas vesicles, the bacterium produces the red pigment prodigiosin and several β-lactam antibiotics (Ramsay et al. 2011). Gas vesicles occur in the stationary phase and allow gaining access to oxygen-rich surface waters by flotation. *Serratia* uses flagella to swim or to swarm. All these traits (swarming motility, prodigiosin, β-lactam antibiotics, gas vesicles) depend on high cell densities and are controlled by the quorum sensing molecule *N*-butanoyl-*L*-homoserine lactone (BHL) (Ramsay and Salmond 2012). Moving passively to the surface might be more effective when the cells float as community as opposed to a single swimming bacterium, as described for filamentous cyanobacteria (Walsby and McAllister 1987; Walsby 1994).

Gas vesicle formation and swarming motility of *Serratia* are oppositely regulated by the global sRNA-binding protein RsmA (Ramsay et al. 2011). An overexpression of RsmA induces gas vesicle formation, whereas the flagella-bound swarming is repressed through repression of flagella and surfactant gene expression (Williamson et al. 2008). The simultaneous synthesis of flagella and gas vesicles is thus excluded. The RsmA regulon is a global regulation system encompassing more than 700 transcripts involved, e.g., in carbon storage and metabolism. Since mutants in *rsmA* sink faster than a strain lacking gas vesicles, RsmA presumably controls additional buoyancy factors. Carbon storage products could act as ballast and counteract floating, as found for cyanobacteria (Ramsay and Salmond 2012). The inverse regulation of gas vesicle formation and the flagella-dependent motility in *Serratia* is in contrast to *Hbt. salinarum,* where gas-vesiculate cells swim with archaella.

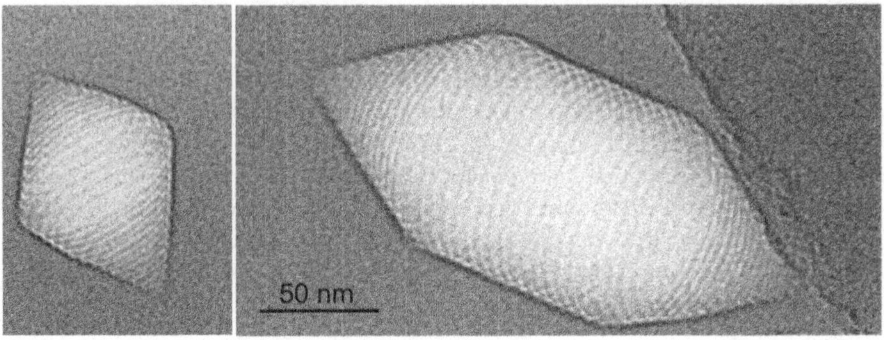

Fig. 3 Cryo-electron micrographs of isolated gas vesicles of *Hbt. salinarum*. Spindle-shaped and cylinder-shaped gas vesicles; the 4.6-nm-wide ribs formed by GvpA are running perpendicular to the long axis of the gas vesicle. The micrographs were taken in cooperation with D. Bollschweiler and H. Engelhardt (MPI Biochemistry, Martinsried, Germany). The bar equals 50μm for both images

3 Gas Vesicle Wall and Proposed Structure of GvpA

The properties of the gas vesicles depend on the proteins constituting the gas vesicle wall. A major component is GvpA forming the ribs of the wall (Fig. 3), and GvpC attaching to the exterior surface like a mesh to stabilize the structure.

3.1 *Gas Permeability and Critical Collapse Pressure*

Gas vesicles are freely permeable by gases such as O_2, N_2, H_2, CO_2, CO, CH_4 and Ar (Walsby 1994). They cannot store gas, and the gas inside is in equilibrium with the gas dissolved in the surrounding. Gas vesicles are permeable even to the large molecule C_4F_8, leading to the assumption that the wall must have pores or holes. Penetration of gas vesicles by C_4F_8 requires a pore diameter of 0.63 nm (Walsby 1982). Water vapour might be able to enter the gas vesicle, but the surface tension at the hydrophobic interior surface excludes the formation of water droplets, because the equilibrium vapour pressure over the convex surface of a water drop is higher than over a planar water surface (Walsby 1994). For gas vesicles of *Microcystis*, the filling time has been calculated as 0.4 μs, and the permeability coefficient of the wall as $\kappa = 32\ \mu ms^{-1}$ (Walsby et al. 1992). The latter value is more than 1000-fold higher than the gas permeability coefficient of a C_{16} lipid monolayer. Recent high-resolution cryo-electron micrographs of gas vesicles indicate regularly spaced holes along the ribs that might allow free exchange of gas molecules (D. Bollschweiler, F. Pfeifer, H. Engelhardt, unpublished results).

Gas vesicles are pressure sensitive and irreversibly collapse at higher hydrostatic pressures. The critical pressure at which isolated gas vesicles collapse varies from

0.09 MPa determined for *Hbt. salinarum* to more than 1 MPa for gas vesicles of *Oscillatoria* (Walsby and Bleything 1988). The critical pressure also correlates inversely with the width of the structure. Gas vesicles with the largest diameter (250 nm for haloarchaea) collapse at the lowest critical pressure, whereas the gas vesicles of *Dactylococcopsis salina, Microcystis* and *Oscillatoria* with diameters of 110 to 62 nm collapse at pressures of 0.3, 0.6 and 1 MPa, respectively (Walsby 1994). The turgor pressure in cyanobacteria ranges from 0.32 to 0.46 MPa, whereas the turgor pressure in *Hbt. salinarum* is negligible. Environmental studies with *Planktothrix rubescens* confirm the inverse correlation of gas vesicle width with the critical collapse pressure, since isolates derived from deeper regions of the lake indeed contain gas vesicles with smaller diameters than those collected closer to the surface (Bright and Walsby 1999; Beard et al. 2000). The differences in the width of the gas vesicles also correlates with the molecular mass of GvpC, and especially with the number of 33-amino acid (aa) repeats of this protein in cyanobacteria. Shorter GvpC variants are found with isolates from deeper water, whereas larger GvpC proteins are present in species collected near the surface water that contain the wider gas vesicles (Beard et al. 1999, 2000; Dunton and Walsby 2005). In vitro experiments demonstrate that GvpCs with reduced numbers of 33-aa repeats strengthen the cyanobacterial gas vesicle (Kinsman et al. 1995). However, a complete removal of GvpC from gas vesicles causes a three-fold decrease in the critical pressure, and the addition of GvpC restores the gas vesicle strength (Buchholz et al. 1993).

3.2 GvpA Protein and the Gas Vesicle Wall

Electron micrographs of collapsed and also of intact haloarchaeal gas vesicles indicate 4.6-nm-wide ribs running perpendicular to the long axis of the gas vesicle (Fig. 3) (Stoeckenius and Kunau 1968; Walsby 1972). The ribs are formed by a helix of low pitch rather than by a stack of hoops, and consist of a single layer of the GvpA protein (Walsby 1972; Offner et al. 1998). The GvpA protein is small (7–8 kDa) (see Table 1) and one of the most hydrophobic proteins known. The protein aggregates in aqueous solutions and is resistant to chaotropic reagents. Isolated gas vesicles cannot be disintegrated by detergents to deduce the protein content by SDS-polyacrylamide gel electrophoresis (SDS-PAGE); they remain in the separating gel. Only the second gas vesicle structural protein, GvpC (42 kDa) (Table 1), is easily washed off and appears as monomer in the gel (Englert et al. 1992b; Englert and Pfeifer 1993). Several additional Gvp proteins are required for gas vesicle formation in minor amounts (Table 1). The *gvp* genes encoding all these proteins in *Hbt. salinarum* PHH1 are clustered in a 10 kbp region (p-vac region) located on plasmid pHH1 (Englert et al. 1992a).

A crystal structure of GvpA is not available, since the aggregates cannot be dissolved without denaturing (Belenky et al. 2004). Secondary structure predictions and solid-state NMR imply a coil–α–β–β–α–coil fold (Sivertsen et al. 2010). X-ray analysis has been performed with layers of collapsed gas vesicles from

Table 1 Gvp proteins encoded by p-vac in *Halobacterium salinarum*

Protein	Size, putative function and remarks
GvpA	8 kDa; major gas vesicle structural protein; hydrophobic proteins of coil-α–β–β–α-coil structure, member of A-J-M class; aggregates to form the helical ribs of the gas vesicle wall. Sequence similarity to GvpJ (50%) and GvpM (48%).
GvpC	42.3 kDa; structural gas vesicle protein attached to the exterior surface forming a mesh; sequence contains 6-7 aa repeats of 32–38 aa of α-helical structure; strengthens gas vesicle wall; determines cylinder shape.
GvpF	24 kDa; essential accessory[a] Gvp; structural similarity to GvpL. Interacts with GvpM and GvpA, forms a putative nucleation complex with other accessory Gvp.
GvpG	10 kDa; essential accessory Gvp; member of a putative nucleation complex.
GvpH	20.8 kDa; accessory Gvp not required to form a gas-filled vesicle; gas vesicles of ΔH transformants are weaker compared to wild type. Prevents GvpM aggregation.
GvpI	16 kDa; accessory Gvp; basic pI of 10.8; not required to form gas vesicle, GV are up to 2.7μm in length. Involved in termination of the enlargement of gas vesicles. Member of the nucleation complex.
GvpJ	11.9 kDa; essential hydrophobic accessory Gvp; member of the A-J-M class; sequence similarity to GvpA (50%) and GvpM (60%). Member of the nucleation complex
GvpK	12.6 kDa; essential accessory Gvp. Member of the nucleation complex.
GvpL	32 kDa; essential accessory Gvp, structural homology to GvpF; interacts with GvpF, GvpG, GvpI, GvpJ and GvpM. Member of the nucleation complex.
GvpM	9.2 kDa; essential hydrophobic accessory Gvp; required for initial step in their synthesis. Member of A-J-M class. Similarity to GvpA (48%) and GvpJ (60%). Interacts with GvpH, GvpJ, GvpL and is member of a nucleation complex.
GvpN	39 kDa; accessory Gvp; contains a NTP binding/AAA+ domain; chaperon; required to enlarge the bicones into cylinder-shaped gas vesicles.
GvpO	13.2 kDa; essential accessory Gvp of unknown function.
GvpD[b]	59.3 kDa; oligomeric regulator protein with repressing function; NTP-binding domain essential for its function; NTP and arginine-rich region are similar to KaiC or RadA. Presence of GvpD leads to the degradation of GvpE.
GvpE[b]	21 kDa; dimer; transcriptional activator of the divergent promoters P_A and P_D; the 20-nt GvpE-responsive elements locate upstream and adjacent to BRE/TATA-box and overlap in the centre. Leucine-zipper domain at the C-terminus.

[a]Accessory Gvp proteins are present in minor amounts; either essential or not essential for gas vesicle formation

[b]The two endogenous regulatory proteins

cyanobacteria and haloarchaea (Blaurock and Walsby 1976; Blaurock and Wober 1976). A stacking periodicity with multiples of 1.95 nm (*Anabaena flos-aquae*) or 2 nm (*Halobacterium*) was found that relates to the thickness of the wall. The subunits along the ribs are regularly spaced at intervals of 1.15 nm, and paired β-chains of GvpA might cross the ribs at an angle of 55 ° to the rib axis as confirmed by atomic force microscopy (McMaster et al. 1996). The data suggest that the β-chains cross much of the full width of the rib (Walsby 1994).

MALDI-TOF mass spectrometry with gas vesicles of *Hbt. salinarum* and *Anabaena flos-aquae* confirm the lack of posttranslational modifications of GvpA (Belenky et al. 2004). Proteases and peptidases were used to determine which

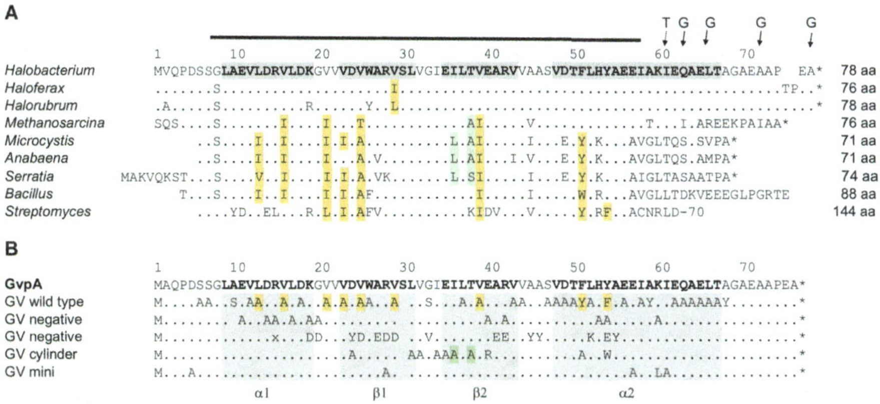

Fig. 4 Alignment of GvpA sequences and summary of the results obtained with GvpA mutations. The GvpA sequence derived from p-vac is presented on top, respectively. The amino acids constituting the helices α1, α2 and the β-strands β1, β2 are in bold and shaded in grey. Dots refer to identical amino acids. (**a**) GvpA sequences of archaea and different bacteria. The bar on top denotes the conserved 50-aa region. Letters and arrows above the *Hbt. salinarum* GvpA sequence (derived from p-vac) indicate sites that are accessible to Trypsin (T; cleaves R-X and K-X) and the endoproteinase Glu-C (G; cleaves D-X and E-X). Amino acid alterations that occur at positions where single substitutions in p-vac GvpA result in wild-type gas vesicles (Fig. 4b) are shaded in yellow, and alterations leading to cylinder-shaped gas vesicles in GvpA are shaded in green. See Fig. 4b for comparison. Further explanations are in the text. The GvpA sequences derive from: *Hbt. salinarum, Hfx. mediterranei, Halorubrum vacuolatum, Methanosarcina barkeri, Microcystis aeruginosa, Anabaena flos-aquae, Serratia* sp. ATCC39006, *Bacillus megaterium, Streptomyces coelicolor*. (**b**) Summary of amino acid substitutions in GvpA derived from the p-vac region of *Hbt. salinarum* and Vac phenotypes. Note that each of the substitutions indicated as letter refers to a single amino acid replacement (after Knitsch et al. 2017)

peptide bonds of GvpA are exposed at the exterior surface. The proteinase Trypsin is able to cleave K59-I60 in some GvpA molecules of *Hbt. salinarum*, whereas all lysines in the conserved central portion are inaccessible (Fig. 4a) (Belenky et al. 2004). The endoproteinase Glu-C from *Staphylococcus aureus* (cleaves D-X and E-X) degrades only the charged C-terminal portion of GvpA, whereas the D-X or E-X bonds in the core region are not affected. Treatment of gas vesicles with Glu-C results in the immediate collapse of the structure (Belenky et al. 2004), suggesting that the C-terminal portion of GvpA is exposed at the exterior side and important for an intact gas vesicle.

The amino acid sequence of GvpA is highly conserved, especially in a central region of 50 amino acids (Fig. 4a). In the case of aquatic archaea and bacteria, GvpA is 71–78 aa long, whereas GvpA sequences of soil bacteria are significantly longer (88–144 aa). The 50-aa core sequence of the bacterial GvpA exhibits characteristic differences compared to the haloarchaeal GvpA, and many of these differences are also found in GvpA of the gas-vesiculate methanogenic archaeon *Methanosarcina barkeri* (Fig. 4a) (Archer and King 1984). A first structural model of GvpA was obtained by high-performance de novo modelling for GvpA derived from *Haloferax*

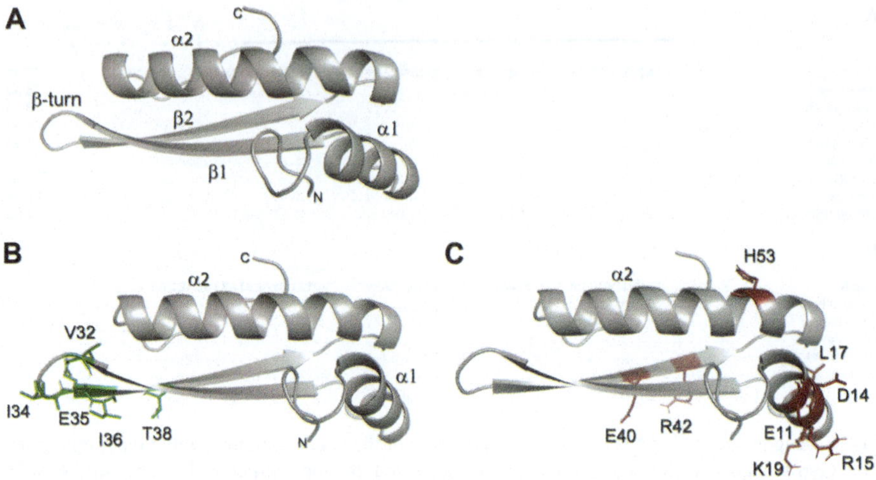

Fig. 5 Model of the GvpA structure and effect of amino acid substitutions. The structure was obtained by *de-novo* modelling (Strunk et al. 2011). (**a**) Side view of GvpA illustrating the α–β–β–α 3D-structure. (**b**) Side view of GvpA indicating substitutions near the β-turn resulting in cylindrical gas vesicles labelled in green. (**c**) Side view of GvpA with alterations leading to a Vac$^-$ phenotype labelled in red

mediterranei (Fig. 5a) and confirmed for the almost identical GvpA of the *Hbt. salinarum* p-vac region (Strunk et al. 2011; Ezzeldin et al. 2012). The model predicts two α-helices (α1, L9-K19 and α2, V48-T67), and two β-chains including a β-turn (β1, V23-L31; β2, E35-V43; β-turn, V32-I34) (Figs. 4a and 5a) (Strunk et al. 2011). The hydrophobic inner surface of the gas vesicles is presumably constituted by the antiparallel β-strands of GvpA.

3.3 *Effect of Point Mutations in GvpA*

The proposed *in silico* structure of GvpA allows pinpointing contact sites to adjacent GvpA, and single amino acids of GvpA were substituted by alanine or another amino acid to test the ability of these GvpA variants to form gas vesicles (Strunk et al. 2011; Knitsch et al. 2017). The 85 individual amino acid substitutions in the *Hbt. salinarum* GvpA were studied in *Haloferax volcanii* ΔA + A_{mut} transformants (containing construct ΔA, harbouring except for *gvpA* all *gvp* genes of the p-vac region, and construct A_{mut}, expressing the mutated *gvpA* reading frame under ferredoxin promoter control in vector pJAS35). The possession of gas vesicles can be assessed by the colony phenotype on agar plates, or by phase-contrast and transmission electron microscopy of the cells.

A single amino acid substitution in GvpA either abolishes gas vesicle formation, alters the gas vesicle shape or has no influence at all (Knitsch et al. 2017). A

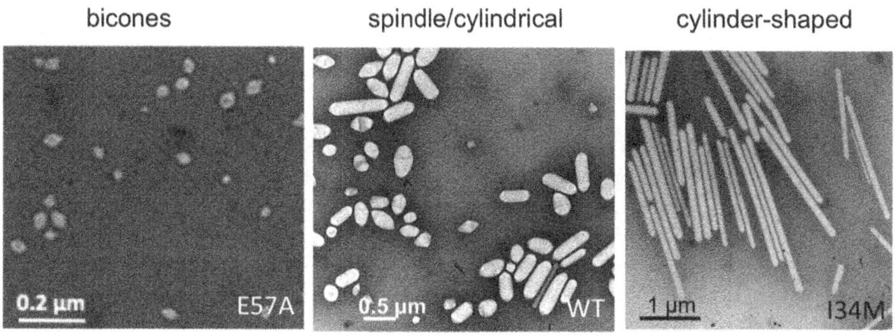

Fig. 6 Electron micrographs of gas vesicles of different shape derived from wild-type or GvpA mutants. The respective wild-type or mutant GvpA leading to these morphologies is indicated on the right side at the bottom

summary of the results is presented in Fig. 4b, and typical gas vesicle morphologies obtained are shown in Fig. 6. Replacement of a non-polar by a charged residue in the β1- or β2-strand of GvpA always results in a Vac⁻ phenotype. Introducing a negative charge in this region might disrupt the hydrophobic interior surface of the gas vesicles. In contrast, replacement of a larger hydrophobic aa by alanine usually results in gas vesicles of wild-type shape (Fig. 4a). However, an alanine replacement of a residue in or near the β-turn (31-LVGIEI-37) yields gas vesicles of cylinder instead of the spindle shape (Figs. 4b and 6). These aa residues constitute a tip in GvpA presumably contacting another GvpA in an adjacent rib (Fig. 5b). Helix α1 located on the opposite side of GvpA might contact a GvpA molecule in another rib. An alanine replacement of one of the charged aa in helix α1 often leads to a Vac⁻ phenotype (Figs. 4b and 5c). All replacements preventing gas vesicle formation localize to aa at the exterior surface of GvpA. They include charged and non-polar residues of helix α1, E40 and R42 presumably forming a salt-bridge between β2 and β2′, and H53-Y54 of α2 (Figs. 4b and 5c). All these residues could offer contact sites between GvpA molecules in the same or in an adjacent rib (Knitsch et al. 2017). In contrast, most replacements in helix α2 near the C-terminus do not affect gas vesicle formation (Fig. 4b). It is possible that α2 contacts other Gvp proteins at the exterior surface of the gas vesicle wall, such as GvpC or GvpN. A comparison of the amino acid positions in GvpA affecting the gas vesicle formation in haloarchaea with the bacterial GvpA sequences yields common features. Many differences in the bacterial GvpA are found at positions where a replacement in the haloarchaeal GvpA results in Vac$^+$ ΔA + A_{mut} transformants (Fig. 4a, b) (Knitsch et al. 2017).

3.4 The Stabilizing Protein GvpC

The second gas vesicle structural protein, GvpC, is attached to the exterior surface of the gas vesicle in archaea and bacteria (Tables 1 and 2). The sequence of GvpC

Table 2 Gas vesicle proteins of *Serratia* sp. ATCC39006

Protein	Size, putative function and remarks (after Ramsay et al. 2011)
GvpA1	8 kDa; major gas vesicle structural protein; 31% identity with GvpA2; 51% identity with GvpA3.
GvpA2	7.1 kDa; essential Gvp; 31% identity with GvpA1; 34% identity with GvpA3. Clusters with bacterial GvpS proteins; distantly related to the halobacterial GvpM/GvpJ.
GvpA3	12 kDa; essential Gvp; 51% identity with GvpA1; 34% identity with GvpA2. Clusters with bacterial GvpJ proteins.
GvpC	18 kDa; second gas vesicle structural protein; 3 degenerate copies of 33 aa repeat.
GvpF1	30 kDa; essential minor structural protein; related to the cyanobacterial GvpF protein.
GvpF2	28.6 kDa; essential minor structural protein GvpF/L.
GvpF3	24.2 kDa; essential minor structural protein; distantly related to bacterial GvpL.
GvpG	15.2 kDa; essential Gvp; nucleotide sequence contains nine near-identical 17 bp repeats.
GvpH	18.4 kDa; not essential gas vesicle protein GvpH
GvpK	12.6 kDa; essential gas vesicle protein GvpK; closely related to haloarchaeal GvpK.
GvpN	37.8 kDa; contains AAA+ ATPase domain; required to enlarge the bicones.
GvpV	11.5 kDa; essential bacterial Gvp protein; required to enlarge bicones.
GvpW	70 kDa; unrelated to known Gvp; not essential. Contains two AAA+ domains.
GvpX	10.9 kDa; unrelated to known Gvp in haloarchaea, not essential.
GvpY	21.1 kDa; unrelated to known Gvp in haloarchaea, not essential.
GvpZ	79.2 kDa; not essential bacterial gas vesicle protein of unknown function; two AAA+ ATPase domains.
GvrA[a]	52 kDa; GV regulator protein, related to NtrC; contains phosphorylation receiver domain, AAA+ ATPase and HTH DNA binding domain.
GvrB[a]	53.4 kDa; GV regulator protein, related to NtrB. Contains PAS sensor and histidine kinase output domain.
GvrC[a]	14 kDa; GV regulator protein; CheY-like receiver domain; 30% ident. to GvrA.

[a]The three endogenous regulatory proteins

contains several 33–34 amino acid repeats (4–5 in cyanobacteria and 6–7 less conserved repeats in haloarchaea) of predicted α-helical structure (Fig. 7). The length calculated for a repeat of 33 aa in α-helical structure is 5 nm which already exceeds the width of the 4.6 nm rib formed by GvpA (Hayes et al. 1988). For the cyanobacterial GvpC it was hypothesized that it could cross a rib at an angle by contacting five adjacent GvpA proteins and connecting them with other packages of five GvpA molecules in adjacent ribs like a stabilizing fibre. The GvpA to GvpC ratio of cyanobacterial gas vesicles has been determined to be 25:1 (Buchholz et al. 1993). In contrast, the haloarchaeal GvpC contains 6–7-aa repeats, 32–38 aa in length and of α-helical structure, and the negatively charged C-terminal domain contains a zinc-finger motif (CCHH) (Fig. 7b) (Englert and Pfeifer 1993). Negative charges at the surface of proteins are a typical adaptation to the high salt (4 M KCl) environment of the haloarchaeal cytoplasm. Cyanobacterial GvpC proteins lack this highly negatively charged C-terminal portion (Fig. 7b).

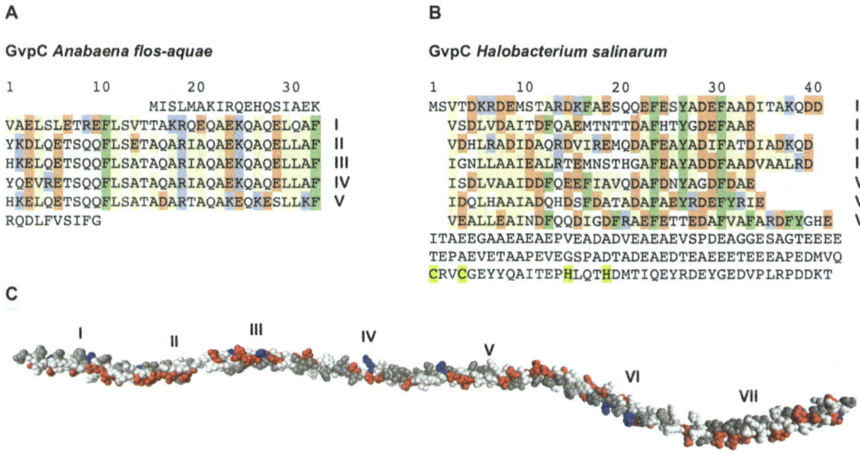

Fig. 7 Sequences of GvpC derived from the p-vac region of *Hbt. salinarum* and of *A. flos-aquae*, and structural prediction of the repeat sequence of the haloarchaeal GvpC. (**a**) The amino acid sequence of the GvpC derived from *A. flos-aquae*, and (**b**) of GvpC derived from the p-vac region of *Hbt. salinarum*, indicating the repeated sequences. Charged, aliphatic and aromatic aa are shaded in colour (red, acidic; blue, basic; green, aromatic; yellow, aliphatic). The zinc-finger motif (CCHH) in the C-terminal domain is labelled in bright yellow. (**c**) 3D-structural prediction of the seven amino acid repeats of the haloarchaeal GvpC using I-Tasser. The seven α-helical structures are labelled I through VII. The acidic C-terminal domain of GvpC is lacking in this structure

GvpC is easily removed from cyanobacterial gas vesicles by treatment with SDS or urea, and the stripped off gas vesicles remain intact and cylindrical, but become weaker (Hayes et al. 1992). A complete removal causes a three-fold decrease in the critical pressure, whereas the addition of GvpC restores their strength (Buchholz et al. 1993). GvpC attaches to both end-cones and the cylinder section of the gas vesicle. In the case of haloarchaea, GvpC is removed by washing the isolated gas vesicles with water. GvpC is not required for the formation of an intact gas vesicle in haloarchaea, since *Hfx. volcanii* ΔC transformants (construct ΔC contains a p-vac region lacking *gvpC)* produce large and pleomorphic gas-filled structures (Offner et al. 1996). However, the gas vesicles formed in the absence of GvpC are altered in shape and exhibit varying diameters throughout a single vesicle, suggesting that GvpC also influences the gas vesicle shape.

4 Gas Vesicle Genes and Proteins of Haloarchaea

The genes encoding the gas vesicle structural proteins GvpA and GvpC were identified in cyanobacteria and halobacteria, and transformation experiments demonstrated that a larger genomic region is required for gas vesicle formation. A total of fourteen *gvp* genes were identified in *Hbt. salinarum* and *Hfx. mediterranei*.

4.1 Gas Vesicle Gene Clusters

The gene encoding GvpA was first identified in the cyanobacterium *Calothrix* (Damerval et al. 1987) and then in the *Hbt. salinarum* strains NRC-1 and PHH1 (DasSarma et al. 1987; Horne et al. 1988). The haloarchaeal *gvpA* is located on large plasmids or mini-chromosomes (pNRC100 and pNRC200 in *Hbt. salinarum* NRC-1, or pHH1 in *Hbt. salinarum* PHH1) and affected by a mutation rate of 1% caused by insertion elements (Pfeifer et al. 1981; DasSarma et al. 1988; Horne et al. 1991). Frequent transpositions are characteristic for *Hbt. salinarum*; similar mutations also affect the bacterioruberin and bacteriorhodopsin synthesis (Pfeifer et al. 1981). Mutations affecting gas vesicle formation are easily observed with the change of the colony phenotype from opaque (Vac$^+$) to translucent (Vac$^-$), and some colonies are even sectored (Fig. 2a). In *Hbt. salinarum* PHH1, the insertions affect a 10-kbp region of plasmid pHH1 comprising the 14 *gvp* genes (p-vac region) (Horne et al. 1991). The genes are arranged as two oppositely oriented clusters, p-*gvpACNO* and p-*gvpDEFGHIJKLM* (Fig. 8) (Englert et al. 1992a). Expression of p-vac throughout growth leads to the spindle-shaped gas vesicles. An additional *gvp* gene cluster, c-vac, harbouring related 14 *gvp* genes in an identical arrangement as in p-vac, is present on another large plasmid in *Hbt. salinarum* PHH1, but these c-*gvp* genes are rarely expressed. However, *Hbt. salinarum* PHH4 and the type strain *Hbt. salinarum* DSM3457 both lack the p-vac region and express the c-vac region,

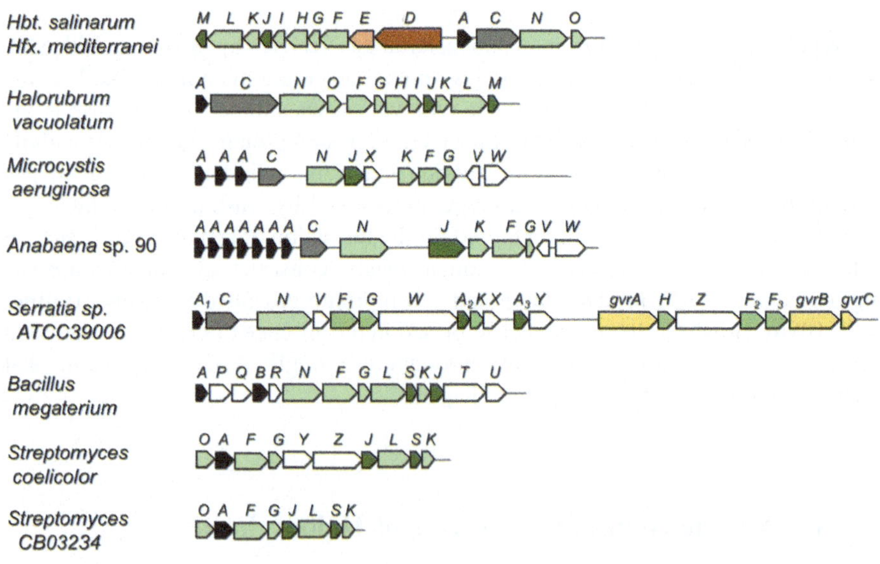

Fig. 8 Genetic maps of various *gvp* gene clusters of archaeal or bacterial gas vesicle producers. The letters designate *gvp* genes. The *gvpA* gene encoding the major gas vesicle protein is marked in black, and *gvpC* in grey. Genes encoding accessory proteins are marked in green. GvpJ and M are homologues of GvpA and indicated in dark green. Regulatory genes are labelled in red (*Halobacterium/Haloferax*) or yellow (*Serratia*)

leading to cylinder-shaped gas vesicles in the stationary phase (Krüger and Pfeifer 1996). *Hbt. salinarum* strain NRC-1 contains two copies of the *gvp1* gene cluster (related to p-vac), one on the 192-kbp pNRC100, and a second one on the 365-kbp pNRC200 (Ng et al. 2000). The latter mini-chromosome carries in addition to *gvp1* also the *gvp2* gene cluster resembling the c-vac region.

The moderately halophilic species *Haloferax mediterranei* contains similar 14 *gvp* genes oppositely arranged as mc-*gvpACNO* and mc-*gvpDEFGHIJKLM* clusters in the chromosome (mc-vac region). The strain produces long, cylinder-shaped gas vesicles in the stationary phase in high-salt media only (Englert et al. 1990, 1992a) (Fig. 1e). A different arrangement of 12 *gvp* genes is found in the haloalkaliphilic archaeon *Halorubrum (Natronobacterium) vacuolatum;* these genes are expressed as single nv-*gvpACNOFGHIJKLM* unit (nv-vac region) (Fig. 8). The two genes *gvpD* and *gvpE* encoding the endogenous gene regulators in the p-vac, c-vac or mc-vac region are missing in *Halorubrum vacuolatum* (Mayr and Pfeifer 1997; Pfeifer 2004). The GvpA sequences encoded by these four haloarchaeal vac regions are >97% identical (see Fig. 4a), the accessory proteins GvpF, GvpG, GvpJ, GvpK, GvpM and GvpN indicate sequence similarities of >79%, whereas similarities between 54 and 74% are found with GvpC, GvpH, GvpI, GvpL and GvpO. Putative functions of the accessory Gvp are described in more detail below (Tables 1 and 2).

The importance of each *gvp* gene for gas vesicle formation was assessed in *Hfx. volcanii* transformants (Offner et al. 2000). Single *gvp* genes of p-vac were deleted, and the remaining genes analyzed in respect to gas vesicle formation in transformants. The complementation was often performed using two fragments of p-vac inserted in two vector plasmids. Complementation of the fragments results in Vac^+ transformants, when all *gvp* genes are present. Also, the lack of *gvpC, gvpD, gvpE, gvpH, gvpI* or *gvpN* yields Vac^+ transformants suggesting that each of the proteins encoded here is not required to form a gas-filled protein vesicle. However, altered gas vesicle shapes are observed for the ΔC, ΔH, ΔI and ΔN transformants lacking the *gvpC, gvpH, gvpI* or *gvpN* gene, respectively (see below). The remaining eight *gvp* genes are essential, since the absence of one of them results in a Vac^- phenotype. The latter *gvp* genes constitute the minimal vac-region *gvpFGJKLM-AO*. Transformants harbouring these genes on a single construct produce minor amounts of gas vesicles (Offner et al. 2000).

Also, an overexpression of single *gvp* genes of the early transcription unit *gvpFGHIJKLM* (Fig. 8) was studied in p-vac + X^{ex} transformants. The proteins encoded by *gvpFGHIJKLM* are all produced in minor amounts and presumably accessory Gvp proteins. Construct X^{ex} contains one of these *gvp* reading frames highly expressed in pJAS35, leading to an overproduction of the respective Gvp protein in the transformant. Vac^- transformants were obtained with X = *gvpG, gvpH*, or *gvpM*, whereas all other transformants were Vac^+ (Tavlaridou et al. 2013). Since p-vac + $GHIJKLM^{ex}$ transformants (overexpressing *gvpGHIJKLM* including *gvpG, gvpH* and *gvpM* in pJAS35) are Vac^+, a balanced amount of these proteins is obviously important. Further studies show that the inhibitory effect of GvpM is neutralized by GvpH, GvpJ or GvpL in the respective p-vac + MX^{ex} transformants

(MX^{ex} = overexpressing *gvpH, gvpJ, or gvpL* in addition to *gvpM*). These results imply that GvpM forms complexes with GvpH, GvpJ and/or GvpL. Such complexes are indeed found in protein–protein interaction studies (Tavlaridou et al. 2014; Winter et al. 2018).

4.2 *Transcription of* gvp *Genes and Putative Functions of Their Gene Products*

The *gvpACNO* gene cluster is transcribed in late exponential and stationary stage, highly induced by the endogenous transcriptional activator GvpE. The unit encodes both gas vesicle structural proteins GvpA and GvpC in addition to GvpN and GvpO. The sizes and characteristics of all Gvp proteins are summarized in Table 1. GvpA forms the ribbed gas vesicle wall and GvpC attaches to the exterior gas vesicle surface (see above for details). GvpN contains a nucleotide-binding motif similar to ATPases of the AAA+ chaperone-like superfamily. ΔN transformants (lacking *gvpN*) produce only tiny bicones implying that further enlargement by the incorporation of GvpA is impaired when GvpN is not present. GvpN might have a chaperone and/or energy delivery function (Fig. 9). GvpO is essential, but its function in not clear (Offner et al. 1996; Englert et al. 1992b).

The *gvpFGHIJKLM* unit of p-vac is transcribed during exponential growth from promoter P_F and independent of GvpE-activation, leading to minor amounts of these accessory Gvp proteins (Offner and Pfeifer 1995). Because of the early transcription, we assume that these Gvp proteins are all required in initial or early steps in gas vesicle formation (Fig. 9). The activity of P_F is much lower compared to the GvpE-activated promoter P_D, driving the expression of the two regulatory genes *gvpDE*, and P_A yielding the structural proteins GvpA and GvpC (Hofacker et al. 2004; Born and Pfeifer 2019). The *gvpFGHIJKLM* mRNA encodes eight accessory Gvp proteins, six of which are essential, i.e. GvpF, GvpG, GvpJ, GvpK, GvpL and GvpM (Offner et al. 2000). Except for GvpF and GvpL, these proteins are small (9.2–16 kDa) (Table 1). GvpJ and GvpM are hydrophobic and related to GvpA

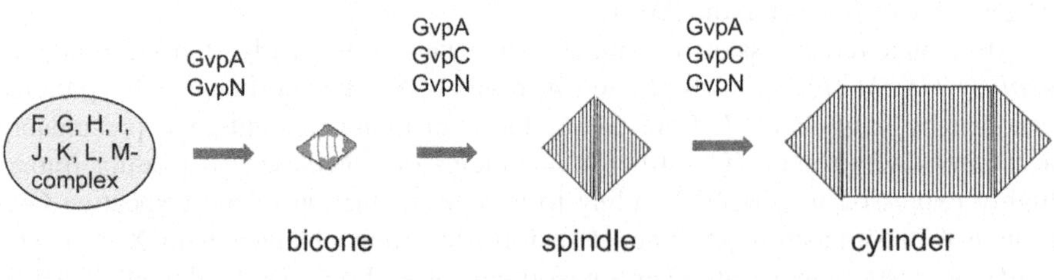

Fig. 9 Model of the gas vesicle formation in haloarchaea. The accessory Gvp constituting the putative nucleation complex are shown on the left. Attraction of GvpA leads to small bicone structures that are enlarged by the addition of GvpA and GvpC to a spindle- and then cylinder-shaped gas vesicle. The chaperone GvpN might provide energy during the assembly. For further explanations, see text

and thus members of the A-J-M class of gas vesicle proteins. GvpF and GvpL exhibit mutually very similar tertiary structures (Winter et al. 2018). A crystal structure of the cyanobacterial GvpF of *Microcystis* is available (Xu et al. 2014), and homology modelling of the haloarchaeal GvpF and GvpL proteins shows the close structural relationship (Winter et al. 2018). In vivo protein–protein interaction studies using split-GFP suggest that GvpL contacts GvpM near the N-terminus, whereas GvpF, GvpH and GvpJ interact with the C-terminal fragment of GvpM (Winter et al. 2018). These proteins might form a nucleation complex required at an early stage of bicone formation. GvpI and GvpH are not required for gas vesicle formation. ΔI transformants produce very long (up to 2.7 μm) cylinder-shaped gas vesicles, suggesting that the highly basic GvpI protein (pI = 10.8) determines the length of the structure by terminating the enlargement. ΔH transformants contain gas vesicles of wild-type size, but they are much weaker and disaggregate into ribs as soon as they are prepared for electron microscopy, implying a stabilizing role of GvpH (Offner et al. 2000).

Similar to GvpA, single amino acid substitutions were tested with the related GvpM in $\Delta M + M_{mut}$ transformants to determine important amino acid positions required for gas vesicle formation. These transformants contain the constructs ΔM (p-vac region lacking *gvpM*) and M_{mut}, expressing a mutated *gvpM* under ferredoxin promoter control in pJAS35 (Tavlaridou et al. 2014). Wild-type GvpM proteins complement ΔM and lead to Vac^+ $\Delta M + M^{ex}$ transformants. Single substitutions in GvpM either had no effect on the gas vesicle formation or resulted in the complete lack of these structures. Neither bicones nor other shape variants are observed, implying that GvpM is required in a very early step. The point mutations might disturb GvpM in such a way that interactions with other Gvps are impossible. These results on GvpM differ from mutations of GvpA, where an aa substitution either yields bicones, spindle- or cylinder-shaped gas vesicles or the complete lack of these structures. Despite the fact that GvpA, GvpJ and GvpM are related proteins, they cannot substitute each other and must have different functions in gas vesicle formation.

Further protein–protein interaction studies of GvpF, GvpG, GvpH, GvpI, GvpJ, GvpK GvpL and GvpM by Ni-NTA chromatography in vitro or split-GFP in vivo suggest that almost all of these proteins interact (Tavlaridou et al. 2014; Winter et al. 2018; Völkner et al. 2020). GvpL has several interaction partners (GvpF, GvpG, GvpI, GvpJ and GvpM), and GvpF contacts besides GvpL also GvpA (Völkner et al. 2020). Some or all of the accessory Gvp proteins might form a nucleation complex located in the tip of the conical end-structure (Fig. 9). GvpA attaches to this complex via GvpF and is forced to aggregate into ribs. A protein complex at the tip of the cones and inside the tip ensures that the ribs formed by GvpA become distant enough to expose their hydrophobic surface, forming a hollow structure rapidly filling with gas. Help of the accessory Gvp might also be required at the transition of the conical end-cones to the cylinder structure. The second structural protein GvpC attaches to the exterior surface already during the formation of the end cones, similar to the chaperone GvpN required for the enlargement of the structure.

4.3 Regulation of Gas Vesicle Formation by GvpDE and Other Factors

GvpD and GvpE regulate the *gvp* gene expression in *Hbt. salinarum* (p-vac and c-vac) and *Hfx. mediterranei* (mc-vac). GvpE is a transcriptional activator, whereas GvpD plays a role in the repression (Englert et al. 1992b, Offner and Pfeifer 1995; Krüger and Pfeifer 1996; Zimmermann and Pfeifer 2003; Hofacker et al. 2004). The 21-kDa GvpE forms a dimer in solution and resembles a basic leucine-zipper protein typical for eukaryotic gene regulators (Krüger et al. 1998; Plösser and Pfeifer 2002). With the exception of P_F and P_O of p-vac, and P_D of c-vac, all other P_D and P_A promoters are induced by GvpE (see Fig. 10a). Archaeal promoters consist of a TATA-box and a BRE element, binding the transcription factors TBP and TFB to attract the multi-component RNA polymerase for transcription initiation (Bell and Jackson 1998). The distance between the oppositely arranged BREs of P_D and P_A in p-vac is 35 bp only, and GvpE binds upstream of BRE. A 20-nt sequence, CATAACGA-n4-TGAAACCA, serves as GvpE-responsive element in P_A of p-vac, and similar 20-nt GvpE-activator sites are found in all other P_A and P_D promoters of c-vac and mc-vac. These UAS sequences overlap by 7 nucleotides in

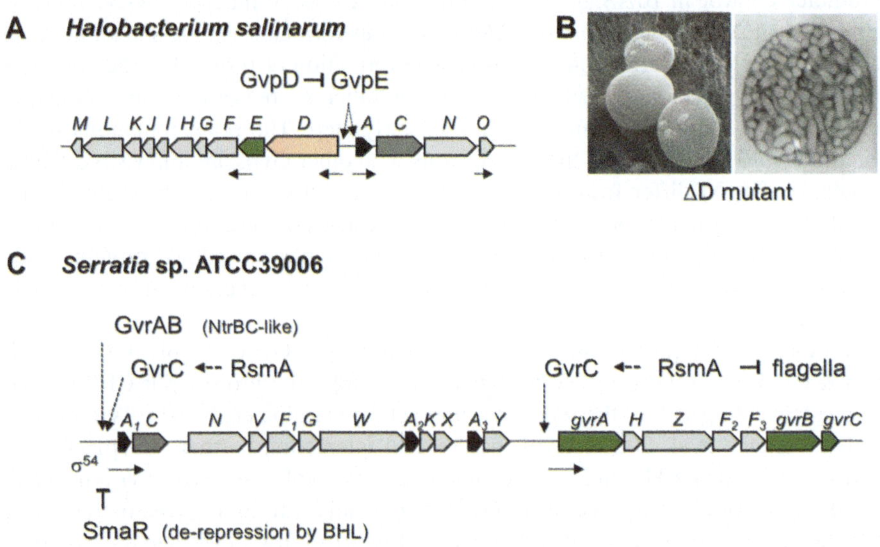

Fig. 10 Regulation of gas vesicle gene expression in *Hbt. salinarum* and *Serratia* ATCC39006. (**a**, **c**) Genetic maps of the two *gvp* gene clusters are shown, with regulatory genes encoding activators coloured in green, and GvpD involved in repression in orange. Arrows mark the start sites and direction of transcription. The action of the endogenous and global regulatory proteins is indicated by arrows. (**b**) Electron micrograph of *Hfx. volcanii* ΔD transformants lacking *gvpD*, with a high-resolution scanning electron micrograph of ΔD transformant cells (left) and transmission electron micrograph of ultrathin sections of these gas vesicle overproducer cells (right). The diameter of the cells is approx. 1.7µm. The micrographs were taken in cooperation with Gerhard Wanner (LMU Munich, Germany)

the center of P_A-P_D (Bauer et al. 2008; Marschaus and Pfeifer 2012). As determined by GFP fluorescence in vivo, the P_A promoter shows the highest basal activity of all p-vac promoters (14 × 10^3 [AU]) and is 13-fold induced by the endogenous GvpE, whereas P_D exhibits a 10-fold weaker basal activity (2 × 10^3 [AU]) and is 8-fold induced by GvpE (Born and Pfeifer 2019).

GvpD is involved in the repression of gas vesicle formation. The deletion of *gvpD* in the mc-vac region (construct ΔD) leads to gas vesicle overproduction in *Hfx. volcanii* ΔD transformants (Englert et al. 1992b). The numerous gas vesicles that appear alter the flat *Haloferax* cells into spheres (Fig. 10b). These ΔD transformants immediately float at the surface. The amount of gas vesicles is reduced to the wild-type level in ΔD + D^{ex} transformants (harbouring the ΔD construct and the second construct D^{ex} expressing mc-*gvpD*) (Pfeifer et al. 1994). GvpD forms larger oligomers and contains a nucleotide binding motif (p-loop) required for its repression function (Pfeifer et al. 2001; Scheuch et al. 2008). GvpE and GvpD are able to interact, and the presence of both proteins leads to a strong reduction in the amount of GvpE followed by the repression of gene expression (Zimmermann and Pfeifer 2003; Scheuch and Pfeifer 2007; Scheuch et al. 2008; Schmidt and Pfeifer 2013). GvpD does not bind to the promoter sequence. Homologues of GvpD and GvpE are not found in bacteria, since the transcription apparatus and promoter sequences, but also the regulatory systems are different.

Environmental factors such as light, temperature, oxygen supply and salt concentration influence transcription of the *gvp* gene clusters in haloarcheaea. Anoxic conditions reduce the expression of *gvpACNO* and *gvpDEFGHIJKLM* ten-fold, and only very few cells contain gas vesicles (Hechler and Pfeifer, 2009). Since ΔD transformants overexpress gas vesicles under aerobic conditions, but are gas vesicle free under anoxic conditions, GvpD is not involved in this repression. With respect to temperature, *Hbt. salinarum* PHH1 grown at 15 °C contains large amounts of gas vesicles due to a two-fold increased expression of *gvpACNO* compared to growth at 37 °C. Also, the *gvpDE*-expression is increased, whereas the transcription of most other household genes is decreased at 15 °C (Bleiholder et al. 2012). Similar results are observed by microarray analysis of strain NRC-1 (DasSarma et al. 2012). With respect to salt concentration, high salt induces the transcription of mc-vac in *Hfx. mediterranei* (Englert et al. 1990; Röder and Pfeifer 1996; Hechler and Pfeifer 2009). Cells grown in 15% salt media are Vac^-, whereas the transcription of *gvpACNO* is seven-fold enhanced when *Haloferax* is grown in 25% salt medium where the cells are Vac^+.

Signal perception and transduction pathways of all these parameters are not yet known. The signals could be transferred to GvpE or GvpD, or to the constituents of the basal transcription apparatus. *Hbt. salinarum* contains several *tbp* and many different *tfb* genes encoding transcription factors that are expressed under different conditions. In addition, a faster degradation of *gvp* mRNAs was observed in *Hfx. mediterranei* grown in <17% salt media (Jäger et al. 2002), implying that differences in mRNA stability contribute to the salt- and growth phase-dependent presence of *gvp* transcripts. Also observed was a negative regulation by the 5′-mRNA leader regions of the *gvpA* and *gvpD* transcripts (Sartorius-Neef and Pfeifer 2004;

Born and Pfeifer 2019). Leaderless transcripts are translated at a much higher rate, suggesting a negative regulatory role of the 5′-untranslated mRNA regions (5′-UTRs). Presumably, translation is reduced since the secondary structures of the 5′-UTR mask the start site of translation. Such mRNA secondary structures could also be stabilized by proteins or sRNA. Taken together, multiple factors influence *gvp* gene expression.

5 Gas Vesicle Genes and Proteins of Bacteria

Gas vesicle gene clusters are present in many genome sequences of bacteria, but are investigated in more detail in a few species only. This chapter focuses on the description of the *gvp* gene clusters of cyanobacteria, *Serratia* sp. ATCC39006, *Bacillus megaterium* and *Streptomyces*. The expression of the *gvp* gene clusters in soil bacteria raises the question on the possible function of gas vesicles in bacteria where flotation in water is not the dominant lifestyle. They might reduce the volume of the cytoplasm and promote diffusion of gases.

5.1 *Gas Vesicle Genes of Cyanobacteria*

The *gvpA* gene was originally identified in the cyanobacterium *Calothrix* (Damerval et al. 1987). Two *gvpA* copies followed by *gvpC* are present, and additional genes related to the haloarchaeal accessory *gvp* genes. The presence of several *gvpA* genes is typical for cyanobacteria. Two copies of *gvpA* are also found in *Calothrix*, *Anabaena flos-aquae*, *Pseudoanabaena* and some *Planktothrix* species (Damerval et al. 1987, 1991; Hayes and Powell 1995; Beard et al. 2000). Three *gvpA* copies are present in the *Microcystis aeruginosa* (Mlouka et al. 2004), and even seven copies are described for *Anabaena* sp. 90 and *Planktothrix* species (Fig. 8) (Beard et al. 2000; Beard et al. 2002). The multiple *gvpA* genes might be ecoparalogs. In contrast, all gas-vesiculate haloarchaea contain only a single *gvpA* gene per cluster.

The *gvp* gene cluster of *Microcystis aeruginosa* consists of the two transcription units $gvpA_1,A_2,A_3CNJX$ and *gvpKFG*, followed by *gvpV* in inverse orientation, and *gvpW* in the same orientation as *gvpKFG* (Fig. 8) *GvpV* and *gvpW* are bacterial *gvp* genes. Similar gene clusters are found in *Anabaena/Nostoc* PCC7120, including *gvpW* and *gvpV* but lacking *gvpX* (Kinsman and Hayes 1997; Mlouka et al. 2004). The gene encoding the essential GvpM of haloarchaea appears to be missing. Analyses to test whether these *gvp* gene clusters are sufficient for gas vesicle formation have not been done so far. Light and also the pH-value of the surroundings influence the *gvp* gene expression in *Microcystis* and other cyanobacteria (Walsby 1994; Gao et al. 2016).

5.2 *The Gas Vesicle Gene Cluster of* Serratia *Sp. ATCC39006*

The *gvp* gene cluster identified in *Serratia* sp. ATCC39006 comprises 19 genes arranged in two consecutive clusters transcribed from two promoters, P_{gvpA1} and P_{gvrA} (Fig. 8) (Ramsay et al. 2011; Tashiro et al. 2016). The region was defined by transposon mutagenesis. Eleven of these genes are homologous to the respective *gvp* genes of haloarchaea (*gvpA1-A3, gvpC, gvpF1-F3, gvpG, gvpH, gvpK, gvpN*), and five additional genes, *gvpV, gvpW, gvpX, gvpY* and *gvpZ,* are typical for bacterial *gvp* gene clusters. The *gvpV* and *gvpZ* genes are conserved in all gas vesicle–containing proteobacteria (Tashiro et al. 2016). The genes *gvpA1-A3* encode proteins of the A-J-M family, with GvpA1 (72 aa) serving as major gas vesicle structural protein, whereas GvpA2 (65 aa) is somewhat related to the more distant GvpM relative GvpS in soil bacteria. GvpA3 (110 aa) is homologous to GvpJ (Table 2). All three *gvpA* genes are required for gas vesicle formation and cannot substitute each other, underlining their distinct functions. The three *gvpF* genes encode proteins related to GvpF and GvpL of the haloarchaea.

Mutational studies (deletion or overexpression of single *gvp* genes) similar to those already described for haloarchaea were performed and analyzed in *E. coli.* In-frame deletions of single *gvp* genes reveal that the proteins GvpA1-A3, GvpF1-F3, GvpG and GvpK are required for gas vesicle formation (Table 2). GvpN and GvpV are essential for the morphogenetic assembly, since only small bicones appear that cannot be enlarged to cylinder-shaped gas vesicles (Tashiro et al. 2016). These results are similar to the results obtained with ΔN transformants in haloarchaea. Also, GvpH is not required in *Serratia*, similar to the results on haloarchaea. The genes *gvpW, gvpX, gvpY* and *gvpZ* are not essential to form a gas-filled vesicle, and might play more subtle roles in gas vesicle formation. Also analyzed was the overexpression of a single essential *gvp* gene in addition to the entire *gvp* gene cluster (Monson et al. 2016). A significant reduction of gas vesicles was observed with an overexpression of *gvpF1* and *gvpF2*, and also with *gvpV* and *gvpA3*. Thus, balanced amounts of the proteins encoded here are required for a successful gas vesicle formation. In the case of an overexpression of *gvpG*, gas vesicles were still present, but the transformants were arrested in cell growth after induction, leading to the speculation on a link between gas vesicle biogenesis and binary fission of the cells. The overexpression of all other *gvp* genes did not affect gas vesicle formation.

With respect to the regulation, the gene cluster contains the three regulatory genes, *gvrA*, *gvrB* and *gvrC* (Fig. 10c). All three are encoded by the second operon. GvrA and GvrB resemble the two-component system NtrBC involved in global regulation of genes involved in nitrogen assimilation in bacteria. GvrA has similarities to NtrC, and GvrB to NtrB (Table 2). GvrA activates the σ^{54}-dependent promoter P_{gvpA1} in the stationary phase, but not P_{gvrA} (Fig. 10c). Thus, an autoregulation of the regulator genes is not observed as found with the GvpE-activated *gvpDE* in haloachaea. Limited oxygen leads to an increase in *gvp* gene expression and gas vesicle formation, whereas *ΔgvrA* mutants show only a basal level of expression under these conditions, implying that GvrAB respond to reduced aeration and transduce the signal to *gvp* gene expression (Ramsay et al. 2011). The third regulator protein GvrC contains a CheY-like phosphorylation receiver domain and stimulates both promoters P_{gvpA1} and P_{gvrA} (Table 2).

Additional signals affect the expression of the *gvp* gene cluster of *Serratia*. The expression depends on high cell density, mediated by the quorum sensing signal BHL (*N*-butanoyl-*L*-homoserin lactone). Addition of exogenous BHL allows the transcription start at P_{gvpA1}, whereas the promoter P_{gvrA} is not influenced (Tashiro et al. 2016). BHL regulates also other traits of *Serratia*, such as swarming motility, the production of prodigiosin and of the broad-spectrum antibiotic carbapenem. BHL binds to the global LuxR-type repressor SmaR to de-repress gene expression (Fig. 10c). Another global regulator is RsmA that connects the flagella gene expression and gas vesicle formation. RsmA is a sRNA-binding protein repressing swarming motility and imposes a strong positive effect on the expression of the *gvp* gene cluster (Ramsay et al. 2011). A DNA-insertion in *rsmA* eliminates the activity of P_{gvpA1} and P_{gvrA}, and complementation by *rsmA* yields highly gas-vacuolated cells. An overexpression of *rsmA* induces gas vesicle formation, whereas the flagella-bound swarming is absent due to a repression of the flagella and surfactant gene expression (Williamson et al. 2008; Ramsay et al. 2011). Thus, flagella expression and gas vesicle expression are oppositely regulated by RsmA in *Serratia*.

5.3 The gvp *Gene Clusters of Soil Bacteria*

The Gram-positive *Bacillus megaterium* and various *Streptomyces, Rhodococcus* and *Frankia* species contain *gvp* gene clusters. *B. megaterium* harbours the *gvpAPQBRNFGLSKJTU* gene cluster with *gvpA* and *gvpB* encoding GvpA-type proteins (Fig. 8) (Li and Cannon 1998). *Escherichia coli* transformants carrying *gvpA-gvpU* (or *gvpB-gvpU*) produce tiny gas vesicles with an average length of 40 nm. GvpR exhibits 44% sequence similarity to GvpO of *Streptomyces coelicolor*, and 39% to the GvpO of haloarchaea. GvpS could be a more distant relative of the GvpA-J-M family of hydrophobic gas vesicle proteins and might be a substitute for GvpM in soil bacteria. The basic GvpQ is related to GvpI of the haloarchaea. The *gvpTU* genes have no archaeal nor *Serratia* homologs, but are essential for gas vesicle formation (Li and Cannon 1998).

Even more astounding is the presence of gas vesicle gene clusters in the mycelium-forming *Streptomyces* and other actinomycetes species (Bentley et al. 2002). The *gvpOAFGYZJLSK* gene cluster of *S. coelicolor* and *S. lividans* contains homologs to all essential haloarchaeal *gvp* genes (Fig. 8) (van Keulen et al. 2005). None of these soil bacteria contain the genes encoding the stabilizer protein GvpC or the chaperone GvpN. Gas vesicle formation was not observed in *S. coelicolor* or *S. lividans,* although the expression of the gene cluster was shown by microarray analysis (Lee et al. 2005). The GvpA of *Streptomyces* is with 124–170 aa almost twice as large compared to the haloarchaeal or other bacterial GvpA (see Fig. 4a) (van Keulen et al. 2005). The sequence of the 50-aa core region is relatively conserved, but some aa positions in GvpA are different and would lead to a gas vesicle negative phenotype in the respective GvpA of haloarchaea (Fig. 4). The additional C-terminal portion of the *Streptomyces*-GvpA does not resemble any known sequence.

Recently, gas vesicles were observed in the *Streptomyces* species CB03234-S (Huang et al. 2019). The strain carries a *gvpOAFGJLSK* gene cluster (*gvp*^{3234}) lacking the *gvpYZ* genes found in *S. coelicolor* (Fig. 8). Strain CB03234-S produces the secondary metabolites tiancimycins (TNMs) exhibiting antibiotic activities, making it interesting for biomedical research. An improved expression of TNMs by substituting the nitrogen source also significantly increased the expression of the *gvp*^{3234} gene cluster. The gas vesicles of CB03234-S observed in the mycelia are of irregular structure and resemble those of ΔC transformants (containing the ΔC construct lacking *gvpC*) of haloarchaea; spindle- or cylinder-shaped morphologies are not observed. The irregular shape could be due to the enlarged GvpA and/or the lack of GvpC in *Streptomyces*. Also observed was that the gas vesicle overproducing strain exhibits a morphological difference by producing fluffy mycelial pellets with longer hyphae compared to the non-gas vesiculate strain (Huang et al. 2019). The authors speculate that these less-compact mycelia could be a valuable trait for industrial *Streptomyces* species.

6 Applications of Gas Vesicles

Gas vesicles of *Hbt. salinarum*, *A. flos-aquae* and *B. megaterium* are used for special applications in biomedical research and clinical diagnostics. These nanoparticles are non-toxic to mammalian cells and self-adjuvanting, and are an effective antigen display system when peptides or proteins of interest are fused to the surface-located GvpC. Such recombinant gas vesicles enable the stimulation of immune reactions in rabbits or mice. In addition, gas vesicles are used as a novel ultrasound contrast agent providing large mechanical response in contrast to the commonly used micro-bubbles. GvpC fusions with peptides or GvpC deletion variants modulate the mechanical, acoustic, surface and targeting properties of gas vesicles and allow multiplex and multimodal ultrasound imaging.

6.1 Antigen Display and Antibody Formation

Haloarchaeal gas vesicle are easy to isolate, and recombinant gas vesicles are used for the production of antibodies (Stuart et al. 2001, 2004). Peptides of 6- to 235-aa length fused to the C-terminal portion of GvpC result in recombinant gas vesicles, displaying these peptides on the surface, and can be used without the application of adjuvant to immunize rabbits or mice. In addition, the toxic lipopolysaccharide of Gram-negative bacteria is lacking. The procedure leads to a long-lived antibody response, and has been applied to raise antibodies against peptides of SIV (Tat, Rev., NefI) (Sremac and Stuart 2008, 2010). Also fused to GvpC were outer membrane proteins of *Chlamydia trachomatis*, and the antibodies raised are able to detect the pathogen in sera derived from *Chlamydia*-positive patients (Childs and Webley 2012). Another protein tested is SopB of *Salmonella enterica* serovar *typhimurium*, a secreted inosine phosphate effector protein injected into the host cell during the pathogenesis of typhoid (DasSarma et al. 2014). The recombinant SopB-gas vesicles were injected in mice immunized with a live attenuated *Salmonella* vaccine strain, and the bacterial burden in mice boosted with SopB-gas vesicles was reduced, suggesting that the SopB-gas vesicles elicit a protective response (DasSarma et al. 2014). Also displayed on gas vesicles was the murine bactericidal permeability-increasing protein, and the recombinant gas vesicles rescue mice from lethal endotoxic shock (Balakrishnan et al. 2016). The display of a *Plasmodium falciparum* enolase peptide was useful as a protective antigen (Dutta et al. 2015).

Another application of gas vesicles might be their use as drug delivery system for intradermal and transdermal permeation (Andar et al. 2017). Gas vesicles injected with microneedles easily travel into the skin as demonstrated by *in vitro* diffusion studies with miniature pig skin. Such gas vesicles could be used to transport protein- or peptide-based drugs into mammalia, but more research is required to apply these procedures for biomedical applications.

6.2 Novel Enhancement Tool in Acoustic Resonance Imaging

Gas vesicles constitute a new class of ultrasound contrast agents (Shapiro et al. 2014a). Ultrasound is commonly used for biomedical imaging, but contrast agents suitable for nanoscale observations are scarce. Usually applied are micro-sized bubbles stabilized by a lipid or protein shell, but these microbubbles have limitations because of their inherent instability. Gas vesicles are very robust, and the rigidity of the gas vesicle wall maintains a hollow, gas-filled core. They generate second harmonic amplitudes of −2 to −6 dB and enable harmonic, multiplex and multimodal ultrasound imaging. Gas vesicles of *A. flos-aquae* (Ana GV), *Hbt. salinarum* (Halo GV) and *B. megaterium* (Mega GV) were tested; the latter ones are the smallest (diameter: 73 ± 14 nm; length: 249 ± 99 nm) compared to those of *A. flos-aquae* (diameter: 136 ± 21 nm; length: 519 ± 160 nm) or *Hbt.* (diameter: 251 ± 51 nm; length: 400 ± 113 nm) (Mukherjee et al. 2017). The shapes and the sizes of gas vesicles affect buoyancy and resistance to external pressure, and these gas vesicles show a differential response to acoustic pressure and collapse at distinct thresholds. Haloarchaeal gas vesicles embedded in a 1% agar matrix collapse when exposed to 6 cycle 20 MHz pulses (Cherin et al. 2017). The acoustic collapse pressure is >550 kPa at all frequencies and thus up to nine-fold higher than the critical hydrostatic pressure determined for these gas vesicles. The difference might be due to the fact that hydrostatic pressure allows the gas to flow outward due to compression of the gas vesicle volume, whereas the acoustic pressure compresses the gas vesicle on a time scale (40 ns at 12.5 MHz) that is shorter than the exponential filling time (μs). The Halo GV produce contrast in vivo in mouse liver and collapse with ultrasound at 6 MHz and pressure of 98 kPa (Cherin et al. 2017). Overall, gas vesicles are of significant advantage over conventional microbubble contrast agents because of their stability and unique physical properties.

Gas vesicles also serve as imaging agent in hyperpolarized xenon magnetic resonance imaging (MRI) and enable non-invasive observations of the anatomy of organisms (Shapiro et al. 2014b; Lakshmanan et al. 2017). Conventional MRI contrast agents, based on super paramagnetic ion oxides or lanthanide chelates, are potentially toxic and require relatively high concentrations (μM range) for detection by MRI. Imaging of gas vesicles filled with hyperpolarized ^{129}Xe takes advantage of non-equilibrium spin polarization and increases the molecular sensitivity significantly (by factors of up to 105). Gas vesicles are detected at picomolar concentrations through hyperpolarized chemical exchange saturation transfer (HyperCEST) pulse sequences (Farhadi et al. 2018). The biochemical and mechanical properties of gas vesicles can be altered by the replacement of GvpC on the surface by recombinant GvpC displaying certain peptides. Such functionalized gas vesicles, carrying, e.g., surface-attached streptavidin antibodies against the HER2 receptor, are able to label a HER2-expressing breast cancer cell line, so that these cells are distinguishable from others by MRI. GvpC has been modified to alter the acoustic properties of Ana GV (Lakshmanan et al. 2016). In this case, His-tagged $GvpC_{WT}$ or smaller GvpC variants lacking the N- or C-terminal portion were added to purified Ana GV stripped of their native GvpC. Gas vesicles devoid of GvpC collapse under a lower acoustic pressure compared to the gas vesicles coated with the GvpC deletion variant, or with $GvpC_{WT}$, similar to hydrostatic pressure application. Thus, the gas vesicle surface properties can be tuned for different zeta potentials (Lakshmanan et al. 2016).

Engineered GvpC also display ligands for reduced or enhanced cellular targeting (Lakshmanan et al. 2016, 2017). Fusion of the lysin-rich protein, LRP, to GvpC enhances the gas vesicle surface charge 100-fold, and gas vesicles with $GvpC_{LRP}$ exhibit a 28 ± 4 mV higher zeta potential. $GvpC_{RGD}$ contains arginyl glycyl aspartic acid (RGD) that efficiently binds integrins, allowing targeting of an integrin-overexpressing human glioblastoma cell line in vivo. Also, the interaction of gas vesicles with macrophages was altered fusing a peptide of the mammalian membrane protein CD47 to GvpC. Decoration of gas vesicles with $GvpC_{CD47}$ results in a reduced uptake by macrophages, whereas decoration with the polycationic peptide polyarginine ($GvpC_{R8}$) leads to gas vesicles taken up more efficiently compared to wild-type gas vesicles. Thus, these GvpC modifications allow cellular labelling. To simplify the functionalization of the gas vesicle surface, the Shapiro laboratory fused GvpC with SpyTag (ST), a 13-residue peptide forming a covalent amide bond with a partner SpyCatcher protein under physiological conditions. $GvpC_{ST}$ binds to gas vesicles with a similar stoichiometry as $GvpC_{WT}$, providing on average 1000 SpyTag functionalities per gas vesicle as determined with SpyCatcher-mNeoGreen.

These fluorescent gas vesicles were used for acoustic and fluorescent imaging; they remain fluorescent after acoustic pressure-induced collapse and allow further histological examinations following ultrasound imaging (Lakshmanan et al. 2016, 2017). Such modifications of GvpC allow modulating the mechanical, acoustic, surface and targeting properties of gas vesicles in mammalian cells, and are milestones for the development of gas vesicles as acoustic reporters.

6.3 Imaging of Microbes in the Mammalian Microbiome

Gas vesicles are also used as reporters to visualize and locate gas vesiculate bacteria in the mammalian gut (Bourdeau et al. 2018). Since the *gvp* gene cluster of *Anabaena* is not expressed in *E. coli* and expression of the *gvp* gene cluster of *B. megaterium* leads to gas vesicles not detectable by ultrasound, *gvp* genes derived from both bacteria were combined. Complementation of the *Bacillus* accessory genes *gvpRNFGLSKJTU* with the *Anabaena gvpAC* genes yielded gas vesicles with robust ultrasound contrast, and application of acoustic impulses with amplitudes above the collapse pressure eliminates this contrast. A truncated variant of GvpC consisting of the first α-helical domain of the five repeating α-helical units forms less stable gas vesicles with smaller hydrostatic and acoustic collapse pressures. Thus, bacteria carrying one of these gas vesicle types can be distinguished by ultrasound. The method was used to image microbes in the gut microbiome of mice. A probiotic *E. coli* strain expressing one of these clusters was easily recognized and localized in a living animal (Bourdeau et al. 2018).

7 Conclusion

The proteinaceous gas vesicles provide buoyancy and are used by aquatic microbes for upward migration to reach more oxygenated areas and light. The rigid gas vesicle wall formed by GvpA allows gases to freely diffuse in and out. GvpC forms a mesh on the outside and stabilizes the wall. GvpC located on the surface can be easily replaced by recombinant versions fused to peptides of interest. Such functionalized gas vesicles are of special interest in medical research and are novel ultrasound contrasting agents. Further work will advance this reporter technology for different biotechnological and clinical applications.

Acknowledgements This work was supported by the Deutsche Forschungsgemeinschaft (DFG; Pf 165). Arnulf Kletzin is thanked for bioinformatic analyses, and Daniel Bollschweiler and Harald Engelhardt (Max-Planck-Institute for Biochemistry in Martinsried, Germany) for the cryo-electron micrograph of gas vesicles in Fig. 3. Kerstin Völkner (Winter), and Alisa Jost are thanked for their contributions to the studies on GvpA and the interaction of accessory Gvp proteins. Gerald Losensky is thanked for critical reading of the manuscript.

References

Andar A, Karan R, Recher W, DasSarma P, Hedrich W, Stinchcomb A et al (2017) Microneedle-assisted skin permeation by non-toxic bioengineerable gas vesicle nanoparticles. Mol Pharm 14:953–958. https://doi.org/10.1021/acs.molpharmaceut.6b00859

Archer D, King N (1984) Isolation of gas vesicles from *Methanosarcina barkeri*. J Gen Microbiol 30:167–172

Balakrishnan A, DasSarma P, Bhattacharjee O, Kim J, DasSarma S, Chakravortty D (2016) Halobacterial nano vesicles displaying murine bactericidal permeability-increasing protein rescue mice from lethal endotoxic shock. Sci Rep 6:33679

Bauer M, Marschaus L, Reuff M, Besche V, Sartorius-Neef S, Pfeifer F (2008) Overlapping activator sequences determined fort wo oppositely oriented promoters in halophilic archaea. Nucleic Acids Res 36:598–606

Beard S, Handley B, Hayes PK, Walsby AE (1999) The diversity of gas vesicle genes in *Planktothrix rubescens* from Lake Zurich. Microbiol 145:2757–2768

Beard S, Davis P, Iglesias-Rodriguez D, Skulberg O, Walsby AE (2000) Gas vesicle genes in *Planktothrix* spp. from Nordic lakes: strains with weak gas vesicles possess a longer variant of GvpC. Microbiol 146:2009–2018

Beard S, Handley B, Walsby AE (2002) Spontaneous mutations in gas vesicle genes of *Planktothrix* spp. affect gas vesicle production and critical pressure. FEMS Microbiol Lett 215:189–195

Belenky M, Meyers R, Herzfeld J (2004) Subunit structure of gas vesicles: a MALDI-TOF mass spectrometry study. Biophys J 86:499–505

Bell SD, Jackson S (1998) Transcription and translation in Archaea: a mosaic of eukaryal and bacterial features. Trends in Microbiol 6:222–228

Bentley S, Chater K, Cerdeno-Tárraga A, Challis G, Thompson N, James K et al (2002) Complete genome sequence of the model actinomycete *Streptomyces coelicolor* A3(2). Nature 417:141–147

Blaurock A, Walsby AE (1976) Crystalline structure of gas vesicle wall from *Anabaena flos-aquae*. J Mol Biol 105:183–199

Blaurock A, Wober W (1976) Structure of wall of *Halobacterium halobium* gas vesicles. J Mol Biol 106:871–888

Bleiholder A, Frommherz R, Teufel K, Pfeifer F (2012) Expression of multiple *tfb* genes in different *Halobacterium salinarum* strains and interaction of TFB with transcriptional activator GvpE. Arch Microbiol 194:269–279

Bolhuis H, te Poele EM, Rodriguez-Valera F (2004) Isolation and cultivation of Walsby's square archaeon. Environ Microbiol 6:1287–1291

Born J, Pfeifer F (2019) Improved GFP variants to study gene expression in haloarchaea. Front Microbiol 10:1200. https://doi.org/10.3389/fmicb.2019.01200

Bourdeau R, Lee-Gosselin A, Lakshmanan A, Farhadi A, Kumar S, Nety S, Shapiro MG (2018) Acoustic reporter genes for noninvasive imaging of microbes in mammalian hosts. Nature 553 (7686):86–90

Bowen C, Jensen T (1965) Blue-green algae: fine structure of the gas vacuoles. Science 147:146–152

Bright D, Walsby AE (1999) The relationship between critical pressure and width of gas vesicles in isolates of *Planktothrix rubescens* from Lake Zurich. Microbiol 145:2769–2775

Buchholz B, Hayes PK, Walsby AE (1993) The distribution of the outer gas vesicle protein, GvpC, on the *Anabaena* gas vesicle, and its ratio to GvpA. J Gen Microbiol 139:2353–2363

Burns D, Camakaris H, Janssen P, Dyall-Smith M (2004) Cultivation of Walsby's square haloarchaeon. FEMS Microbiol Lett 238:469–473

Cherin E, Malis J, Bourdeau R, Yin M, Kochmann D, Foster F, Shapiro MG (2017) Acoustic behavior of *Halobacterium salinarum* gas vesicles in the high frequency range: experiments and modelling. Ultrasound Med Biol 43(5):1016–1030

Childs T, Webley W (2012) *In vitro* assessment of halobacterial gas vesicles as a *Chlamydia* vaccine display and delivery system. Vaccine 30:5942–5948

Clarke A, Walsby AE (1978) The occurrence of gas vacuolate bacteria in lakes. Arch Microbiol 118:223–228

Cui H, Gao X, Li X, Xu X, Zhou Y, Liu H, Zhou P (2010) *Haloplanus vescus* sp nov, an extremely halophilic archaea from a marine solar saltern, and emended description of the genus *Haloplanus*. Int J Syst Evol Microbiol 60:1824–1827

Cui H, Gao X, Yang Y, Xu X (2011) *Haloplanus aerogenes* sp nov, an extremely halophilic archaeon from a marine solar saltern. Int J Syst Evol Microbiol 61:965–968

Damerval T, Houmard J, Guglielmi G, Csiszar K, Tandeau de Marsac N (1987) A developmentally regulated *gvpABC* operon is involved in the formation of gas vesicles in the cyanobacterium *Calothrix*-7601. Gene 54:83–92

Damerval T, Castets AM, Guglielmi G, Houmard J, Tandeau de Marsac N (1991) Gas vesicle synthesis in the cyanobacterium *Pseudoanabaena* sp.: occurrence of a single photoregulated gene. Mol Microbiol 5:657–664

DasSarma S, Damerval T, Jones J, Tandeau de Marsac N (1987) A plasmid-encoded gas vesicle protein gene in a halophilic archaebacterium. Mol Microbiol 1:365–370

DasSarma S, Halladay J, Jones J, Donovan J, Giannasca P, Tandeau de Marsac N (1988) High-frequency mutations in a plasmid-encoded gas vesicle gene in *Halobacterium halobium*. Proc Natl Acad Sci U S A 85:6861–6865

DasSarma P, Zamora R, Müller J, DasSarma S (2012) Genome-wide responses of the model archaeon *Halobacterium* sp. strain NRC-1 to oxygen limitation. J Bacteriol 194:5530–5537

DasSarma P, Negi V, Balakrishnan A, Karan R, Barnes S, Ekulona F, Chakravortty D, DasSarma S (2014) Haloarchaeal gas vesicle nanoparticles displaying Salmonella SopB antigen reduce bacterial burden when administered with live attenuated bacteria. Vaccine 32:4543–4549

Dunton P, Walsby AE (2005) The diameter and critical collapse pressure of gas vesicles in *Microcystis* are correlated with GvpCs of different length. FEMS Microbiol Let 247:37–43

Dutta S, DasSarma P, DasSarma S, Jarori G (2015) Immunogenicity and protective potential of a *Plasmodium* spp. enolase peptide displayed on archaeal gas vesicle nanoparticles. Malaria J 4:406

Elvi Bardavid R, Mana L, Oren A (2007) *Haloplanus natans* gen. Nov., sp. nov., an extremely halophilic gas-vacuolate archaeon from Dead Sea-Red-Sea water mixtures in experimental mesocosms. Int J Syst Evol Microbiol 57:780–783

Englert C, Pfeifer F (1993) Analysis of gas vesicle gene expression in *Haloferax mediterranei* reveals that GvpA and GvpC are both gas vesicle structural proteins. J Biol Chem 268:9329–9336

Englert C, Horne M, Pfeifer F (1990) Expression of the major gas vesicle protein gene in the halophilic archaebacterium *Haloferax mediterranei* is modulated by salt. Mol Gen Genet 222:225–232

Englert C, Krüger K, Offner S, Pfeifer F (1992a) Three different but related gene clusters encoding gas vesicles in halophilic Archaea. J Mol Biol 227:586–592

Englert C, Wanner G, Pfeifer F (1992b) Functional analysis of the gas vesicle gene cluster of the halophilic Archaeon *Haloferax mediterranei* defines the vac-region boundary and suggests a regulatory role for the GvpD gene or its product. Mol Microbiol 6:3543–3550

Ezzeldin H, Klaus J, Solares S (2012) Modeling of the major gas vesicle protein, GvpA: from protein sequence to vesicle wall structure. J Struct Biol 179:18–28

Farhadi A, Ho G, Kunth M, Ling B, Lakshmanan A, Lu G, Bourdeau R, Schröder L, Shapiro MG (2018) Recombinantly expressed gas vesicles as nanoscale contrast agents for ultrasound and hyperpolarized MRI. AICHE J 64(8):2927–2933

Gao H, Zhu T, Xu M, Wang S, Xu X, Kong R (2016) pH-dependent gas vesicle formation in *Microcystis*. FEBS Let 590:3195–3201

Gosink J, Herwig R, Staley J (1997) *Octadecabacter arcticus* gen nov, sp nov, and *O. antarcticus*, sp nov, nonpigmented, psychrophilic gas vacuolate bacteria from polar sea ice and water. Syst Appl Microbiol 20:356–365

Hayes PK, Powell R (1995) The *gvpA/C* cluster of *Anabaena-flos-aquae* has multiple copies of a gene encoding GvpA. Arch Microbiol 164:50–57

Hayes PK, Lazarus C, Bees A, Walker J, Walsby AE (1988) The protein encoded by GvpC is a minor component of gas vesicles isolated from the cyanobacteria *Anabaena-flos-aquae* and *Microcystis* sp. Mol Microbiol 2:545–552

Hayes PK, Buchholz B, Walsby AE (1992) Gas vesicles are strengthened by the outer-surface protein, GvpC. Arch Microbiol 157:229–234

Hechler T, Pfeifer F (2009) Anaerobiosis inhibits gas vesicle formation in halophilic archaea. Mol Microbiol 71:13245

Hofacker A, Schmitz K, Cichonczyk A, Sartorius-Neef S, Pfeifer F (2004) GvpE- and GvpD-mediated transcription regulation of the p-*gvp* genes encoding gas vesicles in *Halobacterium salinarum*. Microbiol 150:1829–1838

Horne M, Englert C, Pfeifer F (1988) Two genes encoding gas vacuole proteins in *Halobacterium halobium*. Mol Gen Genet 213:459–464

Horne M, Englert C, Wimmer C, Pfeifer F (1991) A DNA region of 9 kbp contains all genes necessary for gas vesicle synthesis in halophilic archaebacteria. Mol Microbiol 5:1159–1174

Houwink A (1956) Flagella, gas vacuoles and cell-wall structure in *Halobacterium halobium*; an electron microscopy study. J Gen Microbiol 15:146–150

Huang R, Lin J, Gao D, Zhang F, Yi L, Huang Y, Yan X, Duan Y, Zhu X (2019) Discovery of gas vesicles in *Streptomyces* sp. CB03234-S and potential effects of gas vesicle gene overexpression on morphological and metabolic changes in streptomycetes. Appl Microbiol Biotechnol 103:5751–5761

Irgens RL, Suzuki I, Staley J (1989) Gas vacuolate bacteria obtained from marine waters of Antarctica. Curr Microbiol 18:261–265

Jäger A, Samorski R, Pfeifer F, Klug G (2002) Individual *gvp* transcript segments in *Haloferax mediterranei* exhibit varying half-lives, which are differentially affected by salt concentration and growth phase. Nucleic Acids Res 30:5436–5443

Jung D, Achenbach L, Karr E, Takaichi S, Madigan M (2004) A gas vesiculate planktonic strain of the purple non-sulfur bacterium *Rhodoferax antarcticus* isolated from Lake Fryxell, dry valleys, Antarctica. Arch Microbiol 182:236–243

Karr E, Sattley W, Jung D, Madigan M, Achenbach L (2003) Remarkable diversity of phototrophic purple bacteria in a permanently frozen Antarctic lake. Appl Environ Microbiol 69:4910–4914

Kinsman R, Hayes PK (1997) Genes encoding proteins homologous to halobacterial Gvps N, J, K, F & L are located downstream of *gvpC* in the cyanobacterium *Anabaena flos-aquae*. DNA Seq 7:97–106

Kinsman R, Walsby AE, Hayes PK (1995) GvpCs with reduced numbers of repeating sequence elements bind to and strengthen cyanobacterial gas vesicles. Mol Microbiol 17:147–154

Klebahn H (1895) Gasvakuolen, ein Bestandteil der Zellen der wasserblütenbildenden Phycochromaceen. Flora (Jena) 80:241

Knitsch R, Schneefeld M, Weitzel K, Pfeifer F (2017) Mutations in the major gas vesicle protein GvpA and impacts on gas vesicle formation in *Haloferax volcanii*. Mol Microbiol 106:530–542

Konopka A, Lara J, Staley J (1977) Isolation and characterization of gas vesicles from *Microcyclus aquaticus*. Arch Microbiol 112:133–140

Krüger K, Pfeifer F (1996) Transcript analysis of the c-vac region and differential synthesis of the two regulatory gas vesicle proteins GvpD and GvpE in *Halobacterium salinarium* PHH4. J Bacteriol 178:4012–4009

Krüger K, Hermann T, Armbruster V, Pfeifer F (1998) The transcriptional activator GvpE for the halobacterial gas vesicle genes resembles a basic region leucine-zipper regulatory protein. J Mol Biol 279:761–771

Lakshmanan A, Farhadi A, Nety S, Lee-Gosselin A, Bourdeau R, Maresca D, Shapiro MG (2016) Molecular engineering of acoustic protein nanostructures. ACS Nano 10:7314–7322
Lakshmanan A, Lu G, Farhadi A, Nety S, Kunth M, Lee-Gosselin A, Maresca D, Bourdeau R, Yin M, Yan J, Witte C, Malounda D, Foster F, Schröder L, Shapiro MG (2017) Preparation and noninvasive imaging of biogenic gas vesicle nanostructures. Nat Protoc 12(10):2050–2080
Larsen H, Omang S, Steensland H (1967) On the gas vacuoles of the halobacteria. Arch Microbiol 59:197–203
Lee E, Karoonutharisiri N, Kim H, Park J, Cha C, Kao C, Roe J (2005) A master regulator σ^B governs osmotic and oxidative response as well as differentiation via a network of sigma factors in *Streptomyces coelicolor*. Mol Microbiol 57:1252–1264
Li N, Cannon M (1998) Gas vesicle genes identified in *Bacillus megaterium* and functional expression in *Escherichia coli*. J Bacteriol 180:2450–2458
Marschaus L, Pfeifer F (2012) A dual promoter region with overlapping activator sequences drives the expression of gas vesicle protein genes in haloarchaea. Microbiol 158:2815–2825
Mayr A, Pfeifer F (1997) The characterization of the nv-*gvpACNOFGH* gene cluster involved in gas vesicle formation in *Natronobacterium vacuolatum*. Arch Microbiol 168:24–32
McMaster T, Miles M, Walsby AE (1996) Direct observation of protein secondary structure in gas vesicles by atomic force microscopy. Biophys J 70:2432–2436
Mlouka A, Comte K, Castets A, Bouchier C, Tandeau de Marsac N (2004) The gas vesicle gene cluster from *Microcystis aeruginosa* and DNA rearrangements that lead to loss of cell buoyancy. J Bacteriol 186:2355–2365
Monson RE, Tashiro Y, Salmon G (2016) Overproduction of individual gas vesicle proteins perturbs flotation, antibiotic production and cell division in the enterobacterium Serratia sp. ATCC 39006. Microbiol 162:1595–1607
Montalvo-Rodriguez R, Vreeland R, Oren A, Kessel M, Betancourt C, López-Garriga J (1998) *Halogeometricum borinquense* gen nov, sp nov, a novel halophilic archaeon from Puerto Rico. Int J Syst Bacteriol 48:1305–1312
Mukherjee A, Davis H, Ramesh P, Lu G, Shapiro MG (2017) Biomolecular MRI reporters: evolution of new mechanisms. Prog Nucl Magn Reson Spectrosc 102-103:32–42
Mwatha W, Grant W (1993) *Natronobacterium vacuolatum* sp nov, a haloalkaliphilic archaeon isolated from Lake Magadi, Kenya. Int J Syst Bacteriol 43:401–404
Ng W, Kennedy S, Mahairas G, Berquist B, Pan M, Shukla H, Lasky S, Baliga N, Thorsson V, Sbrogna J et al (2000) Genome sequence of *Halobacterium* species NRC-1. Proc Natl Acad Sci U S A 97:12176–12181
Oesterhelt D (1998) The structure and mechanism of the family of retinal proteins from halophilic archaea. Curr Opin Struct Biol 8:489–500
Offner S, Pfeifer F (1995) Complementation studies with the gas vesicle-encoding p-vac region of *Halobacterium salinarium* PHH1 reveal a regulatory role for the p-*gvpDE* genes. Mol Microbiol 16:9–19
Offner S, Wanner G, Pfeifer F (1996) Functional studies of the *gvpACNO* operon of *Halobacterium salinarium* reveal that the GvpC protein shapes gas vesicles. J Bacteriol 178:2071–2078
Offner S, Ziese U, Wanner G, Typke D, Pfeifer F (1998) Structural characteristics of halobacterial gas vesicles. Microbiol 144:1331–1342
Offner S, Hofacker A, Wanner G, Pfeifer F (2000) Eight of fourteen *gvp* genes are sufficient for formation of gas vesicles in halophilic archaea. J Bacteriol 182:4328–4336
Oliver R, Walsby AE (1984) Direct evidence for the role of light-mediated gas vesicle collapse in the buoyancy regulation of *Anabaena flos-aquae* (cyanobacteria). Limnol Oceanogr 29:879–886
Oren A (2013) The function of gas vesicles in halophilic archaea and bacteria: theories and experimental evidence. Life 3:1–20
Overmann J, Lehmann S, Pfennig N (1991) Gas vesicle formation and buoyancy regulation in *Pelodictyon phaeoclathratiforme* (green Sulphur bacteria). Arch Microbiol 157:29–37

Parkes K, Walsby AE (1981) Ultrastructure of a gas-vacuolate square bacterium. J Gen Microbiol 126:503–506
Petter H (1931) On bacteria of salted fish. Proc Natl Acad Sci Amsterdam 34:1417–1423
Pfeifer F (2004) Gas vesicle genes in halophilic archaea and bacteria. In: Ventosa A (ed) Halophilic microorganisms. Springer, Berlin, Heidelberg, pp 229–239
Pfeifer F (2012) Distribution, formation and regulation of gas vesicles. Nature Rev Microbiol 10:705–715
Pfeifer F (2015) Haloarchaea and the formation of gas vesicles. Life 5:385–402
Pfeifer F, Weidinger G, Goebel W (1981) Genetic variability in *Halobacterium halobium*. J Bacteriol 145:375–381
Pfeifer F, Offner S, Krüger K, Ghahraman P, Englert C (1994) Transformation of halophilic archaea and investigation of gas vesicle synthesis. System Appl Microbiol 16:569–577
Pfeifer F, Zotzel J, Kurenbach B, Röder R, Zimmermann P (2001) A p-loop motif and two basic regions in the regulatory protein GvpD are important for the repression of gas vesicle formation in the archaeon *Haloferax mediterranei*. Microbiol 147:63–73
Plösser P, Pfeifer F (2002) A bZIP protein from halophilic archaea: structural features and dimer formation of cGvpE from *Halobacterium salinarum*. Mol Microbiol 45:511–520
Ramsay JP, Salmond G (2012) Quorum sensing-controlled buoyancy through gas vesicles. Comm & Intregrat Biology 5:96–98
Ramsay JP, Williamson NR, Spring DR, Salmond G (2011) A quorum-sensing molecule acts as a morphogen controlling gas vesicle organelle biogenesis and adaptive flotation in an enterobacterium. Proc Natl Acad Sci USA 6:14932–14937
Röder R, Pfeifer F (1996) Influence of salt on the transcription of the gas-vesicle genes of *Haloferax mediterranei* and identification of the endogenous transcriptional activator gene. Microbiol 142:1715–1723
Sartorius-Neef S, Pfeifer F (2004) *In vivo* studies on putative Shine-Dalgarno sequences of the halophilic archaeon *Halobacterium salinarum*. Mol Microbiol 51:579–588
Scheuch S, Pfeifer F (2007) GvpD-induced breakdown of the transcriptional activator GvpE of halophilic archaea required a functional p-loop and an arginine-rich region in GvpD. Microbiol 153:947–958
Scheuch S, Marschaus L, Sartorius-Neef S, Pfeifer F (2008) Regulation of *gvp* genes encoding gas vesicle proteins in halophilic archaea. Arch Microbiol 190:333–340
Schmidt I, Pfeifer F (2013) Use of GFP-GvpE fusions to quantify the GvpD-mediated reduction of the transcriptional activator GvpE in haloarchaea. Arch Microbiol 195:403–412
Shapiro MG, Goodwill P, Neogy A, Yin M, Foster F, Schaffer D, Conolly S (2014a) Biogenic gas nanostructures as ultrasonic molecular reporters. Nat Nanotechnol 9:311–316
Shapiro MG, Ramirez R, Sperling L, Sun G, Sun J, Pines A, Schaffer D, Bajaj V (2014b) Genetically encoded reporters for hyperpolarized xenon magnetic resonance imaging. Nature Chem 6:629–634
Sivertsen A, Bayro M, Belenky M, Griffin R, Herzfeld J (2010) Solid-state NMR characterization of gas vesicle structure. Biophys J 99:1932–1939
Sremac M, Stuart E (2008) Recombinant gas vesicles from *Halobacterium* sp displaying SIV peptides demonstrate biotechnology potential as a pathogen delivery vehicle. BMC Biotechnol 8:9
Sremac M, Stuart E (2010) SIVsm tat, rev, and Nef1: functional characteristics of r-GV internalization on isotypes, cytokines, and intracellular degradation. BMC Biotechnol 10:54
Staley J (1968) *Prosthecomicrobium* and *Ancalomicrobium*: new prosthecate freshwater bacteria. J Bacteriol 95:1921–1942
Staley J, Irgens R, Brenner D (1987) *Enhydrobacter aerosarcus* gen nov, sp nov, a gas-vacuolated, facultatively anaerobic, heterotrophic rod. Int J Syst Bacteriol 37:289–291
Staley J, Irgens R, Herwig R (1989) Gas vacuolate bacteria from the sea ice of Antarctica. Appl Environ Microbiol 55:1033–1036

Stoeckenius W, Kunau W (1968) Further characterization of particulate fractions from lysed cell envelopes of *Halobacterium halobium* and isolation of gas vacuole membranes. J Cell Biol 38:337–357

Strunk T, Hamacher K, Hoffgaard F, Engelhardt H, Zillig M, Faist K, Pfeifer F (2011) Structural model of the gas vesicle protein GvpA and analysis of GvpA mutants *in vivo*. Mol Microbiol 81:56–68

Stuart E, Morshed F, Sremac M, DasSarma S (2001) Antigen presentation using novel particulate organelles from halophilic archaea. J Biotechnol 88:119–128

Stuart E, Morshed F, Sremac M, DasSarma S (2004) Cassette-based presentation of SIV epitopes with recombinant gas vesicles from halophilic archaea. J Biotechnol 114:225–237

Tashiro Y, Monson R, Ramsay J, Salmond G (2016) Molecular genetic and physical analysis of gas vesicles in buoyant enterobacteria. Environ Microbiol 18:1264–1276

Tavlaridou S, Faist K, Weitzel K, Pfeifer F (2013) Effect of an overproduction of accessory Gvp proteins on gas vesicle formation in *Haloferax volcanii*. Extremophiles 17:277–287

Tavlaridou S, Winter K, Pfeifer F (2014) The accessory gas vesicle protein GvpM of haloarchaea and its interaction partners during gas vesicle formation. Extremophiles 18:693–706

Thomas R, Walsby AE (1985) Buoyancy regulation on a strain of *Microcystis*. J Gen Microbiol 131:799–809

Van Ert M, Staley J (1971) A new gas vacuolated heterotrophic rod from freshwaters. Arch Microbiol 80:70–77

Van Keulen G, Hopwood D, Dijkhuizen L, Sawers G (2005) Gas vesicles in actinomycetes: old buoys in novel habitats? Trends Microbiol 13:350–354

Völkner K, Jost A, Pfeifer F (2020) Accessory Gvp proteins form a complex during gas vesicle formation of haloarchaea. Front Microbiol. https://doi.org/10.3389/fmicb.2020.610179

Walsby AE (1972) Structure and function of gas vacuoles. Bacteriol Rev 36:1–32

Walsby AE (1982) Permeability of gas vesicles to perfluorocyclobutane. J Gen Microbiol 128:1679–1684

Walsby AE (1994) Gas vesicles. Microbiol Rev 58:94–144

Walsby AE (2005) Archaea with square cells. Trends Microbiol 13:193–195

Walsby AE, Bleything A (1988) The dimensions of cyanobacterial gas vesicles in relation to their efficiency in providing buoyancy and withstanding pressure. J Gen Microbiol 134:2635–2645

Walsby AE, Hayes PK (1988) The minor cyanobacterial gas vesicle protein, GvpC, is attached to the outer surface of the gas vesicle. J Gen Microbiol 134:2647–2657

Walsby AE, McAllister G (1987) Buoyancy regulation of *Microcystis* in Lake Okaro. New Zeal J Mar Fresh Res 21:521–524

Walsby AE, Revsbech NP, Griffel DH (1992) The gas-permeability coefficient of the cyanobacterial gas vesicle wall. J Gen Microbiol 138:837–845

Williamson N, Fineran P, Ogawa W, Woodley L, Salmond G (2008) Integrated regulation involving quorum sensing, a two-component system, a GGDEF/EAL domain protein and a post-transcriptional regulator controls swarming and RhlA-dependent surfactant biosynthesis in Serratia. Environ Microbiol 10:1202–1217

Winter K, Born J, Pfeifer F (2018) Interaction of haloarchaeal gas vesicle proteins determined by split-GFP. Front Microbiol 9:1897

Xu B, Dai Y, Zhou K, Liu Y, Sun Q, Ren Y, Chen Y, Zhou C (2014) Structure of the gas vesicle protein GvpF from the cyanobacterium *Microcystis aeruginosa*. Acta Crystallogr D Biol Crystallogr 70:3013–3022

Yao V, Spudich J (1992) Primary structure of an archaebacterial transducer, a methyl-accepting protein associated with sensory rhodopsin I. Proc Natl Acad Sci U S A 89:11915–11919

Zhang W, Broun A, Mueller M, Alam M (1996) The primary structure of the archaeon *Halobacterium salinarum* blue light receptor sensory rhodopsin II and its transducer, a methyl-accepting protein. Proc Natl Acad Sci U S A 93:8230–8235

Zimmermann P, Pfeifer F (2003) Regulation of the expression of gas vesicle genes in *Haloferax mediterranei*: interaction of the two regulatory proteins GvpD and GvpE. Mol Microbiol 49:783–794

The Anammoxosome Organelle: The Power Plant of Anaerobic Ammonium-Oxidizing (Anammox) Bacteria

Laura Claret Fernández, Rob Mesman, and Laura van Niftrik

Contents

1 Introduction ... 108
2 Anammoxosome Structure ... 110
 2.1 Membrane Curvature ... 111
 2.2 Tubule-Like Structures ... 112
 2.3 Iron-Containing Particles and Encapsulins ... 112
3 Anammoxosome Function: Energy Metabolism ... 113
 3.1 Anammox Reaction 1: Nitrite Reduction to Nitric Oxide ... 114
 3.2 Anammox Reaction 2: Hydrazine Synthesis from Nitric Oxide and Ammonium ... 115
 3.3 Hydrazine Oxidation to Dinitrogen Gas ... 116
 3.4 Electron Cycling and Energy Conversion ... 117
4 Anammox Membrane Lipids: Ladderanes ... 118
5 Concluding Remarks ... 119
References ... 119

Abstract Anaerobic ammonium-oxidizing (anammox) bacteria are important players in both the environment and industry where they contribute substantially to nitrogen removal from natural and man-made (eco)systems. Apart from their ecological value, anammox bacteria are extremely interesting from both a physiological and cell biological perspective. Cells of anammox bacteria contain a major membrane-bound compartment, the anammoxosome, which is dedicated to the anammox reaction and energy conversion. The anammox reaction converts the substrates ammonium and nitrite to the product dinitrogen gas via two reactive intermediates: nitric oxide and hydrazine. First, nitrite is converted to nitric oxide by a nitrite reductase, nitric oxide and ammonium are then combined by hydrazine synthase to form hydrazine, and, finally, hydrazine is oxidized to dinitrogen gas by hydrazine dehydrogenase. The hydrazine synthase and hydrazine dehydrogenase enzymes are biochemical novelties. Electrons released from hydrazine oxidation are

L. Claret Fernández · R. Mesman · L. van Niftrik (✉)
Department of Microbiology, Institute for Water & Wetland Research, Faculty of Science, Radboud University, Nijmegen, The Netherlands
e-mail: l.vanniftrik@science.ru.nl

D. Jendrossek (ed.), *Bacterial Organelles and Organelle-like Inclusions*, Microbiology Monographs 34, https://doi.org/10.1007/978-3-030-60173-7_5

proposed to be shuttled to an electron transport chain in the anammoxosome membrane resulting in the establishment of a proton motive force and subsequent ATP synthesis. The anammoxosome compartment has a highly curved membrane and contains tubule-like structures and electron-dense, iron-containing particles. Finally, anammox bacteria contain unique ladderane membrane lipids that are postulated to render anammox membranes less permeable than conventional membranes. In this chapter, the cell biology, physiology and biochemistry of the anammox reaction and anammoxosome compartment are discussed.

1 Introduction

Anammox bacteria combine ammonium and nitrite anaerobically to form dinitrogen gas with nitric oxide and hydrazine as intermediates (Kartal et al. 2011). The first indications for this anaerobic ammonium oxidation (anammox) process were obtained in the early 1990s, when, in an anoxic fluidized-bed bioreactor at the Gist-Brocades yeast factory in the Netherlands, ammonium was found to be converted to dinitrogen gas at the expense of either nitrate or nitrite (Mulder et al. 1995). The existence of this process was already predicted as early as the 1940s (Broda 1977, Hamm and Thompson 1941, Richards 1966), as thermodynamical calculations and field observations indicated that far less ammonium accumulated in anoxic water bodies than expected from Redfield stoichiometry (Redfield 1958). The organisms responsible for the anammox process were enriched a few years later and identified as bacteria belonging to the *Planctomycetes* phylum (van de Graaf et al. 1995; Strous et al. 1999). Anammox bacteria form a distinct and deep branching phylogenetic group within this phylum and are assigned the order *Brocadiales* and family *Brocadiaceae* (Jetten et al. 2010). Up to date, five different anammox genera have been described: "*Candidatus* Brocadia", "*Ca.* Kuenenia", "*Ca.* Jettenia", "*Ca.* Anammoxoglobus" and "*Ca.* Scalindua". The so-called type strain is "*Ca.* Kuenenia stuttgartiensis"—the first anammox species for which a planktonic enrichment culture (Fig. 1) has been obtained (van der Star et al. 2008). Anammox bacteria grow very slowly with typical generation times ranging from one to several weeks depending on the growth conditions. They are cultivated in bioreactor systems (Fig. 1) as enrichment cultures of up to 95% purity (hence the "*Candidatus*" status).

In nature, the chemolithoautotrophic anammox bacteria are found in anoxic or oxygen-limited terrestrial (Penton et al. 2006; Clark et al. 2008; Hu et al. 2011), marine (Dalsgaard et al. 2005; Schmid et al. 2007; Woebken et al. 2008) and freshwater (Schubert et al. 2006; Zhang et al. 2007) environments and have been shown to play a crucial role in the biological nitrogen cycle where they are estimated to produce approximately half of the dinitrogen gas present in the atmosphere (Francis et al. 2007). Besides their presence in nature, all genera of anammox

Fig. 1 Membrane bioreactor system with a 95% enrichment culture of planktonic *K. stuttgartiensis* anammox bacteria. Note the bright red colour of anammox bacteria which is derived from their numerous haem-containing proteins

bacteria have been detected in conventional wastewater treatment plants (Kartal et al. 2007, 2008; Schmid et al. 2000, 2003; Strous et al. 1999, 2006).

Since the anaerobic and autotrophic anammox bacteria do not require organic matter as carbon source or oxygen for their metabolism, they have been established as an attractive alternative to conventional nitrogen removal from wastewater (Kartal et al. 2010). The long generation time of anammox bacteria is also a benefit here, since the accumulation of sludge is greatly reduced. The first 75 m^3 anammox wastewater treatment plant (WWTP) in Rotterdam (the Netherlands) has been in operation from 2002 onwards. In a survey in 2014, 114 full-scale anammox WWTPs were reported worldwide (Lackner et al. 2014), and this number has been steadily increasing.

The coccoid anammox bacteria are approximately 500–800 nm in diameter and divide by binary fission, though several of the canonical genes for this process are missing from the genome (van Niftrik et al. 2009). The cell envelope (Fig. 2) follows that of Gram-negative bacteria with an outer membrane and a periplasmic space which contains peptidoglycan (van Teeseling et al. 2015, 2018). In freshwater species, the outer membrane has been shown to be covered by a glycosylated surface protein (S-)layer which was characterized in the type strain *K. stuttgartiensis* (van Teeseling et al. 2014, 2016). The major exception to the standard bacterial cell plan

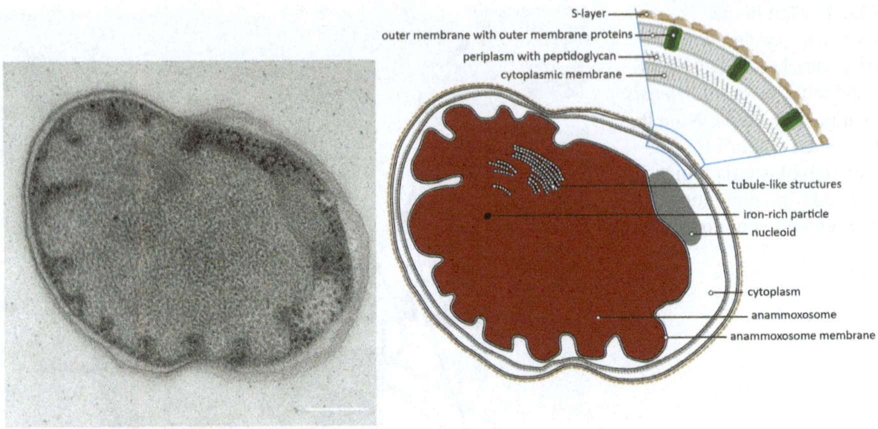

Fig. 2 Transmission electron micrograph (left) and schematic model (right) of a cross-section of an anammox cell. Scale bar; 200 nm

is that the cytoplasm of anammox bacteria is dominated by an additional membrane-bound compartment (Fig. 2): the anammoxosome (van Niftrik et al. 2004). The anammoxosome has been shown to house the energy metabolism of the cell (Neumann et al. 2014, de Almeida et al. 2015) and can thus be considered a true "prokaryotic organelle" with a specific cellular function: energy conversion. In addition to the unique cell plan, anammox bacteria have unique membrane lipids called ladderanes (Sinninghe Damsté et al. 2002). In this chapter we review the ins and outs of the structure and function of the unique anammoxosome compartment as well as ladderane lipids of anammox bacteria.

2 Anammoxosome Structure

The anammoxosome (Fig. 2) is a major compartment in the anammox cell (Strous et al. 1999; Lindsay et al. 2001). It occupies about 61% of the cell volume in both *K. stuttgartiensis* and *B. fulgida* and 51% and 66% in the slightly larger *Scalindua* sp. and *A. propionicus* cells, respectively (van Niftrik et al. 2008a). The condensed DNA (nucleoid) of anammox bacteria is associated with, and maybe even physically connected to, the anammoxosome (van Niftrik et al. 2004; Fuerst 2005). This is clearly visible in both light (DAPI staining) and transmission electron microscopy, where the DNA is crescent-shaped as a result of its attachment to the anammoxosome. During the entire process of cell division, the DNA stays attached to the anammoxosome. Strikingly, the DNA even stays attached to the outside of the organelle upon anammoxosome isolation (Neumann et al. 2014). This putative connection between DNA and anammoxosome might be functionally linked to the coordinated and equal partitioning of both the hereditary material and organelle to both daughter cells upon binary fission.

Anammoxosomes of planktonic anammox cells (i.e. *K. stuttgartiensis*) can be isolated from the cell at relatively high yield by using a combination of chemical and physical disruption followed by density gradient centrifugation (Sinninghe Damsté et al. 2002; Neumann et al. 2014). The fact that intact anammoxosomes can be isolated from the cell indicates that anammoxosomes are independent organelles without permanent membrane connections to the other cell compartments. Whether temporary, nonstructural membrane connections are present throughout the anammox life cycle remains a topic for future study. Anammoxosome isolation from cells growing in clusters is challenging and up to now has only been achieved with very low yield (Sinninghe Damsté et al. 2002). This is caused by the extensive extracellular polymeric substance (EPS) matrix which first has to be removed resulting in damage and subsequent lysis of many cells. Most isolated anammoxosomes lose their membrane curvature (see below) even though the bilayer membrane is seemingly intact. In addition, to maintain osmotic equilibrium and prevent lysis, isolated anammoxosomes have to be kept in buffer containing 0.25 M sucrose. Isolated anammoxosomes were shown to be able to perform the anammox reaction by measuring the production of $^{29}N_2$ from ^{15}N-ammonium and ^{14}N-nitrite in activity assays (Neumann et al. 2014). However, hydrazine was needed as external electron source indicating that the electron flow in isolated anammoxosomes is most likely impaired. The impaired electron flow could be caused by the absence of electron acceptors (such as $NAD(P)^+$ and the carbon fixation pathway) that are located in the cytoplasm of intact anammox cells.

2.1 Membrane Curvature

The anammoxosome is bounded by a single, bilayer membrane. This membrane is highly curved as shown by thin sections and 3D electron tomography models of anammox cells (van Niftrik et al. 2008b). The curvature is irregular and differs greatly between individual cells ranging from an almost completely oval anammoxosome to anammoxosomes with extensive and deep protrusions into the interior of the anammoxosome. The curvature remains during cell division although the anammoxosome is often stretched straight on one side which might be linked to the attachment of the DNA at that location. The membrane curvature is hypothesized to be linked to generating an increased membrane surface available for membrane-bound metabolic processes and subsequently an enhanced metabolic rate, analogous to mitochondrial cristae (see Sect. 3). How the anammoxosome membrane is folded is still unknown. It could be the result of a passive folding mechanism, i.e. the osmotic pressure folding the membrane inwards to reach osmotic equilibrium. This is substantiated by the fact that isolated anammoxosomes tend to lose their membrane curvature. Alternatively, the membrane could be folded by an active mechanism such as changes in local lipid composition, cytoskeletal or microtubule motor activity, direct protein-protein interactions and indirect protein-protein interactions (such as in the case of ATP synthase assembly and cristae formation (Cui 2018)) (McMahon and Gallop 2005).

2.2 Tubule-Like Structures

The anammoxosome contains tubule-like structures (Fig. 2) (Lindsay et al. 2001, van Niftrik et al. 2004, 2008b). When cross-sectioned, the tubules appear hexagonal and composed of three identical units with a width of about 9.4 nm each (van Niftrik et al. 2008b). The tubules always appear in packed arrays and often stretch the full length of the anammoxosome in the long plane of cell division (pole to pole). Membrane connections have not been observed. Immunogold labeling of key metabolic enzymes showed that nitrite oxidoreductase (NXR), which is proposed to convert nitrite to nitrate to replenish the electron pool (van de Graaf et al. 1996, Schouten et al. 2004, Hu et al. 2019) (see Sect. 3), is located at the tubule-like structures (de Almeida et al. 2015). This indicates that the tubules contain, or are, the NXR enzyme or that the enzyme is linked to the tubules. The present hypothesis is that the NXR enzyme forms tubule-like structures in order to facilitate high local concentrations of this protein complex.

2.3 Iron-Containing Particles and Encapsulins

One other striking feature in anammox cells is the presence of electron-dense particles within the anammoxosome (Fig. 2) (van Niftrik et al. 2008b). The particles range in diameter from 16 to 25 nm and have a heterogeneous electron density, without any obvious scattering associated with semi-crystalline structures. Energy dispersive X-ray analysis showed increased levels of both iron and phosphorus in these particles compared to the surrounding material (van Niftrik et al. 2008b), indicating a possible role in iron metabolism, storage or detoxificaton.

Many bacteria have intracellular reserves of iron, stored within iron storage proteins (Andrews 1998), of which the prokaryotic non-haem ferritins and haem-containing bacterioferritins are the most well-known. These proteins form large multimeric complexes (~120 Å in diameter for ferritin and 105 Å for bacterioferritin) with a large hollow core which can accommodate up to 2000 iron atoms, complexed as ferric hydroxyphosphate. Initially it was proposed that the iron particles within the anammoxosome could be bacterioferritins (van Niftrik et al. 2008b; van Niftrik and Jetten 2012; van Teeseling et al. 2013). However, the two bacterioferritin genes encoded in the *K. stuttgartiensis* genome (kuste3640 and kuste4480) lack a distinctive signal peptide needed to be imported into the anammoxosome (Ferousi et al. 2017). Furthermore, there is a marked discrepancy in the sizes reported for bacterioferritins and the observed iron particles. Other proteins that potentially could store iron, such as some members of the Dps protein family (Bozzi et al.

1997), have also been detected in the *K. stuttgartiensis* genome (kuste2684, kustd2167). However, the reported size of the Dps multimeric complexes is even smaller than those of bacterioferritins, making it less likely that these are the iron particles.

The genome of *K. stuttgartiensis* also harbours an encapsulin gene (kuste2478) (Radford 2014; Giessen and Silver 2017). Like (bacterio)ferritins, these proteins form nanocompartments, which depending on their cargo proteins can have different functions (Giessen and Silver 2017; Nichols et al. 2017). Using small ferritin molecules as cargo, encapsulins can effectively store ten times more iron than (bacterio)ferritin complexes (McHugh et al. 2014; He et al. 2016). The larger size of the iron-containing core (260 Å for the iron-storing encapsulin from *Myxococcus xanthus*) also matches better with the observed size of the iron particles in anammox bacteria. Encapsulins encapsulate specific cargo proteins which are targeted to the encapsulin via targeting peptides (TPs) present in their C-terminus (Rurup et al. 2014; Giessen and Silver 2017).

The proposed cargo for *K. stuttgartiensis* encapsulin is an HAO-like protein (kuste2479) (Giessen and Silver 2017) coded in line with the encapsulin protein. In some cases, secondary cargos can also be found further away from the encapsulin gene (Radford 2014; Giessen and Silver 2017). The *K. stuttgartiensis* genome encodes several small ferritin-like proteins which could be potential cargo of the encapsulin and also match the reported structure of the encapsulated ferritin from *R. rubrum* (He et al. 2016), though a clear encapsulation sequence is missing. Future research may provide the evidence needed to elucidate the identity, structural components and function of the anammox iron particles.

3 Anammoxosome Function: Energy Metabolism

The anammox energy metabolism takes place inside the anammoxosome (Fig. 3). It is heavily dependent on haem-containing proteins (Kartal et al. 2011; van Niftrik and Jetten 2012; Neumann et al. 2014; Kartal and Keltjens 2016). The presence of haems, both as part of the enzymes performing the unique anammox reaction and as putative electron shuttles, gives these bacteria their characteristic red colour (Fig. 1).

The anammox reaction converts nitrite and ammonium into dinitrogen gas, with hydrazine and nitric oxide as intermediates (Fig. 3). The anammox reaction is carried out in three consecutive redox reactions and involves three main enzymes: nitrite reductase (NIR), hydrazine synthase (HZS) and hydrazine dehydrogenase (HDH) (Strous et al. 2006). The electrons released in the anammox redox reactions are proposed to be shuttled to an electron transport chain in the anammoxosome membrane for energy conversion and subsequent ATP synthesis (Kartal et al. 2013; Kartal and Keltjens 2016).

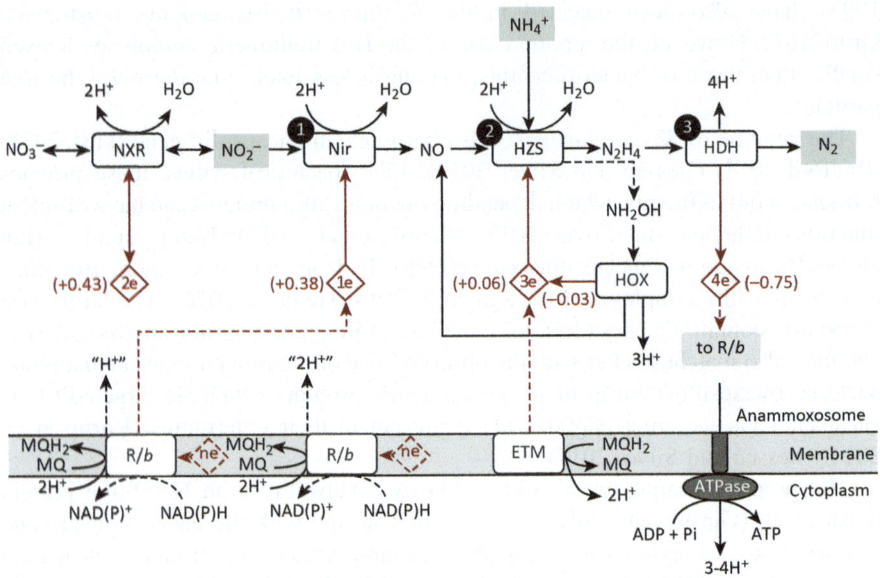

Fig. 3 Model for anaerobic ammonium oxidation and energy conversion within the anammoxosome compartment of anammox bacteria. (1) Nitrite reduction to nitric oxide by a nitrite reductase enzyme. (2) The condensation of ammonium with nitric oxide to form hydrazine by the hydrazine synthase enzyme. (3) Hydrazine oxidation to dinitrogen gas by the hydrazine dehydrogenase enzyme. NXR, nitrite oxidoreductase; Nir, nitrite reductase; HZS, hydrazine synthase; HDH, hydrazine dehydrogenase; HOX, hydroxylamine oxidase; R/b, Rieske-haem *b* complex (*bc*1 complex); ETM, electron transfer module. Modified with permission from Kartal and Keltjens (2016)

3.1 Anammox Reaction 1: Nitrite Reduction to Nitric Oxide

The first step in the anammox reaction is the reduction of nitrite to nitric oxide, consuming one electron in the process (Fig. 3). The current hypothesis states that a nitrite reductase (NIR) performs this reaction (Strous et al. 2006; Kartal et al. 2011; de Almeida et al. 2016). Although there is metagenomic evidence that (different) anammox bacteria encode different types of NIR enzymes, it still has to be experimentally validated which protein is responsible for this reaction in each specific anammox species (Kartal et al. 2013; Kartal and Keltjens 2016). In *K. stuttgartiensis* nitrite reduction to nitric oxide is proposed to be carried out by a cd_1-NIR (NirS), encoded by gene cluster kuste4136-4140. However, the mRNA expression levels of this enzyme are very low compared to those of other anammox key metabolic enzymes. *Scalindua profunda* encodes for the same NirS, but in this case the mRNA expression levels are strikingly high in comparison with *K. stuttgartiensis* (van de Vossenberg et al. 2013; Kartal et al. 2013). *Jettenia caeni* strain KSU-1, on the other hand, encodes for a Cu-NIR (NirK) enzyme instead of a NirS (Hira et al. 2012). Finally, the metagenomes of *Brocadia* sp. do not encode for any known nitrite reductase (Kartal et al. 2013; Speth et al. 2016; Peeters and van Niftrik 2019). Alternatively, it was proposed that a hydroxylamine oxidoreductase (HAO), kustc0458, could function as alternative nitrite reductase in at least *K. stuttgartiensis* and *Brocadia* sp. but perhaps also other anammox bacteria (Hu et al. 2019).

3.2 Anammox Reaction 2: Hydrazine Synthesis from Nitric Oxide and Ammonium

After the reduction of nitrite to nitric oxide, nitric oxide is taken up by the hydrazine synthase (HZS), a unique enzyme that bears no homology with other known enzymes (Dietl et al. 2015). HZS combines nitric oxide and ammonium to form hydrazine, consuming three electrons in the process (Fig. 3). *K. stuttgartiensis* genome analysis suggested a set of three genes to code for HZS (kuste2859-2860-2861) that are part of a larger gene cluster (kuste2854-61). HZS accounts for up to 20% of the *K. stuttgartiensis* protein content, and its purification indicated that it indeed catalyses the proposed reaction (Kartal et al. 2011; Maalcke et al. 2014). At least three other genes in the HZS gene cluster might encode a multicomponent enzyme system for proper HZS functioning (Fig. 3: ETM). The integral membrane protein kuste2856, with two haem b and one quinone binding site, would transfer the electrons yielded from quinol oxidation to the membrane-anchored multihaem cytochrome protein kuste2855. Next, the soluble tetrahaem protein kuste2854 would donate the electrons to HZS.

Based on the HZS crystal structure (Fig. 4), a mechanism for hydrazine synthesis was proposed (Dietl et al. 2015). HZS is a dimer of heterotrimers ($\alpha_2\beta_2\gamma_2$). The α-subunit contains a six-bladed β-propeller and two c-type haems: one near the protein surface and one representing a catalytic site deep inside the protein. The β-subunit contains a seven-bladed β-propeller. The γ-subunit contains two c-type haems: one on the protein surface likely involved in electron transfer and one representing a catalytic site deep inside the protein. The α- and γ-subunits are

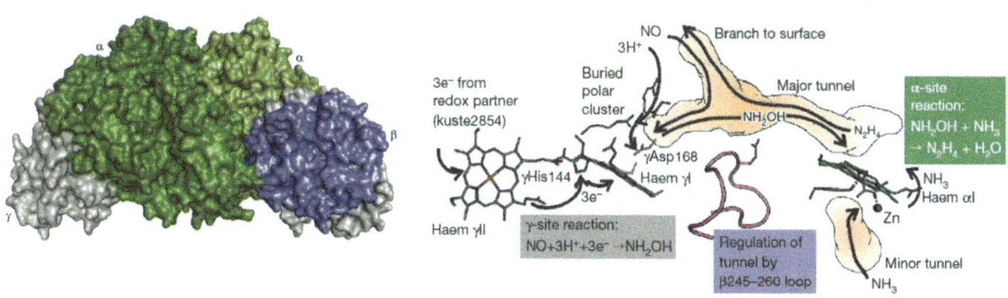

Fig. 4 Proposed mechanism of hydrazine synthesis (right) by the *K. stuttgartiensis* hydrazine synthase (HZS) enzyme based on its crystal structure (left). HZS is a dimer of heterotrimers. Adapted with permission from Dietl et al. (2015)

connected by tunnels between their active sites, leading to the protein surface. The latter might allow entry and exit of substrate and product. Nitric oxide would enter the tunnel and travel to the catalytic site of the γ-subunit. Here, the three-electron reduction of nitric oxide to hydroxylamine takes place where the electrons are obtained from the tetrahaem cytochrome *c* protein kuste2854. Next, hydroxylamine would be transferred through the major tunnel to the catalytic haem of the α-subunit and combined with ammonium, which enters through the minor tunnel, to form hydrazine. Hydrazine can leave the enzyme from the same place where NO entered. The β-subunit could play a role in modulating transport through the tunnel.

3.3 Hydrazine Oxidation to Dinitrogen Gas

The last step of the anammox reaction involves the oxidation of hydrazine to dinitrogen gas (Fig. 3) and is performed by another biochemical novelty: the hydrazine dehydrogenase (HDH) enzyme (Kartal et al. 2013; Maalcke et al. 2016).

The multihaem *c*-type cytochrome *K. stuttgartiensis* HDH enzyme is a soluble protein complex and encoded by kustc0694 (de Almeida et al. 2015; Maalcke et al. 2016). It provides low potential electrons that drive the energy metabolism, and these electrons thus have to be transported via the anammoxosome plasma to the quinone pool in the anammoxosome membrane (Fig. 3). The process of electron shuttling must be tightly regulated as care has to be taken that these high-energy electrons are not simply dumped on the next available electron acceptor. The 2.8 Å resolution crystal structure (Fig. 5) shed some light on how this might be achieved (Akram et al. 2019). Hydrazine dehydrogenase shares homology with HAO proteins (Kartal et al. 2011). However, the homotrimers of HDH form octamers in solution, whilst the HAOs from AOB and the other HAOs from *K. stuttgartiensis* are

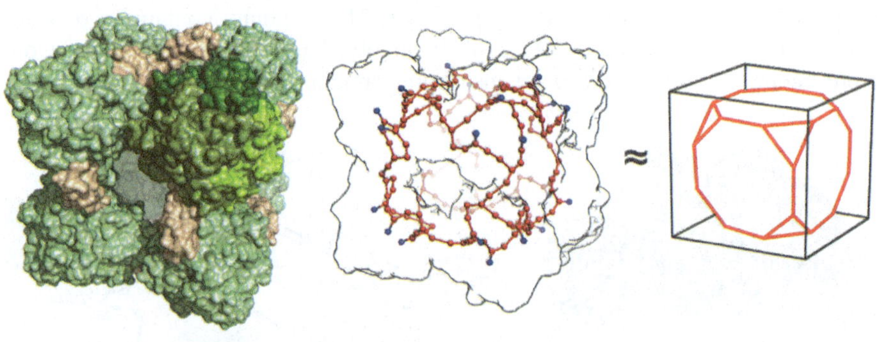

Fig. 5 The 2.8 Å resolution crystal structure (left) of the hydrazine dehydrogenase (HDH) enzyme from *K. stuttgartiensis*. HDH contains a 192-haem electron transfer network (right), which is proposed to store and release the low potential electrons from hydrazine oxidation. HDH is an octamer of trimers with an additional protein (in pink) at the vertices of the cube-like enzyme which might function as an assembly factor. Adapted from Akram et al. (2019)

homotrimeric proteins. The HDH octamer of trimers is cube-shaped and contains 24 copies of the HDH monomer. In addition, a 10 kDa protein (kustc1130) is located at the vertices of the cube which is proposed to act as an assembly factor (Akram et al. 2019). The total 1.7 Mda HDH complex thus consists of $(\alpha_3)_8\beta_{12}$. The HDH complex contains no less than 192 haem groups and is proposed to function as an electron transport and storage system to safeguard the high potential electrons. Acceptor proteins might specifically dock onto the HDH complex, and the electrons are proposed to be able to be transferred to the acceptor proteins through all 24 exit sites. The holes in the assembly structure thus function as a selectivity filter to prevent transfer to the wrong acceptor.

Hydrazine dehydrogenase is inhibited by hydroxylamine (Maalcke et al. 2016). Another HAO, the hydroxylamine oxidase (HOX), is proposed to function as a safety mechanism by oxidizing any hydroxylamine, which might leak out of the hydrazine synthase, back to nitric oxide (Fig. 3). Nitric oxide can be fed back directly in the energy metabolism as substrate for hydrazine synthase.

3.4 *Electron Cycling and Energy Conversion*

It is proposed that the low potential electrons released from hydrazine oxidation are shuttled via soluble cytochromes to an electron transport chain in the anammoxosome membrane and ultimately fed back to the nitrite reductase and hydrazine synthase (Strous et al. 2006; Kartal et al. 2013; Kartal and Keltjens 2016). In the anammoxosome membrane, electrons would flow via Rieske-haem *b* complexes (bc1 complexes) and a quinone pool at the expense of NAD(P)H, resulting in the establishment of a proton motive force. The proton motive force could be used by anammoxosome membrane-bound ATPases (van Niftrik et al. 2010; Karlsson et al. 2014) to produce ATP in the cytoplasm. In addition, a complexome analysis showed that *K. stuttgartiensis* expresses respiratory complexes that could utilize sodium instead of protons and anammox bacteria may therefore also have the potential to use a sodium motive force (de Almeida et al. 2016).

In order to sustain growth, it is proposed that some of the electrons from hydrazine oxidation are shuttled towards carbon fixation via the acetyl coenzyme A (CoA) pathway (Schouten et al. 2004; Strous et al. 2006). As mentioned before, the oxidation of nitrite to nitrate by NXR (Fig. 3) is proposed to replenish the electron pool by feeding electrons to the nitrite reductase (Hu et al. 2019).

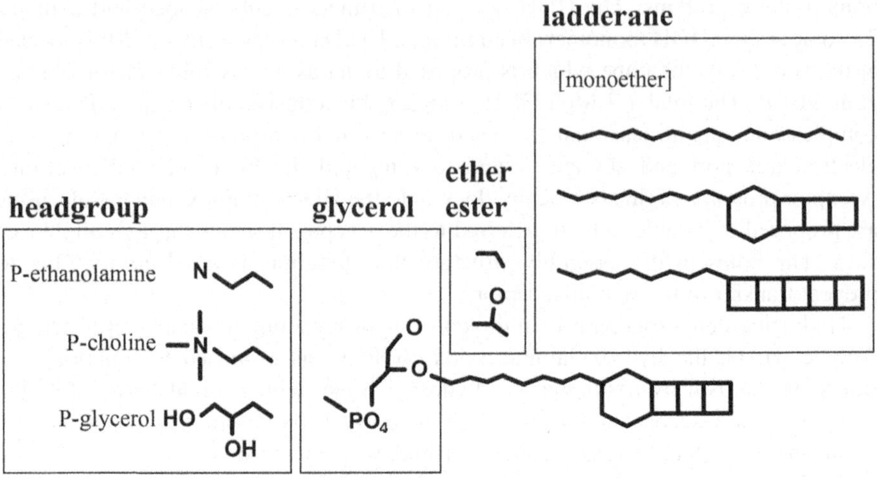

Fig. 6 Composition of anammox ladderane lipids. Anammox bacteria contain ladderane fatty acids with either three or five linearly concatenated cyclobutane rings. The hydrocarbon tails are bound to a polar head group by either ester or ether bonds. The polar head groups are phosphocholine, phosphoethanolamine or phosphoglycerol. Modified with permission from Jetten et al. (2009)

4 Anammox Membrane Lipids: Ladderanes

Anammox bacteria contain very unusual membrane lipids that are called ladderanes (Fig. 6) (Sinninghe Damsté et al. 2002). Ladderane lipids are present in all three anammox membranes (Neumann et al. 2014) and constitute up to at least 22–63% of total lipids depending on the species (Rattray et al. 2008). These percentages were determined from 75 to 90% anammox enrichment cultures so the actual percentage in the anammox cells is most likely higher. The distribution of the ladderane lipids versus other membrane lipids (such as straight chain and branched fatty acids and hopanoids) is unknown and could vary per anammox membrane and species. Anammox bacteria contain C_{18} and C_{20} ladderane fatty acids with either three or five linearly concatenated cyclobutane rings (Fig. 6) (Rattray et al. 2008). C_{22} ladderane fatty acids were also detected (Rattray et al. 2008). The hydrocarbon tails are bound to a polar head group by either ester or ether bonds. The polar head groups are phosphocholine (PC), phosphoethanolamine (PE) or phosphoglycerol (PG) (Boumann et al. 2006; Rattray et al. 2008). Ladderane lipids form a much denser, less permeable membrane compared to conventional, non-ladderane membrane lipids (Sinninghe Damsté et al. 2002; Boumann et al. 2009). Synthetic ladderane-only membranes were reported to have low permeability for protons but not for hydrazine (Moss III et al. 2018). This corroborates the hypothesis that the slow-growing anammox bacteria require a denser, less permeable membrane to limit the loss of protons by passive diffusion (Sinninghe Damsté et al. 2002).

How ladderane lipids are synthesized is still unknown (Javidpour et al. 2016). Four gene clusters were identified in *K. stuttgartiensis*, containing 34 genes that might be involved in ladderane biosynthesis (Rattray et al. 2009). The proposed biosynthetic pathway involves desaturation of acyl-ACPs to form polyunsaturated intermediates. Next, a radical cascade mechanism would perform polycyclization to form the ladderane structure.

5 Concluding Remarks

Anammox bacteria are of scientific interest for multiple reasons. In the environment, they are important players in the global nitrogen cycle; in industry, they are applied for the removal of nitrogen from wastewater streams; and on top of that, they have many unique properties. With regard to the structure and function of the anammoxosome compartment, a major question for future research is how the anammox reaction is coupled to energy conversion and, more specifically, which components play a role in the electron transport chain in the anammoxosome membrane and how are electrons shuttled between the membrane and the soluble NIR, HZS and HDH enzymes. In addition, it is still unknown which enzyme (s) perform(s) nitrite reduction in specific anammox species, and the proposed ladderane biosynthetic pathway awaits experimental validation. Even though these questions are not easily solved without a pure culture, standard cultivation techniques and no genetic system, they are of major interest to the scientific community both from a fundamental and an applied perspective.

References

Akram M, Dietl A, Mersdorf U, Prinz S, Maalcke W, Keltjens J, Ferousi C, de Almeida NM, Reimann J, Kartal B, Jetten MSM, Parey K, Barends TRM (2019) A 192-heme electron transfer network in the hydrazine dehydrogenase complex. Sci Adv 5:eaav4310

Andrews SC (1998) Iron storage in bacteria. In: Poole RK (ed) Advances in microbial physiology. Academic, pp 281–351

Boumann HA, Hopmans EC, van de Leemput I, Op den Camp HJM, van de Vossenberg J, Strous M, Jetten MSM, Sinninghe Damsté JS, Schouten S (2006) Ladderane phospholipids in anammox bacteria comprise phosphocholine and phosphoethanolamine headgroups. FEMS Microbiol Lett 258:297–304

Boumann HA, Longo ML, Stroeve P, Poolman B, Hopmans EC, Stuart MCA, Sinninghe Damsté JS, Schouten S (2009) Biophysical properties of membrane lipids of anammox bacteria: I. Ladderane phospholipids form highly organized fluid membranes. BBA Biomembranes 1788:1444–1451

Bozzi M, Mignogna G, Stefanini S, Barra D, Longhi C, Valenti P, Chiancone E (1997) A novel non-heme iron-binding ferritin related to the DNA-binding proteins of the Dps family in *Listeria innocua*. J Biol Chem 272:3259–3265

Broda E (1977) Two kinds of lithotrophs missing in nature. Z Für Allg Mikrobiol 17:491–493

Clark I, Timlin R, Bourbonnais A, Jones K, Lafleur D, Wickens K (2008) Origin and fate of industrial ammonium in anoxic ground water—15N evidence for anaerobic oxidation (anammox). Groundw Monit Remediat 28:73–82

Cui Q (2018) Membrane-mediated interaction drives mitochondrial ATPase assembly and cristae formation. J Gen Physiol 150:777–780

Dalsgaard T, Thamdrup B, Canfield DE (2005) Anaerobic ammonium oxidation (anammox) in the marine environment. Res Microbiol 156:457–464

de Almeida NM, Neumann S, Mesman RJ, Ferousi C, Keltjens JT, Jetten MSM, Kartal B, van Niftrik L (2015) Immunogold localization of key metabolic enzymes in the anammoxosome and on the tubule-like structures of *Kuenenia stuttgartiensis*. J Bacteriol 197:2432–2441

de Almeida NM, Wessels HJ, de Graaf RM, Ferousi C, Jetten MSM, Keltjens JT, Kartal B (2016) Membrane-bound electron transport systems of an anammox bacterium: a complexome analysis. BBA Bioenerget 1857:1694–1704

Dietl A, Ferousi C, Maalcke WJ, Menzel A, de Vries S, Keltjens JT, Jetten MSM, Kartal B, Barends TRM (2015) The inner workings of the hydrazine synthase multiprotein complex. Nature 527:394–397

Ferousi C, Lindhoud S, Baymann F, Kartal B, Jetten MSM, Reimann J (2017) Iron assimilation and utilization in anaerobic ammonium oxidizing bacteria. Curr Opin Chem Biol 37:129–136

Francis CA, Beman JM, Kuypers MMM (2007) New processes and players in the nitrogen cycle: the microbial ecology of anaerobic and archaeal ammonia oxidation. ISME J 1:19–27

Fuerst JA (2005) Intracellular compartmentation in planctomycetes. Annu Rev Microbiol 59:299–328

Giessen TW, Silver PA (2017) Widespread distribution of encapsulin nanocompartments reveals functional diversity. Nat Microbiol 2:17029

Hamm RE, Thompson TG (1941) Dissolved nitrogen in the sea water of the northeast Pacific with notes on the total carbon dioxide and the dissolved oxygen. J Marine Res 4:11–27

He D, Hughes S, Vanden-Hehir S, Georgiev A, Altenbach K, Tarrant E, Mackay CL, Waldron KJ, Clarke DJ, Marles-Wright J (2016) Structural characterization of encapsulated ferritin provides insight into iron storage in bacterial nanocompartments. ELife 5:e18972

Hira D, Toh H, Migita CT, Okubo H, Nishiyama T, Hattori M, Furukawa K, Fujii T (2012) Anammox organism KSU-1 expresses a NirK-type copper-containing nitrite reductase instead of a NirS-type with cytochrome cd1. FEBS Lett 586:1658–1663

Hu B, Rush D, van der Biezen E, Zheng P, van Mullekom M, Schouten S, Sinninghe Damsté JS, Smolders AJP, Jetten MSM, Kartal B (2011) New anaerobic, ammonium-oxidizing community enriched from peat soil. Appl Environ Microbiol 77:966–971

Hu Z, Wessels HJCT, van Alen T, Jetten MSM, Kartal B (2019) Nitric oxide-dependent anaerobic ammonium oxidation. Nat Commun 10:1244

Javidpour P, Deutsch S, Mutalik VK, Hillson NJ, Petzold CJ, Keasling JD, Beller HR (2016) Investigation of proposed ladderane biosynthetic genes from anammox bacteria by heterologuous expression in *E. coli*. PLoS One 11:e0151087

Jetten MSM, van Niftrik L, Strous M, Kartal B, Keltjens JT, Op den Camp HJM (2009) Biochemistry and molecular biology of anammox bacteria. Crit Rev Biochem Mol Biol 44:65–84

Jetten MSM, Op den Camp HJM, Kuenen JG, Strous M (2010) Family I. "*Candidatus* Brocadiaceae" fam. nov. In: Krieg NR, Staley JT, Brown DR, Hedlund B, Paster BJ, Ward N, Ludwig W, Whitman WB (eds) Bergey's manual of systematic bacteriology, Springer, pp 596–602

Karlsson R, Karlsson A, Bäckman O, Johansson BR, Hulth S (2014) Subcellular localization of an ATPase in anammox bacteria using proteomics and immunogold electron microscopy. FEMS Microbiol Lett 354:10–18

Kartal B, Keltjens JT (2016) Anammox biochemistry: a tale of heme *c* proteins. Trends Biochem Sci 41:998–1011

Kartal B, Rattray J, van Niftrik LA, van de Vossenberg J, Schmid MC, Webb RI, Schouten S, Fuerst JA, Sinninghe Damsté JS, Jetten MSM, Strous M (2007) *Candidatus* "Anammoxoglobus propionicus" a new propionate oxidizing species of anaerobic ammonium oxidizing bacteria. Syst Appl Microbiol 30:39–49

Kartal B, van Niftrik L, Rattray J, van de Vossenberg JLCM, Schmid MC, Sinninghe Damsté JS, Jetten MSM, Strous M (2008) *Candidatus* "Brocadia fulgida": an autofluorescent anaerobic ammonium oxidizing bacterium. FEMS Microbiol Ecol 63:46–55

Kartal B, Kuenen JG, van Loosdrecht MCM (2010) Sewage treatment with anammox. Science 328:702–703

Kartal B, Maalcke WJ, de Almeida NM, Cirpus I, Gloerich J, Geerts W, Op den Camp HJM, Harhangi HR, Janssen-Megens EM, Francoijs K-J, Stunnenberg HG, Ketltjens JT, Jetten MSM, Strous M (2011) Molecular mechanism of anaerobic ammonium oxidation. Nature 479:127–130

Kartal B, de Almeida NM, Maalcke WJ, Op den Camp HJ, Jetten MSM, Keltjens JT (2013) How to make a living from anaerobic ammonium oxidation. FEMS Microbiol Rev 37:428–461

Lackner S, Gilbert EM, Vlaeminck SE, Joss A, Horn H, van Loosdrecht MCM (2014) Full-scale partial nitritation/anammox experiences – an application survey. Water Res 55:292–303

Lindsay MR, Webb RI, Strous M, Jetten MSM, Butler MK, Forde RJ, Fuerst JA (2001) Cell compartmentalisation in planctomycetes: novel types of structural organisation for the bacterial cell. Arch Microbiol 175:413–429

Maalcke WJ, Dietl A, Marritt SJ, Butt JN, Jetten MSM, Keltjens JT, Barends TRM, Kartal B (2014) Structural basis of biological NO generation by octaheme oxidoreductases. J Biol Chem 289:1228–1242

Maalcke WJ, Reimann J, de Vries S, Butt JN, Dietl A, Kip N, Mersdorf U, Barends TRM, Jetten MSM, Keltjens JT, Kartal B (2016) Characterization of anammox hydrazine dehydrogenase, a key N_2-producing enzyme in the global nitrogen cycle. J Biol Chem 291:17077–17092

McHugh CA, Fontana J, Nemecek D, Cheng N, Aksyuk AA, Heymann JB, Winkler DC, Lam AS, Wall JS, Steven AC, Hoiczyk E (2014) A virus capsid-like nanocompartment that stores iron and protects bacteria from oxidative stress. EMBO J 33:1896–1911

McMahon HT, Gallop JL (2005) Membrane curvature and mechanisms of dynamic cell membrane remodelling. Nature 438:590–596

Moss FR III, Shuken SR, JAM M, Cohen CM, Weiss TM, Boxer SG, Burns NZ (2018) Ladderane phospholipids form a densely packed membrane with normal hydrazine and anomalously low proton/hydroxide permeability. Proc Natl Acad Sci USA 115:9098–9103

Mulder A, van de Graaf AA, Robertson LA, Kuenen JG (1995) Anaerobic ammonium oxidation discovered in a denitrifying fluidized bed reactor. FEMS Microbiol Ecol 16:177–183

Neumann S, Wessels HJ, Rijpstra WIC, Sinninghe Damsté JS, Kartal B, Jetten MSM, van Niftrik L (2014) Isolation and characterization of a prokaryotic cell organelle from the anammox bacterium *Kuenenia stuttgartiensis*. Mol Microbiol 94:794–802

Nichols RJ, Cassidy-Amstutz C, Chaijarasphong T, Savage DF (2017) Encapsulins: molecular biology of the shell. Crit Rev Biochem Mol Biol 52:583–594

Peeters SH, van Niftrik L (2019) Trending topics and open questions in anaerobic ammonium oxidation. Curr Opin Chem Biol 49:45–52

Penton CR, Devol AH, Tiedje JM (2006) Molecular evidence for the broad distribution of anaerobic ammonium-oxidizing bacteria in freshwater and marine sediments. Appl Environ Microbiol 72:6829–6832

Radford DR (2014) Understanding the encapsulins: prediction and characterization of phage capsid-like nanocompartments in prokaryotes. Dissertation, University of Toronto

Rattray JE, van de Vossenberg J, Hopmans EC, Kartal B, van Niftrik L, Rijpstra WIC, Strous M, Jetten MSM, Schouten S, Sinninghe Damsté JS (2008) Ladderane lipids distribution in four genera of anammox bacteria. Arch Microbiol 190:51–66

Rattray JE, Strous M, Op den Camp HJM, Schouten S, Jetten MSM, Sinninghe Damsté JS (2009) A comparative genomics study of genetic products potentially encoding ladderane lipid biosynthesis. Biol Direct 4:8

Redfield AC (1958) The biological control of chemical factors in the environment. Am Sci 46:205–221

Richards FA (1966) Chemical observations in some anoxic, sulfide-bearing basins and fjords. S.l.: Dept. of Oceanography, University of Washington

Rurup WF, Snijder J, Koay MST, Heck AJR, Cornelissen JJLM (2014) Self-sorting of foreign proteins in a bacterial nanocompartment. J Am Chem Soc 136:3828–3832

Schmid M, Twachtmann U, Klein M, Strous M, Juretschko S, Jetten MSM, Metzger JW, Schleifer KH, Wagner M (2000) Molecular evidence for genus level diversity of bacteria capable of catalyzing anaerobic ammonium oxidation. Syst Appl Microbiol 23:93–106

Schmid M, Walsh K, Webb R, Rijpstra WIC, van de Pas-Schoonen K, Verbruggen MJ, Hill T, Moffett B, Fuerst JA, Schouten S, Sinninghe Damsté JS, Harris J, Shaw P, Jetten MSM, Strous M (2003) *Candidatus* "Scalindua brodae", sp. nov., *Candidatus* "Scalindua wagneri", sp. nov., two new species of anaerobic ammonium oxidizing bacteria. Syst Appl Microbiol 26:529–538

Schmid MC, Risgaard-Petersen N, van de Vossenberg J, Kuypers MMM, Lavik G, Petersen J, Hulth S, Thamdrup B, Canfield D, Dalsgaard T, Rysgaard S, Sejr MK, Strous M, Op den Camp HJM, Jetten MSM (2007) Anaerobic ammonium-oxidizing bacteria in marine environments: widespread occurrence but low diversity. Environ Microbiol 9:1476–1484

Schouten S, Strous M, Kuypers MMM, Rijpstra WIC, Baas M, Schubert CJ, Jetten MSM, Sinninghe Damsté JS (2004) Stable carbon isotopic fractionations associated with inorganic carbon fixation by anaerobic ammonium-oxidizing bacteria. Appl Environ Microbiol 70:3785–3788

Schubert CJ, Durisch-Kaiser E, Wehrli B, Thamdrup B, Lam P, Kuypers MMM (2006) Anaerobic ammonium oxidation in a tropical freshwater system (Lake Tanganyika). Environ Microbiol 8:1857–1863

Sinninghe Damsté JS, Strous M, Rijpstra WIC, Hopmans EC, Geenevasen JAJ, van Duin ACT, van Niftrik L, Jetten MSM (2002) Linearly concatenated cyclobutane lipids form a dense bacterial membrane. Nature 419:708–712

Speth DR, Guerrero-Cruz S, Dutilh BE, Jetten MSM (2016) Genome-based microbial ecology of anammox granules in a full-scale wastewater treatment system. Nat Commun 7:11172

Strous M, Fuerst JA, Kramer EHM, Logemann S, Muyzer G, van de Pas-Schoonen KT, Webb R, Kuenen JG, Jetten MSM (1999) Missing lithotroph identified as new planctomycete. Nature 400:446–449

Strous M, Pelletier E, Mangenot S, Rattei T, Lehner A, Taylor MW, Horn M, Daims H, Bartol-Mavel D, Wincker P, Barbe V, Fonknechten N, Vallenet D, Segurens B, Schenowitz-Truong C, Médigue C, Collingro A, Snel B, Dutilh BE, Op den Camp HJM, van der Drift C, Cirpus I, van de Pas-Schoonen KT, Harhangi HR, van Niftrik L, Schmid M, Keltjens J, van de Vossenberg J, Kartal B, Meier H, Frishman D, Huynen MA, Mewes H-W, Weissenbach J, Jetten MSM, Wagner M, Le Paslier D (2006) Deciphering the evolution and metabolism of an anammox bacterium from a community genome. Nature 440:790–794

van de Graaf AA, Mulder A, de Bruijn P, Jetten MSM, Robertson LA, Kuenen JG (1995) Anaerobic oxidation of ammonium is a biologically mediated process. Appl Environ Microbiol 61:1246–1251

van de Graaf AA, de Bruijn P, Robertson LA, Jetten MSM, Kuenen JG (1996) Autotrophic growth of anaerobic ammonium-oxidizing microorganisms in a fluidized bed reactor. Microbiology 142:2187–2196

van der Star WRL, Miclea AI, van Dongen UGJM, Muyzer G, Picioreanu C, van Loosdrecht MCM (2008) The membrane bioreactor: A novel tool to grow anammox bacteria as free cells. Biotechnol Bioeng 101:286–294

van de Vossenberg J, Woebken D, Maalcke WJ, Wessels HJCT, Dutilh BE, Kartal B, Janssen-Megens EM, Roeselers G, Yan J, Speth D, Gloerich J, Geerts W, van der Biezen E, Pluk W, Francoijs KJ, Russ L, Lam P, Malfatti SA, Tringe SG, Haaijer SC, Op den Camp HJ, Stunnenberg HG, Amann R, Kuypers MMM, Jetten MSM (2013) The metagenome of the marine anammox bacterium '*Candidatus* Scalindua profunda' illustrates the versatility of this globally important nitrogen cycle bacterium. Environ Microbiol 15:1275–1289 van Niftrik L, Jetten MSM (2012) Anaerobic ammonium-oxidizing bacteria: unique microorganisms with exceptional properties. Microbiol Mol Biol Rev 76:585–596

van Niftrik LA, Fuerst JA, Sinninghe Damsté JS, Kuenen JG, Jetten MSM, Strous M (2004) The anammoxosome: an intracytoplasmic compartment in anammox bacteria. FEMS Microbiol Lett 233:7–13

van Niftrik L, Geerts WJC, van Donselaar EG, Humbel BM, Webb RI, Fuerst JA, Verkleij AJ, Jetten MSM, Strous M (2008a) Linking ultrastructure and function in four genera of anaerobic ammonium-oxidizing bacteria: cell plan, glycogen storage and localization of cytochrome c proteins. J Bacteriol 190:708–717

van Niftrik L, Geerts WJC, van Donselaar EG, Humbel BM, Yakushevska A, Verkleij AJ, Jetten MSM, Strous M (2008b) Combined structural and chemical analysis of the anammoxosome: a membrane-bounded intracytoplasmic compartment in anammox bacteria. J Struct Biol 161:401–410

van Niftrik L, Geerts WJC, van Donselaar EG, Humbel BM, Webb RI, Harhangi HR, Op den Camp HJM, Fuerst JA, Verkleij AJ, Jetten MSM, Strous M (2009) Cell division ring, a new cell division protein and vertical inheritance of a bacterial organelle in anammox planctomycetes. Mol Microbiol 73:1009–1019

van Niftrik L, van Helden M, Kirchen S, van Donselaar EG, Harhangi HR, Webb RI, Fuerst JA, Op den Camp HJM, Jetten MSM, Strous M (2010) Intracellular localization of membrane-bound ATPases in the compartmentalized anammox bacterium "*Candidatus* Kuenenia stuttgartiensis". Mol Microbiol 77:701–715

van Teeseling MCF, Neumann S, van Niftrik L (2013) The anammoxosome organelle is crucial for the energy metabolism of anaerobic ammonium oxidizing bacteria. J Mol Microbiol Biotechnol 23:104–117

van Teeseling MCF, de Almeida NM, Klingl A, Speth DR, Op den Camp HJM, Rachel R, Jetten MSM, van Niftrik L (2014) A new addition to the cell plan of anammox bacteria: "*Candidatus* Kuenenia stuttgartiensis" has a protein surface layer as the outermost layer of the cell. J Bacteriol 196:80–89

van Teeseling MCF, Mesman RJ, Kuru E, Espaillat A, Cava F, Brun YV, VanNieuwenhze MS, Kartal B, van Niftrik L (2015) Anammox Planctomycetes have a peptidoglycan cell wall. Nat Commun 6:6878

van Teeseling MCF, Maresch D, Rath CB, Figl R, Altmann F, Jetten MSM, Messner P, Schäffer C, van Niftrik L (2016) The S-layer protein of the anammox bacterium *Kuenenia stuttgartiensis* is heavily O-glycosylated. Front Microbiol 7:1721

van Teeseling MCF, Benz R, de Almeida NM, Jetten MSM, Mesman RJ, van Niftrik L (2018) Characterization of the first planctomycetal outer membrane protein identifies a channel in the outer membrane of the anammox bacterium *Kuenenia stuttgartiensis*. BBA Biomembranes 1860:767–776

Woebken D, Lam P, Kuypers MMM, Naqvi SWA, Kartal B, Strous M, Jetten MSM, Fuchs BM, Amann R (2008) A microdiversity study of anammox bacteria reveals a novel *Candidatus* Scalindua phylotype in marine oxygen minimum zones. Environ Microbiol 10:3106–3119

Zhang Y, Ruan X-H, Op den Camp HJM, Smits TJM, Jetten MSM, Schmid MC (2007) Diversity and abundance of aerobic and anaerobic ammonium-oxidizing bacteria in freshwater sediments of the Xinyi River (China). Environ Microbiol 9:2375–2382

Bacterial Microcompartments

Sabine Heinhorst and Gordon C. Cannon

Contents

1 Introduction 125
2 Shell Structure and Function 128
3 BMC Superloci 132
4 Organization of the β-Carboxysome Interior 137
5 Positioning and Segregation of Carboxysomes 137
6 Concluding Remarks 139
References 140

Abstract Bacterial microcompartments (BMCs) are polyhedral metabolic organizers in bacteria that consist of a thin, selectively permeable protein boundary and a set of encapsulated enzymes. Most BMCs participate in the catabolism of specific carbon sources; the carboxysome is the only anabolic BMC and promotes efficient fixation of CO_2 by ribulose 1,5-bisphosphate carboxylase/oxygenase (RubisCO) in many autotrophic bacteria. Despite the diverse functions of BMCs, commonalities among them are the oxygen sensitivity of one of the cargo enzymes and the volatility of intermediates that are generated in their interior. The BMC shell prevents the escape of these metabolites, thereby ensures their availability for downstream reactions within the BMC lumen and averts toxic effects on other cell components. In this chapter, we discuss recent advances that have led to new insights into BMC structure, function and cellular interactions.

S. Heinhorst (✉) · G. C. Cannon
The University of Southern Mississippi, Hattiesburg, MS, USA
e-mail: Sabine.heinhorst@usm.edu; Gordon.cannon@usm.edu

D. Jendrossek (ed.), *Bacterial Organelles and Organelle-like Inclusions*,
Microbiology Monographs 34, https://doi.org/10.1007/978-3-030-60173-7_6

1 Introduction

Many bacteria segregate selected metabolic steps into intracellular compartments, termed bacterial microcompartments (BMCs), that are separated from the cytosol by a semipermeable molecular boundary. Functionally analogous to eukaryotic organelles, BMCs, however, lack a lipid bilayer-based permeability barrier and are instead delimited by a thin (2–4 nm) protein shell that determines their polyhedral shape (Fig. 1) and serves to enhance metabolic efficiency of the enzymatic steps within their interior in part through control of metabolite access to and exit from the lumen. Genes for BMC proteins exist in approximately a fifth to a fourth of all sequenced bacterial genomes. Apart from the anabolic carboxysomes of photosynthetic cyanobacteria and some chemoautotrophs, all other known BMCs, often termed metabolosomes, harbour enzymes that aid in the catabolism of specialty carbon sources (Table 1 and reviewed in (Kirst and Kerfeld 2019; Kerfeld et al. 2018)).

The protein shell of BMCs constitutes a permeability barrier for small, volatile metabolites that are formed in their lumen, such as the aldehydes generated in the catabolic BMCs. Limiting their diffusion into the cytosol allows these intermediates to be efficiently processed further within the BMC interior and prevents them from exerting undesirable, often toxic effects on the cell. Likewise, the shell of the only known anabolic BMC, the carboxysome, ensures that CO_2, the substrate of ribulose 1,5-bisphosphate carboxylase/oxygenase (RubisCO) that is generated in the lumen by the carboxysomal carbonic anhydrase (CA), remains trapped in the organelle and thereby supports effective fixation by the otherwise catalytically inefficient RubisCO (Dou et al. 2008).

All BMC shells are constructed from the same, structurally conserved basic building modules: three types of small proteins that readily assemble into disc-shaped oligomers of characteristic stoichiometries (Fig. 2). Hexamers of highly

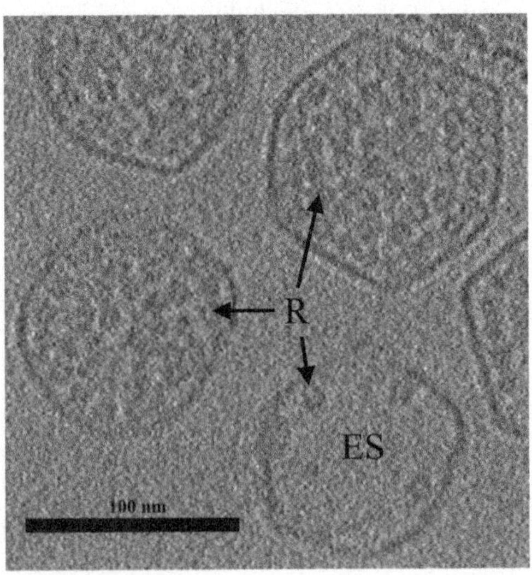

Fig. 1 Cryo electron tomogram of purified *H. neapolitanus* carboxysomes. R = individual RubisCO holoenzyme molecules; ES = empty carboxysome shell with a few RubisCO holoenzymes aligned on the inner periphery. The image was provided by Cristina Iancu and Grant J. Jensen

Table 1 Bacterial microcompartments

BMC type	BMC-entering substrate	Signature enzyme(s)	References
Anabolic bacterial microcompartments			
α-carboxysome	HCO_3^-	RubisCO form IA[a] and β-CA CsoSCA	Shively et al. (1973)
β-carboxysome	HCO_3^-	RubisCO form IB[a], β-CA CcaA and/or γ-CA CcmM	Price and Badger (1991)
Catabolic bacterial microcompartments			
Gre (*G*lycyl *r*adical *e*nzyme-associated)	Choline	Choline lyase	Craciun and Balskus (2012), Craciun et al. (2014, 2016)
	1,2 propanediol	1,2 propanediol dehydratase	Petit et al. (2013), Zarzycki et al. (2017), Schindel et al. (2019)
	L-fuculose-P or L-rhamnulose-P	Fuculose-P aldolase	Petit et al. (2013), Erbilgin et al. (2014)
Etu (*Et*hanol *u*tilization)	Ethanol	Ethanol dehydrogenase	Heldt et al. (2009), Seedorf et al. (2008)
Eut (*E*thanolamine *ut*ilization)	Ethanolamine	Ethanolamine ammonia lyase	Kofoid et al. (1999), Stojiljkovic et al. (1995)
Pdu (*P*ropane*d*iol *u*tilization)	1,2 propanediol	Adenosylcobalamin-dependent 1,2 propanediol dehydratase	Shively et al. (1998), Bobik et al. (1997, 1999) Havemann and Bobik (2003)
RMM (*Rhodococcus* and *Mycobacterium* *m*icrocompartment); **Aut** (1-amino-2-propanol/1-amino-2-propanone *ut*ilization)	1-amino-2-propanol or 1-amino-2-propanone	1-amino-2-propanol dehydrogenase or aminotransferase	Urano et al. (2011), Ravcheev et al. (2019), Mallette and Kimber (2017)
Xau (*x*anthine *a*naerobic *u*tilization)	Xanthine	Xanthine hydrolase	Ravcheev et al. (2019)

Listed here are BMCs whose function is known, or has been inferred. In addition to the references in the table that pertain to specific BMCs, Axen et al. (2014) and Kerfeld et al. (2018) list additional gene clusters for BMCs of presently unknown function
[a]RubisCO Form IA and IB classification based on Tabita et al. (2007, 2008)

abundant pfam00936 proteins (BMC-H) tile into the single layers that form the bulk of the polyhedral BMC facets (Kerfeld et al. 2005; Tanaka et al. 2008; Tsai et al. 2007; Crowley et al. 2008; Mallette and Kimber 2017; Takenoya et al. 2010; Tanaka et al. 2010). Interspersed within the hexamer layer are trimers (Sagermann et al. 2009; Crowley et al. 2010; Heldt et al. 2009; Pang et al. 2011; Tanaka et al. 2010; Takenoya et al. 2010) or stacked dimers of trimers (Cai et al. 2013; Klein et al. 2009; Mallette and Kimber 2017) formed by paralogs (BMC-T) with tandem pfam00936 domains. The vertices of the polyhedral BMC shell are closed by pentamers of pfam03319 proteins (BMC-P) (Tanaka et al. 2008; Cai et al. 2009; Wheatley et al.

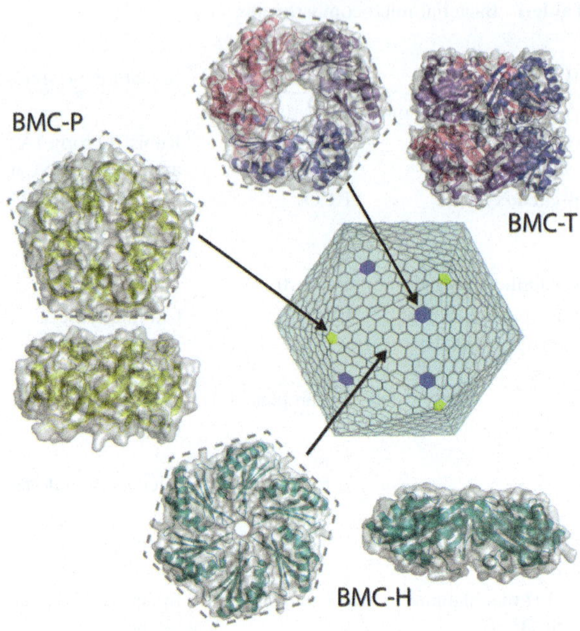

Fig. 2 Schematic of a generic BMC shell and its protein components. BMC-H (teal) = pfam00936 protein hexamers forming the BMC facets; BMC-P (yellow) = pfam03319 protein pentamers at the BMC vertices; BMC-T (purple) = tandem pfam00936 protein stacked dimers of trimers interspersed within the BMC-H hexamers. The image was generated by Markus Sutter and Cheryl Kerfeld

2013; Keeling et al. 2014; Mallette and Kimber 2017). Additional shell-associated proteins exist that are BMC type-specific and play a role in assembly and/or function of the organelle (see BMC Superloci).

Since BMCs have been the subject of several recent reviews (Heinhorst et al. 2014; Kirst and Kerfeld 2019; Kerfeld et al. 2018; Sommer et al. 2017; Yeates et al. 2013), Table 1 provides an overview of BMCs whose functions are either known or have been inferred bioinformatically or through metabolic studies. The remainder of this chapter focuses on recent advances that have significantly enhanced our understanding of BMC structure, function and interaction with other cell components.

2 Shell Structure and Function

Metabolite Transfer Since the structure of the BMC shell proteins was first elucidated and shell models were established (Kerfeld et al. 2005; Tanaka et al. 2008, 2010), the central pores in the oligomeric assemblies of BMC proteins have been viewed as the most likely candidates for mediating and/or regulating metabolite exchange across the shell. Pore size, geometry and surface charge vary between oligomers of different paralogs and orthologs and thereby provide selectivity for the metabolite(s) that are able to traverse the shell of the various BMC types. Lending indirect support to the proposed role of the pores as conduits for small metabolites are the anions (Tsai et al. 2007; Tanaka et al. 2009; Takenoya et al. 2010) and substrate molecules (Pang et al. 2012) that were shown to occupy the pores in various BMC protein crystals. Furthermore, molecular dynamics simulations of substrate (1,2 propanediol) and product (propionaldehyde) diffusion across the pores of the PduA hexamers of the Pdu (*p*ropane*d*iol *u*tilization) BMC predict a higher energy barrier for the toxic aldehyde intermediate than for the substrate (Park et al. 2017).

To address the role of the pores in metabolite transfer across the BMC shell in vivo, Choudhury and colleagues (2015) created a series of *Salmonella enterica* mutants in which PduA was altered to yield hexamers with occluded or less polar pore openings. These changes affect the diffusion rate of 1,2-propanediol into and/or that of propionaldehyde out of the Pdu BMC. A point mutation that changes the pore lining and renders the PduA pore characteristics similar to those of the Pdu BMC of lactobacilli also allows passage of the substrate glycerol (metabolized in the Pdu BMC of lactoballi but not by that of *S. enterica*) into the Pdu BMC of the resulting *S. enterica* mutant. Slininger Lee et al. (2017) expanded these mutant studies by creating a library of 15 PduA pore mutants in which a single pore residue at a key position is replaced with a different amino acid. The authors were able to correlate the resulting mutant growth phenotypes with the characteristics of the substituent amino acid and identified side chain charge and hydrophobicity as crucial determinants for the retention of propionaldehyde in the lumen of the Pdu BMC.

Recombinant, chimeric Pdu BMCs in which the Eut (*e*thanolamine *ut*ilization) BMC shell protein EutM substitutes for PduA appear to be morphologically and compositionally similar to wild-type Pdu BMCs. However, despite their structural similarity, PduA and EutM have different pores: lysine residues line the central pore of PduA hexamers, and glutamine is found in EutM pores. The chimeric Pdu BMC shows improved utilization of 1,2 propanediol, which the authors attribute to more effective retention of luminal metabolites and their increased availability for downstream reactions within the organelle (Slininger Lee et al. 2017).

Similar mutant studies that assess pore permeability have not yet been reported for carboxysomes; however, Mahintchichaichan et al. (Mahinthichaichan et al. 2018) used molecular dynamics simulations and free energy calculations to assess the permeability for HCO_3^-, CO_2 and O_2 of hexamer pores formed by the α- and β-carboxysome shell proteins CsoS1A from *Halothiobacillus neapolitanus* and CcmK4 from *Synechocystis sp.* PCC 6803, respectively. For both pore types, they found preferred interactions between the anion and the positively charged pore surface, corroborating the previously observed selective permeability of the carboxysome shell for bicarbonate over CO_2 (Dou et al. 2008). Retention of CO_2, which is generated from HCO_3^- by the carboxysomal CA and released into the lumen, ensures that RubisCO is supplied with ample amounts of its substrate. Likewise, exclusion of O_2 from the carboxysome interior is believed to prevent the unproductive conversion of this competing RubisCO substrate.

Collectively, theoretical and experimental evidence lends strong support to BMC-H shell pores serving as the main conduits for small metabolites between BMC lumen and the exterior. Furthermore, pore properties are specific for the substrates, intermediates and products that are relevant for a particular BMC type, and pore characteristics can be genetically engineered without compromising the structural integrity of the BMC shell. This flexibility opens up the possibility of designing and engineering specific passages for small metabolites that could enhance and/or broaden the range of BMC functions (Cai et al. 2015b; Kirst and Kerfeld 2019; Kerfeld et al. 2018).

In addition to the abundant BMC-H proteins that form the majority of the single-layer shell facets, the trimers and stacked dimers of trimers of BMC-T proteins are also integral components of the BMC shell (Roberts et al. 2012; Cai et al. 2013). Their pores are generally larger than those of the BMC-H hexamers and have unique geometric and charge properties. The double-layered dimers of trimers seen in some BMC-T protein crystals contain an internal channel or nanocompartment with either one or both pores in the closed conformation (Klein et al. 2009; Mallette and Kimber 2017; Cai et al. 2013; Larsson et al. 2017), suggesting that these proteins may have the ability to bind, temporarily store and gate the transfer of larger metabolites. However, to date experimental evidence for substrate traffic though BMC-T proteins is lacking. Considering their very low abundance in the carboxysome shell (Roberts et al. 2012), at least for the carboxysome a role in substrate transfer does not seem likely.

BMC-T protein assemblies might control entry and/or exit of larger regulatory metabolites that affect the reactions within the BMC and serve to align processes in the BMC lumen with the metabolic status of the cell, or the BMC-T pores may serve as gated conduits to replenish larger cofactors (e.g. NAD(P)H, ATP, cobalamin) (Mallette and Kimber 2017; Larsson et al. 2017). The EutL trimers crystallize with either a closed (Sagermann et al. 2009; Takenoya et al. 2010) or an open (Tanaka et al. 2010) central pore. Exposure of closed-pore EutL crystals to Zn^{2+} induces a conformational change to the open-pore state (Takenoya et al. 2010) that is suggestive of a gating mechanism. The observation that cobalamin, a required cofactor for the catabolism of 1,2-propanediol and ethanolamine in Pdu and Eut BMCs, respectively, binds to *Clostridium perfringens* EutL assemblies lends support to the proposed role of BMC-T proteins in cofactor replenishment (Thompson et al. 2014). However, the binding of only one ligand per trimer towards the periphery, rather than in the pore, casts doubt on the physiological significance of this association and points to the need for in vivo experiments that can help elucidate the contributions of EutL and other BMC-T proteins to BMC function.

Electron and Proton Permeability The electron and proton impermeability of the lipid bilayers that bound eukaryotic organelles begs the question whether BMC protein shells display a similar selectivity. Several observations support the possibility that electrons or redox-active metabolites may be able to cross the BMC boundary.

Several shell protein oligomers of catabolic BMCs are associated with FeS clusters (Crowley et al. 2010; Parsons et al. 2008; Pang et al. 2011; Thompson et al. 2014; Ferlez et al. 2019). The BMC-T protein PduT of the Pdu BMC and the BMC-H protein GrpU found in several glycyl radical propanediol (Grp) BMCs have spectral properties that indicate the presence of a 4Fe-4S cluster, a cofactor in 1,2 propanediol metabolism. Modelling results predict a location of the ligand in the central pore. Although the exact mode of action of the redox moiety has not been experimentally ascertained, likely scenarios are a shuttling of electrons or of the entire 4Fe-4S cluster across the BMC shell to support cofactor regeneration within the BMC lumen (Crowley et al. 2010; Parsons et al. 2008; Pang et al. 2011; Thompson et al. 2014).

Penrod and Roth (2006) had proposed that the maintenance of a pH differential between BMC interior and its surroundings could constitute a potential mechanism by which all BMCs are able to enhance the metabolic reactions that take place in their lumen. A lumenal pH that is lower than that of the cytosol would promote catabolism of volatile metabolites in catabolic BMCs and, in the carboxysome, would favour the conversion of bicarbonate to CO_2 by the carboxysomal CA. However, work by Menon et al. (Menon et al. 2010) convincingly showed that the α-carboxysome shell is permeable to protons. Because the shells of the various BMCs share identical design principles, this property is thought to apply to all BMC types.

Orientation of Shell Protein Oligomers The first published model of a BMC shell (Tanaka et al. 2008) predicted that the hexamers of the abundant α-carboxysome shell proteins are all oriented in the same direction; however, the model was unable to resolve the relative orientations of their concave and convex sides. In the CsoS1A hexamer, only the pore portion that faces the concave side is positively charged; the one on the convex side has a negative surface potential and presents a higher energy barrier for entry of HCO_3^- into the funnel leading towards the narrowest pore diameter (Tanaka et al. 2008). Based on those characteristics, the authors predicted a shell protein orientation in which the concave side of the hexamer faces outward and facilitates the transfer of HCO_3^- into the carboxysome interior. However, size and composition heterogeneity of naturally occurring BMCs have rendered them refractory to crystallization, and to date, the orientation of their shell proteins has not been resolved experimentally.

Using a combination of genes that encode BMC-H, BMC-T and BMC-P shell proteins from *Haliangium ochraceum*, Sutter et al. (2017) recombinantly expressed uniformly sized synthetic BMCs in *E. coli*. Their crystal structure corroborates the previously predicted orientation of the carboxysomal protein pentamers (Tanaka et al. 2008) that seal the icosahedral carboxysome shell at the vertices of its flat facets (Cai et al. 2009). Most importantly, the synthetic BMC structure revealed that the concave surface of the hexameric protein assemblies that form the BMC shell facets faces the cytosol. This finding has important implications for the structure and metabolite transfer properties of all BMC types.

BMC-H Heterohexamers The large majority of gene clusters and operons that encode the various BMC types contain multiple paralogs of BMC-H and BMC-P shell proteins (Axen et al. 2014), raising questions about functional redundancy or complementarity between individual paralogs. Informed by structural characterization of homo-oligomeric crystals of these proteins (Kerfeld et al. 2005; Tanaka et al. 2008; Tsai et al. 2007, 2009; Mallette and Kimber 2017; Cai et al. 2015b), current models of BMC shells are based solely on homo-oligomeric shell proteins. However, the existence of hetero-oligomers in BMC shells has long been considered to be a means by which bacteria could modulate pore properties, broaden the range of metabolites that can cross the shell, and/or regulate small-molecule traffic into and out of the organelle.

Sommer et al. (2019) investigated two BMC-H shell protein paralogs of β-carboxysomes, CcmK3 and CcmK4. Their genes always co-occur in a satellite locus distant from the main *ccm* gene cluster in the ~90% of cyanobacterial genomes in which they exist (Sommer et al. 2017), which suggests that they are not redundant but functionally connected. Mutant studies in *Synechococcus elongatus* PCC 7942 and *Synechocystis* sp. PCC 6803 revealed that CcmK4 is essential for efficient photoautotrophic growth in air (Zhang et al. 2004; Sommer et al. 2019). Both proteins are carboxysome components, but unlike CcmK4 (Kerfeld et al. 2005; Cai et al. 2015b), recombinant CcmK3 does not form homohexamers (Sommer et al. 2019). However, subjecting differentially tagged CcmK protein combinations to sequential affinity purification yielded stable CcmK3/CcmK4 heterohexamers for the orthologs from *Halothece* PCC 7418 and *S. elongatus* PCC 7942. Using a similar approach, Garcia-Alles and colleagues (2019) isolated tagged recombinant *Synechocystis* sp. PCC 6803 CcmK3/CcmK4, as well as CcmK1/CcmK2 heterohexamers. Interestingly, the CcmK3/CcmK4 hexamers also assemble into dodecamers in a pH-dependent manner (Sommer et al. 2019). Because of the poor conservation of residues lining the hetero-hexamer edges and the at-best very low abundance of CcmK3 in purified β-carboxysomes, Sommer et al. (2019) proposed that heterohexamers might be transiently associating with the outer shell surface, forming concave-to-concave side dodecamers with existing hexamers in the shell facets and changing their permeability properties temporarily.

Double-layered Shell Components Clearly, ultrastructural evidence has established a BMC shell thickness of 2–4 nm, which corresponds to the thickness of many BMC-H hexamers and strongly supports the single-layer nature of the shell facets (reviewed in (Heinhorst et al. 2014)). However, the possibility that some BMC-H hexamers assemble into double-layered dodecamers under certain conditions has been raised (Sommer et al. 2019; Samborska and Kimber 2012). The pH-dependent stacking of CcmK3/CcmK4 heterohexamers and the stacked dimers of trimers formed by some BMC-T proteins suggest that localized double-layered assemblies may exist transiently in response to changing intracellular conditions or as permanent shell features.

3 BMC Superloci

Aside from the major shell and cargo proteins of the various bacterial microcompartments (Cannon and Shively 1983; Havemann and Bobik 2003; Axen et al. 2014), additional proteins are associated with the organelles and/or are relevant to organelle function, regulation and/or integration of BMC-associated enzymatic steps with the metabolic state of the cell. The genes for these proteins often cluster in the vicinity of the main locus that encodes shell and cargo proteins but can also be located at greater distance. Collectively, these genes constitute what has been referred to as the BMC superlocus (Roberts et al. 2012; Kirst and Kerfeld 2019; Kerfeld et al. 2018; Axen et al. 2014; Pitts et al. 2012).

Structures and functions of shell and lumen proteins are well known for most BMCs, have been covered in several reviews (Heinhorst et al. 2014; Kerfeld et al. 2010, 2018; Yeates et al. 2013; Kirst and Kerfeld 2019) and are therefore not mentioned here. Proteins encoded by the accessory genes in the various BMC superloci play a role in the assembly of a functional organelle, are components of membrane transporters, regulatory components, or may participate in BMC positioning and partitioning (Pitts et al. 2012, Mallette and Kimber 2017). Since aside from their identification by bioinformatics approaches (Axen et al. 2014) not much is known about the role of accessory proteins in the superloci of the diverse catabolic BMC, this section focuses on selected proteins of α- and β-carboxysome superloci.

Pcd-like Protein The gene for the Pterin-4a-carbinolamine dehydratase (Pcd)-like protein is present in the genome of all α-carboxysome-forming bacteria (Roberts et al. 2012) and encodes a small polypeptide that is structurally very similar to bona fide Pcd but lacks enzymatic activity (Wheatley et al. 2014). Co-expression studies in *E. coli* with combinations of Pcd-like protein, RubisCO and GroESL, suggest that the Pcd-like protein functions as a RubisCO assembly chaperone in the heterologous host (Wheatley et al. 2014). Recombinant carboxysomes expressed in *E. coli* from the *H. neapolitanus cso* operon that lacks the terminal *pcd* gene, on the other hand, contain enzymatically active RubisCO (Bonacci et al. 2012). Preliminary studies in our lab of a *H. neapolitanus pcd* knockout mutant (D. Del Valle, unpublished) indicate that the Pcd-like protein affects the ability of *H. neapolitanus* to grow efficiently at ambient CO_2 levels. However, carboxysomes purified from mutant cells are very similar, if not identical, to wild-type organelles in appearance and composition, and the CO_2 fixation kinetics of their carboxysomal RubisCOs are near identical. The exact role of the Pcd-like protein in α-carboxysome assembly and/or function in vivo remains to be elucidated, and it is still unclear whether the protein is associated with the organelle.

CbbO and CbbQ When present in α-carboxysome-forming bacteria, the genes for the putative RubisCO activase CbbQ and its interaction partner, the Van Willebrand Factor A domain-containing protein CbbO, occur together. Sutter et al. (2015) determined the structure and activity of *H. neapolitanus* CbbQ. The protein, which belongs to the MoxR-type AAA+ ATPase family, forms a hexamer that interacts with one copy of CbbO and has ATPase activity. The CbbQ protein is tightly associated with the shell of *H. neapolitanus* carboxysomes but does not seem to play a crucial structural or functional role in the organelle. Carboxysomes purified from a *cbbQ* knockout mutant are structurally and functionally indistinguishable from wild-type organelles, and the mutant does not require elevated CO_2 levels for efficient growth.

Although annotated as a putative RubisCO activase, unlike its orthologs from *Acidithiobacillus ferrooxidans* that stimulate reactivation of inhibited non-carboxysomal form I and form II RubisCO, respectively (Tsai et al. 2015), the CbbQ/O complex encoded in the *H. neapolitanus* α-carboxysome superlocus does not appear to function as an activase of the carboxysomal RubisCO (Sutter et al. 2015). However, it is possible that the carboxysomal CbbQ/O is required to re-activate carboxysomal RubisCO under certain metabolic conditions that have yet to be identified.

CbbX The gene for the putative RubisCO activase CbbX (Mueller-Cajar et al. 2011) is part of the superlocus in many α-carboxysome-forming cyanobacteria (Roberts et al. 2012) but absent from the genomes of chemoautotrophs. The CbbX polypeptide also belongs to the AAA+ superfamily. The ortholog from *Rhodobacter sphaeroides* has been shown to reactivate inhibited red-type RubisCO of that organism (Mueller-Cajar et al. 2011), but the role of CbbX in assembly and/or activation of the cyanobacterial Form IA carboxysomal RubisCO remains to be addressed experimentally.

RbcX The *rbcX* gene of most β-carboxysome-forming cyanobacteria is either co-localized with the genes for the carboxysomal Form IB RubisCO (*rbcL, rbcS*) (Tabita et al. 2007, 2008) in a small operon or is part of the *ccmKLN* gene cluster (Huang et al. 2019). Onizuka et al. (2004) observed a marked reduction in the amount of RubisCO subunits present and in RubisCO activity in a *Synechococcus* sp. PCC 7002 frameshift mutant that produces partially inactive RbcX protein. Direct stimulation of RubisCO assembly by RbcX, however, has so far only been shown for recombinant Form IB RubisCO expressed in the heterologous host *E. coli* (Onizuka et al. 2004; Li and Tabita 1997).

Notable exceptions to the colocalization of the genes encoding RubisCO and its presumed chaperone are *S. elongatus* PCC 7942 and *S. elongatus* PCC 6301; in these cyanobacteria, the *rbcX* gene is located at a considerable distance from *rbcL* and *rbcS*. Interruption (Emlyn-Jones et al. 2006b) or deletion (Huang et al. 2019) of *rbcX* in *S. elongatus* PCC 7942 does not affect cell growth and RubisCO activity. However, Huang et al. (2019) recently showed that the complete absence of RbcX has a marked effect on the amount of RubisCO present and on size, number and intracellular positioning of carboxysomes. Since the majority of fluorescently labelled RubisCO and RbcX colocalizes to carboxysomes, the collective evidence suggests that, rather than promoting RubisCO assembly, RbcX instead plays a role in carboxysome assembly and distribution within the *S. elongatus* PCC 7942 cell. Proteomic analysis of purified *S. elongatus* PCC 7942 carboxysomes (Faulkner et al. 2017) did not reveal any RbcX protein, which suggests that the RubisCO activase is present in the organelle in very small amounts, or that its association with the β-carboxysome is transient.

McdA and McdB Two proteins encoded by carboxysome superlocus genes are McdA (also previously annotated as ParF or ParA) and its interaction partner McdB (previously annotated as a hypothetical protein), which mediate β-carboxysome positioning in the cell and partitioning among the two daughter cells (Savage et al. 2010; Jain et al. 2012; MacCready et al. 2018). Their mechanism of action is discussed in greater detail in the section Carboxysome Positioning and Segregation.

CsoS2 The CsoS2 protein is encoded in the canonical *cso* operon that contains the majority of α-carboxysome structural genes. The protein is the largest, highly abundant, and most unique α-carboxysome shell component whose function had long remained unknown. Unlike all other known α-carboxysome proteins, CsoS2 has a very high isoelectric point in most bacteria. The protein exists in two isoforms that are present in approximately equimolar amounts in carboxysomes of some but not all bacteria (Baker et al. 1999; Roberts et al. 2012; Cannon and Shively 1983). The shorter isoform (CsoS2A) represents a C-terminally truncated version of the full-length polypeptide (CsoS2B) (Cai et al. 2015a) and is the product of programmed ribosomal frameshifting to the −1 reading frame at an intrinsic slippery sequence (Chaijarasphong et al. 2016). The presence of such a motif in the *csoS2* coding sequence is predicted to correlate with the occurrence of two CsoS2 isoforms.

Three discrete regions can be distinguished in the CsoS2 primary sequence (Cai et al. 2015a): an N-terminal and a middle sequence, each with characteristic amino acid repeat motifs (Cannon et al. 2003), and a C-terminal region. Based on experimental and modelling evidence, CsoS2 likely has an elongated shape and is intrinsically disordered. In vitro binding studies revealed specific interactions of the protein with RubisCO and CsoS1 shell proteins and identified the location of several binding sites in the CsoS2 primary sequence (Cai et al. 2015a).

The essential role of the CsoS2 protein for α-carboxysome assembly is demonstrated by the phenotype of an *H. neapolitanus csoS2* knockout mutant that does not produce CsoS2 protein. The mutant does not grow in air and does not produce carboxysomes (Cai et al. 2015a). Based on the collective data, a model was put forth in which CsoS2 is predicted to recruit shell proteins and RubisCO to the α-carboxysome biogenesis site. Insertion of the CsoS2 C-terminus into the developing shell facets and organization of RubisCO in the emerging lumen space occur simultaneously (Cai et al. 2015a). The model is consistent with the carboxysome biosynthesis intermediates that were observed in *H. neapolitanus* cells by cryo electron tomography (Iancu et al. 2010). Recent expression studies of wild-type and mutant *H. neapolitanus cso* operons in *E. coli* identified the small subunit (CbbS) of the carboxysomal RubisCO as the interaction partner of CsoS2 and the N-terminal domain of the large subunit (CbbL) as the portion of the enzyme that contacts all three BMC-H paralogs (Liu et al. 2018). Those results are consistent with the well-documented single layer of RubisCO holoenzyme molecules that line the inside of the carboxysome shell (Iancu et al. 2010; Schmid et al. 2006; Iancu et al. 2007) and remain shell-associated when carboxysomes are broken (Holthuijzen et al. 1986). The observed requirement for full-length CsoS2B to yield recombinant carboxysomes in *E. coli* (Chaijarasphong et al. 2016) confirms the crucial role of the protein's C-terminal region in α-carboxysome assembly.

CcmM The CcmM protein is encoded within the main β-carboxysome gene cluster alongside the genes for several shell proteins and CcmN (Sommer et al. 2017). The protein was shown to be crucial for β-carboxysome assembly and/or function;

mutants that do not express functional CcmM protein are devoid of carboxysomes and have a strict high CO_2-requiring (*hcr*) phenotype (Ludwig et al. 2000; Woodger et al. 2005; Berry et al. 2005; Emlyn-Jones et al. 2006a). The CcmM protein features an N-terminal domain that is homologous to γ-CA, and a C-terminal portion that contains three to five repeats with weak, albeit detectable similarity to the RubisCO small subunit (RbcS) sequence (Ludwig et al. 2000). The crystal structure of CcmM from *Thermosynechococcus elongatus* BP-1 (Peña et al. 2010) revealed that the N-terminal CA domain assembles into a trimer. Notably, the protein only acquires enzymatic activity after a key disulphide bond in this domain is formed (Peña et al. 2010), presumably in the oxidizing environment of the β-carboxysome interior (Chen et al. 2013; Peña et al. 2010).

In addition to full-length CcmM (CcmM58), β-carboxysomes also contain a truncated form (CcmM35), which consists of only the C-terminal portion of the protein (Long et al. 2007, 2010) and is generated by translation from an internal initiation site upstream of the C-terminal repeat region. Cameron and colleagues (2013) showed that the CcmM protein is needed during β-carboxysome biogenesis to condense RubisCO molecules into the core to which shell components are subsequently recruited. The reduced form of *S. elongatus* PCC 7942 CcmM35 rapidly condenses purified RubisCO holoenzyme into a liquid droplet network; when oxidized, its affinity for RubisCO is reduced. Mutants in which CcmM lacks one or more of the cysteines that form disulphide bonds in its C-terminal domain have an *hcr* phenotype and contain fewer and abnormally shaped (elongated) carboxysomes (Wang et al. 2019), emphasizing the crucial role of the CcmM protein and its oxidized form in β-carboxysome biogenesis that had been deduced earlier from mutant and carboxysome assembly studies.

Because of sequence and structural homology between CcmM35 and RbcS, as well as the reported discrepancy in cellular abundance between RubisCO large and small subunits (Long et al. 2011), replacement of RbcS by CcmM35 in the RubisCO holoenzyme had been proposed as a possible means of modulating the activity of the CO_2 fixing enzyme (Long et al. 2011). However, the RbcS-like domain of CcmM lacks the essential RbcL binding elements of RbcS (Ferlez et al. 2019). Furthermore, co-crystals of *T. elongatus* RubisCO holoenzyme and one RbcS-like CcmM35 domain (Ryan et al. 2019) revealed that CcmM35 binds at a separate site and does not induce release of endogenous RbcS, even if the holoenzyme contains severely truncated RbcS subunits. The authors suggested that CcmM35 binds peripherally to a cleft between two RbcL subunits in the holoenzyme complex. This binding mode was subsequently confirmed by cryo electron tomography analysis of *S. elongatus* PCC 7942 CcmM-RubisCO complexes (Wang et al. 2019).

4 Organization of the β-Carboxysome Interior

Considering the abundance of individual proteins in fractions highly enriched in β-carboxysomes, as well as resulting from in vivo and in vitro interaction studies (Long et al. 2007, 2010, 2011; Cot et al. 2008; Kinney et al. 2012; McGurn et al. 2016), a model for the molecular organization of the β-carboxysome interior has emerged in which the CcmN protein, which is essential for β-carboxysome assembly, tethers a peripheral layer of full-length CcmM58 trimers to the shell. The C-terminal domain of CcmN contacts CcmK2 and probably other BMC-H shell components through a conserved peptide motif. The protein interacts with the CcmM trimer through its N-terminal domain. In those cyanobacteria that also contain the β-CA CcaA, the N-terminal catalytic domains of the CcmM trimer form a complex with a hexamer of three CcaA dimers (McGurn et al. 2016). The C-terminal RbcS-like domains of CcmM bind a layer of RubisCO that abuts the shell. The truncated CcmM35 isoform, which is devoid of the N-terminal domain that mediates interactions with CcaA and CcmN, is thought to organize RubisCO into a paracrystalline array in the interior (Long et al. 2007, 2010, 2011; Cot et al. 2008; Peña et al. 2010). This model is attractive, because it considers the concentration gradient of bicarbonate from the shell towards the centre of the organelle and proposes efficient generation and trapping of the RubisCO substrate CO_2 at the periphery immediately upon bicarbonate entry into the organelle lumen. However, high-resolution imaging studies (Niederhuber et al. 2017) have prompted a revision of the prevailing model. Using fluorescently tagged CcmM35 and CcmM58, the authors observed simultaneous assembly of both isoforms during carboxysome biogenesis. Furthermore, both isoforms appear to be distributed throughout the carboxysome lumen, suggesting that both C-terminal portion and full-length CcmM play a role in condensing the RubisCO core.

5 Positioning and Segregation of Carboxysomes

Early electron microscopic observations (reviewed in (Heinhorst et al. 2014)) frequently noted the close association of carboxysomes with the DNA fibrils of the bacterial nucleoid, a finding that was corroborated by Jain et al. (Jain et al. 2012). Using differential fluorescent staining, the authors showed that the multiple chromosome copies of *S. elongatus* PCC 7942 are regularly interspersed with carboxysomes along the longitudinal cell axis, suggesting that positioning and partitioning of DNA and carboxysomes are connected through shared molecular components.

Since movement of cell constituents is frequently brought about by components of the cytoskeleton and native, as well as recombinant BMCs were reported to be associated with filamentous molecular structures (Savage et al. 2010; Parsons et al. 2010), filaments of cytoskeleton proteins were implicated in the segregation and intracellular distribution of BMCs. A homolog of the DNA segregation protein ParA was shown to play an important role in β-carboxysome spacing and segregation during cell division in *S. elongatus* PCC 7942 (Savage et al. 2010; Jain et al. 2012). Savage and colleagues (2010) observed intracellular oscillations of fluorescently labelled ParA, which appears to be interspersed with carboxysomes, and found evidence for ParA filaments. Deletion of the protein leads to aberrantly distributed carboxysomes without, however, affecting the arrangement of the multiple chromosome copies in the cell (Jain et al. 2012).

MacCready et al. (2018) corroborated the *S. elongatus* PCC 7942 ParA oscillations observed previously (Savage et al. 2010) but found no evidence of filament formation. The authors showed that, like ParA in other bacteria, the cyanobacterial protein, which they re-named McdA (*M*aintenance of *c*arboxysome *d*istribution A), is a non-sequence-specific DNA binding protein when complexed with ATP. However, the genome of carboxysome-containing bacteria does not encode a recognizable ortholog of the known ParA partner protein, ParB. In a screen of deletion mutants of several nearby candidate genes, the researchers identified McdB, which has no homology to any known protein, as an McdA interaction partner that stimulates the ATPase activity of the latter. Schumacher et al. (2019) subsequently solved the crystal structures of McdA and McdB and showed interaction of purified McdA and McdB proteins from *Cyanothece* PCC 7424. Fluorescently tagged McdB co-localizes with carboxysomes in the cell, and two-hybrid screens showed that McdB, but not McdA, can directly interact with several carboxysome shell proteins (CcmK2, CcmK3, CcmK4 and CcmO). In addition, CcmK3 and CcmK4 deletion mutants of *S. elongatus* PCC 7942 contain aberrantly positioned carboxysomes (Rae et al. 2012). In a mutant that does not form carboxysomes, McdB exhibits a diffuse localization pattern, and oscillations of McdA were not detected. Collectively, these results suggest that McdB, when recruited to carboxysomes, affects the dynamics of McdA organization in the cell.

Since the findings of MacCready et al. (2018) rule out a cytoskeletal filament-based carboxysome movement and positioning in the cell, the authors proposed an alternative, Brownian ratchet mechanism. Supported by the observed movement of fluorescently tagged carboxysomes towards nucleoid areas, which are rich in McdA, in time lapse images, the ratchet model posits that DNA-bound McdA, upon stimulation of its ATPase activity by carboxysome-bound McdB, is released from the nucleoid, thus creating an McdA depletion zone in the vicinity of carboxysomes. McdB moves its carboxysome cargo along the McdA concentration gradient towards higher concentrations of nucleoid-bound McdA•ATP. Once a certain distance is reached between two carboxysomes, the nucleoid area between them repopulates with McdA•ATP. The authors further showed that the elongated shape of the nucleoid in *S. elongatus* PCC 7942 is responsible for the alignment of carboxysomes along the longitudinal cell axis in this cyanobacterium.

In several bacteria, particularly under physiological conditions that promote an increase in the number of carboxysomes, clustering of the organelles into groups has been observed (Sun et al. 2016) and reviewed in (Heinhorst et al. 2014; Sun et al. 2016). Whether this phenomenon reflects a physiological advantage, such as trapping of CO_2 that has escaped from the lumen of a carboxysome by a neighbouring organelle (Ting et al. 2007), or is simply the result of molecular crowding (Iancu et al. 2010) was not resolved. MacCready et al. (2018) present evidence that carboxysome clustering or hexagonal packaging correlates with the number of carboxysomes per nucleoid surface. In addition, modelling of carboxysome distribution further predicts clustering above a certain threshold number and supports the Brownian ratchet mechanism as the driver of β-carboxysome positioning in the cell. Interestingly, *mcdA* and *mcdB* knockout and overexpression mutants, in addition to showing defects in carboxysome distribution, also display changes in organelle size and shape that suggest a possible role of McdA and McdB in organelle biogenesis.

6 Concluding Remarks

The BMC field has enjoyed a tremendous expansion in the past 10 years, thanks to a flurry of interest following the realization that BMCs constitute a strategy to compartmentalize key metabolic reactions that is widespread among bacteria. Furthermore, because of their small size and relative simplicity, their tremendous synthetic biology potential is increasingly being realized. Informed by the known structure and identity of individual BMC protein components and a growing understanding of their contributions to BMC biogenesis and function, impressive results have been achieved in designing BMCs with novel properties.

Paving the way for BMC engineering has been the development of heterologous expression systems for entire BMC operons (Bonacci et al. 2012; Liu et al. 2018; Parsons et al. 2008; Graf et al. 2018; Baumgart et al. 2017). The culmination of those efforts to date is the successful transfer of α- and β-carboxysome genes into higher plants and the targeting of their protein products to the chloroplast (Lin et al. 2014; Long et al. 2018), where the bacterial RubisCO is active in CO_2 fixation (Long et al. 2018). Furthermore, empty BMC shells have been generated with specifically engineered architectures, sizes and properties that emphasize the structural and compositional flexibility of BMC shell protein-derived assemblies (reviewed in (Plegaria and Kerfeld 2018, Kirst and Kerfeld 2019).

Capitalizing on known mechanisms of protein encapsulation through short helical N- or C-terminal peptide sequences or utilizing endogenous BMC lumen proteins as vehicles (e.g. (Kinney et al. 2012; Fan et al. 2010; Fan and Bobik 2011; Menon et al. 2008, 2009, 2010); reviewed in (Plegaria and Kerfeld 2018)), orthologs of cargo proteins as well as entirely unrelated proteins have been targeted to the BMC lumen. In several cases, the resulting engineered BMCs are endowed with novel metabolic properties (Liang et al. 2017; Lawrence et al. 2014; Wagner et al. 2017) that demonstrate the ability of BMCs to accommodate substantial variations in their protein cargos.

Despite the scientific advances that have answered many questions related to BMC assembly, composition, structure and function, and the impressive developments in designing synthetic BMCs for specialty purposes, several basic questions remain that need to be addressed before a comprehensive understanding of BMC biology can be reached and the full potential of BMC engineering can be realized.

The most pressing questions in this regard relate to the permeability properties of the shell. At least one of the enzymes encapsulated in most, if not all BMCs is sensitive to oxygen, such as the enzymes of catabolic BMCs that rely on an oxygen-sensitive cofactor or operate through an oxygen-sensitive reaction mechanism. Likewise, for efficient CO_2 fixation RubisCO must avoid O_2, its competing, non-productive substrate. Although the shell is widely assumed to exclude oxygen from the BMC interior, experimental evidence for such selectively is lacking.

Despite the experimental evidence that supports the role of the pores in the BMC protein shell as the conduits through which substrates, products, possibly cofactors and regulatory metabolites enter and exit BMCs, the molecular mechanisms of their transport across the shell are not known. Additional open questions are the range of BMC-H proteins that are capable of forming heterohexamers, the dynamics of their associations with the array of hexamers in the BMC shell, and their significance in mediating and/or regulating metabolite traffic. Likewise, the proposed gating of metabolite transport through BMC-T pores and the properties that are ascribed to their internal nanocompartments constitute attractive potential regulatory strategies that await experimental support. Finally, establishing a complete inventory of pore permeabilities for the oligomers of the various BMC types will be the prerequisite for a more in-depth understanding of metabolic crosstalk between organelle lumen and the cytosol and for engineering efforts to manipulate the shell for the development of BMC-based nanoreactors with specific properties for specialty applications.

Acknowledgements First and foremost, we would like to acknowledge the generous financial support for our carboxysome research throughout the years from the National Science Foundation of the USA. Thanks go to our collaborators, Cristina Iancu and Grant J. Jensen, for providing us with the cryo electron tomogram image for Fig. 1, and to Markus Sutter and Cheryl Kerfeld for generating yet another beautiful BMC schematic (Fig. 2) for us. Their help is greatly appreciated.

References

Axen SD, Erbilgin O, Kerfeld CA (2014) A taxonomy of bacterial microcompartment loci constructed by a novel scoring method. PLoS Comput Biol 10:e1003898

Baker SH, Lorbach SC, Rodriguez-BUEY M, Williams DS, Aldrich HC, Shively JM (1999) The correlation of the gene *csoS2* of the carboxysome operon with two polypeptides of the carboxysome in *Thiobacillus neapolitanus*. Arch Microbiol 172:233–239

Baumgart M, Huber I, Abdollahzadeh I, Gensch T, Frunzke J (2017) Heterologous expression of the *Halothiobacillus neapolitanus* carboxysomal gene cluster in *Corynebacterium glutamicum*. J Biotechnol 258:126–135

Berry S, Fischer JH, Kruip J, Hauser M, Wildner GF (2005) Monitoring cytosolic pH of carboxysome-deficient cells of *Synechocystis sp.* PCC 6803 using fluorescence analysis. Plant Biol (Stuttg) 7:342–347

Bobik TA, Xu Y, Jeter RM, Otto KE, Roth JR (1997) Propanediol utilization genes (*pdu*) of *Salmonella typhimurium*: three genes for the propanediol dehydratase. J Bacteriol 179:6633–6639

Bobik TA, Havemann GD, Busch RJ, Williams DS, Aldrich HC (1999) The propanediol utilization (*pdu*) operon of *Salmonella enterica* serovar typhimurium LT2 includes genes necessary for formation of polyhedral organelles involved in coenzyme B(12)-dependent 1, 2-propanediol degradation. J Bacteriol 181:5967–5975

Bonacci W, Teng PK, Afonso B, Niederholtmeyer H, Grob P, Silver PA, Savage DF (2012) Modularity of a carbon-fixing protein organelle. Proc Natl Acad Sci U S A 109:478–483

Cai F, Menon BB, Cannon GC, Curry KJ, Shively JM, Heinhorst S (2009) The pentameric vertex proteins are necessary for the icosahedral carboxysome shell to function as a CO_2 leakage barrier. PLoS One 4:e7521

Cai F, Sutter M, Cameron JC, Stanley DN, Kinney JN, Kerfeld CA (2013) The structure of CcmP, a tandem bacterial microcompartment domain protein from the beta-carboxysome, forms a subcompartment within a microcompartment. J Biol Chem 288:16055–16063

Cai F, Dou Z, Bernstein SL, Leverenz R, Williams EB, Heinhorst S, Shively J, Cannon GC, Kerfeld CA (2015a) Advances in understanding carboxysome assembly in *Prochlorococcus* and *Synechococcus* implicate CsoS2 as a critical component. Life (Basel) 5:1141–1171

Cai F, Sutter M, Bernstein SL, Kinney JN, Kerfeld CA (2015b) Engineering bacterial microcompartment shells: chimeric shell proteins and chimeric carboxysome shells. ACS Synth Biol 4:444–453

Cameron JC, Wilson SC, Bernstein SL, Kerfeld CA (2013) Biogenesis of a bacterial organelle: the carboxysome assembly pathway. Cell 155:1131–1140

Cannon GC, Shively JM (1983) Characterization of a homogenous preparation of carboxysomes from *Thiobacillus neapolitanus*. Arch Microbiol 134:52–59

Cannon GC, Baker SH, Soyer F, Johnson DR, Bradburne CE, Mehlman JL, Davies PS, Jiang QL, Heinhorst S, Shively JM (2003) Organization of carboxysome genes in the thiobacilli. Curr Microbiol 46:115–119

Chaijarasphong T, Nichols RJ, Kortright KE, Nixon CF, Teng PK, Oltrogge LM, Savage DF (2016) Programmed ribosomal frameshifting mediates expression of the alpha-carboxysome. J Mol Biol 428:153–164

Chen AH, Robinson-Mosher A, Savage DF, Silver PA, Polka JK (2013) The bacterial carbon-fixing organelle is formed by shell envelopment of preassembled cargo. PLoS One 8:e76127

Chowdhury C, Chun S, Pang A, Sawaya MR, SINHA S, Yeates TO, Bobik TA (2015) Selective molecular transport through the protein shell of a bacterial microcompartment organelle. Proc Natl Acad Sci U S A 112:2990–2995

Cot SS-W, So AK-C, Espie GS (2008) A multiprotein bicarbonate dehydration complex essential to carboxysome function in cyanobacteria. J Bacteriol 190:936–945

Craciun S, Balskus EP (2012) Microbial conversion of choline to trimethylamine requires a glycyl radical enzyme. Proc Natl Acad Sci U S A 109:21307–21312

Craciun S, Marks JA, Balskus EP (2014) Characterization of choline trimethylamine-lyase expands the chemistry of glycyl radical enzymes. ACS Chem Biol 9:1408–1413

Craciun S, Marks JA, Balskus EP (2016) Correction to characterization of choline trimethylamine-Lyase expands the chemistry of Glycyl radical enzymes. ACS Chem Biol 11:2068

Crowley CS, Sawaya MR, Bobik TA, Yeates TO (2008) Structure of the PduU shell protein from the Pdu microcompartment of *Salmonella*. Structure 16:1324–1332

Crowley CS, Cascio D, Sawaya MR, Kopstein JS, Bobik TA, Yeates TO (2010) Structural insight into the mechanisms of transport across the *Salmonella enterica* Pdu microcompartment shell. J Biol Chem 285:37838–37846

Dou Z, Heinhorst S, Williams EB, Murin CD, Shively JM, Cannon GC (2008) CO_2 fixation kinetics of *Halothiobacillus neapolitanus* mutant carboxysomes lacking carbonic anhydrase suggest the shell acts as a diffusional barrier for CO_2. J Biol Chem 283:10377–10384

Emlyn-Jones D, Woodger FJ, Andrews TJ, Price GD, Whitney SM (2006a) A *Synechococcus* PCC 7942 ΔCcmM (cyanophyceae) mutant pseudoreverts to air growth without regaining carboxysomes. J Phycol 42:769–777

Emlyn-Jones D, Woodger FJ, Price GD, Whitney SM (2006b) RbcX can function as a rubisco chaperonin, but is non-essential in *Synechococcus* PCC7942. Plant Cell Physiol 47:1630–1640

Erbilgin O, Mcdonald KL, Kerfeld CA (2014) Characterization of a planctomycetal organelle: a novel bacterial microcompartment for the aerobic degradation of plant saccharides. Appl Environ Microbiol 80:2193–2205

Fan C, Bobik TA (2011) The N-terminal region of the medium subunit (PduD) packages adenosylcobalamin-dependent diol dehydratase (PduCDE) into the Pdu microcompartment. J Bacteriol 193:5623–5628

Fan C, Cheng S, Liu Y, Escobar CM, Crowley CS, Jefferson RE, Yeates TO, Bobik TA (2010) Short N-terminal sequences package proteins into bacterial microcompartments. Proc Natl Acad Sci U S A 107:7509–7514

Faulkner M, Rodriguez-Ramos J, Dykes GF, Owen SV, Casella S, Simpson DM, Beynon RJ, Liu LN (2017) Direct characterization of the native structure and mechanics of cyanobacterial carboxysomes. Nanoscale 9:10662–10673

Ferlez B, Sutter M, Kerfeld CA (2019) Glycyl radical enzyme-associated microcompartments: redox-replete bacterial organelles. mBio 10:e02327–e02318

Garcia-Alles LF, Root K, Maveyraud L, Aubry N, Lesniewska E, Mourey L, Zenobi R, Truan G (2019) Occurrence and stability of hetero-hexamer associations formed by beta-carboxysome CcmK shell components. PLoS One 14:e0223877

Graf L, Wu K, Wilson JW (2018) Transfer and analysis of *Salmonella pdu* genes in a range of gram-negative bacteria demonstrate exogenous microcompartment expression across a variety of species. Microb Biotechnol 11:199–210

Havemann GD, Bobik TA (2003) Protein content of polyhedral organelles involved in coenzyme B12-dependent degradation of 1,2-propanediol in *Salmonella enterica* serovar typhimurium LT2. J Bacteriol 185:5086–5095

Heinhorst S, Cannon GC, Shively JM (2014) Carboxysomes and their structural organization in prokaryotes. In: BARTON LL, Bazylinski DA, Xu, H. (eds) Nanomicrobiology. Springer, New York, pp 75–101

Heldt D, Frank S, Seyedarabi A, Ladikis D, Parsons JB, Warren MJ, Pickersgill RW (2009) Structure of a trimeric bacterial microcompartment shell protein, EtuB, associated with ethanol utilization in *Clostridium kluyveri*. Biochem J 423:199–207

holthuijzen YA, Breemen JFL, Konings WN, Bruggen EFJ (1986) Electron microscopic studies of carboxysomes of *Thiobacillus neapolitanus*. Arch Microbiol 144:258–262

Huang F, Vasieva O, Sun Y, Faulkner M, Dykes GF, Zhao Z, Liu L-N (2019) Roles of RbcX in carboxysome biosynthesis in the cyanobacterium *Synechococcus elongatus* PCC7942. Plant Physiol 179:184–194

Iancu CV, Ding HJ, Morris DM, Dias DP, Gonzales AD, Martino A, Jensen GJ (2007) The structure of isolated *Synechococcus* strain WH8102 carboxysomes as revealed by electron cryotomography. J Mol Biol 372:764–773

Iancu CV, Morris DM, Dou Z, Heinhorst S, Cannon GC, Jensen GJ (2010) Organization, structure, and assembly of α-carboxysomes determined by electron cryotomography of intact cells. J Mol Biol 396:105–117

Jain IH, Vijayan V, O'Shea EK (2012) Spatial ordering of chromosomes enhances the fidelity of chromosome partitioning in cyanobacteria. Proc Natl Acad Sci U S A 109:13638–13643

Keeling TJ, Samborska B, Demers RW, Kimber MS (2014) Interactions and structural variability of beta-carboxysomal shell protein CcmL. Photosynth Res 121:125–133

Kerfeld CA, Sawaya MR, Tanaka S, Nguyen CV, Phillips M, Beeby M, Yeates TO (2005) Protein structures forming the shell of primitive bacterial organelles. Science 309:936–938

Kerfeld CA, Heinhorst S, Cannon GC (2010) Bacterial microcompartments. Ann Rev Microbiol 64:391–408

Kerfeld CA, Aussignargues C, Zarzycki J, Cai F, SUTTER M (2018) Bacterial microcompartments. Nat Rev Microbiol 16:277–290

Kinney JN, Salmeen A, Cai F, Kerfeld CA (2012) Elucidating essential role of conserved carboxysomal protein CcmN reveals common feature of bacterial microcompartment assembly. J Biol Chem 287:17729–17736

Kirst H, Kerfeld CA (2019) Bacterial microcompartments: catalysis-enhancing metabolic modules for next generation metabolic and biomedical engineering. BMC Biol 17:79

Klein MG, Zwart P, Bagby SC, Cai F, Chisholm SW, Heinhorst S, Cannon GC, Kerfeld CA (2009) Identification and structural analysis of a novel carboxysome shell protein with implications for metabolite transport. J Mol Biol 392:319–333

Kofoid E, Rappleye C, Stojiljkovic I, Roth J (1999) The 17-gene ethanolamine (*eut*) operon of *Salmonella typhimurium* encodes five homologues of carboxysome shell proteins. J Bacteriol 181:5317–5329

Larsson AM, Hasse D, Valegard K, Andersson I (2017) Crystal structures of beta-carboxysome shell protein CcmP: ligand binding correlates with the closed or open central pore. J Exp Bot 68:3857–3867

Lawrence AD, Frank S, Newnham S, Lee MJ, Brown IR, Xue WF, Rowe ML, Mulvihill DP, Prentice MB, Howard MJ, Warren MJ (2014) Solution structure of a bacterial microcompartment targeting peptide and its application in the construction of an ethanol bioreactor. ACS Synth Biol 3:454–465

Li LA, Tabita FR (1997) Maximum activity of recombinant ribulose 1,5-bisphosphate carboxylase/ oxygenase of *Anabaena sp.* strain CA requires the product of the *rbcX* gene. J Bacteriol 179:3793–3796

Liang M, Frank S, Lünsdorf H, Warren MJ, Prentice MB (2017) Bacterial microcompartment-directed polyphosphate kinase promotes stable polyphosphate accumulation in *E coli.* Biotechnol J 68:1600415

Lin MT, Occhialini A, Andralojc PJ, Devonshire J, Hines KM, Parry MA, Hanson MR (2014) Beta-Carboxysomal proteins assemble into highly organized structures in *Nicotiana* chloroplasts. Plant J 79:1–12

Liu Y, He X, Lim W, Mueller J, Lawrie J, Kramer L, Guo J, Niu W (2018) Deciphering molecular details in the assembly of alpha-type carboxysome. Sci Rep 8:15062

Long BM, Badger MR, Whitney SM, Price GD (2007) Analysis of carboxysomes from *Synechococcus* PCC7942 reveals multiple RubisCO complexes with carboxysomal proteins CcmM and CcaA. J Biol Chem 282:29323–29335

Long BM, Tucker L, Badger MR, Price GD (2010) Functional cyanobacterial β-carboxysomes have an absolute requirement for both long and short forms of the CcmM protein. Plant Physiol 153:285–293

Long BM, Rae BD, Badger MR, Price GD (2011) Over-expression of the beta-carboxysomal CcmM protein in *Synechococcus* PCC7942 reveals a tight co-regulation of carboxysomal carbonic anhydrase (CcaA) and M58 content. Photosynth Res 109:33–45

Long BM, Hee WY, Sharwood RE, Rae BD, Kaines S, Lim YL, Nguyen ND, Massey B, Bala S, von Caemmerer S, Badger MR, Price GD (2018) Carboxysome encapsulation of the CO_2-fixing enzyme rubisco in tobacco chloroplasts. Nat Commun 9:3570

Ludwig M, Sültemeyer D, Price GD (2000) Isolation of *ccmKLMN* genes from the marine cyanobacterium *Synechococcus* sp. PCC7002 (cyanophyceae), and evidence that CcmM is essential for carboxysome assembly. J Phycol 36:1109–1119

MacCready JS, Hakim P, Young EJ, Hu L, Liu J, Osteryoung KW, Vecchiarelli AG, Ducat DC (2018) Protein gradients on the nucleoid position the carbon-fixing organelles of cyanobacteria. elife 7:e39723

Mahinthichaichan P, Morris DM, Wang Y, Jensen GJ, Tajkhorshid E (2018) Selective permeability of carboxysome shell pores to anionic molecules. J Phys Chem B 122:9110–9118

Mallette E, Kimber MS (2017) A complete structural inventory of the mycobacterial microcompartment shell proteins constrains models of global architecture and transport. J Biol Chem 292:1197–1210

McGurn LD, Moazami-Goudarzi M, White SA, Suwal T, Brar B, Tang JQ, Espie GS, Kimber MS (2016) The structure, kinetics and interactions of the beta-carboxysomal beta-carbonic anhydrase, CcaA. Biochem J 473:4559–4572

Menon BB, Dou Z, Heinhorst S, Shively JM, Cannon GC (2008) *Halothiobacillus neapolitanus* carboxysomes sequester heterologous and chimeric RubisCO species. PLoS One 3:e3570

Menon BB, Dou Z, Milam J, Shively JM, Heinhorst S, Cannon GC (2009) Phenotypic analysis of a *Halothiobacillus neapolitanus* mutant harboring beta-cyanobacterial form IB RubisCO. Amercian Society for Microbiology 109th General Meeting, Philadelphia, PA

Menon BB, Heinhorst S, Shively JM, Cannon GC (2010) The carboxysome shell is permeable to protons. J Bacteriol 192:5881–5886

Mueller-Cajar O, Stotz M, Wendler P, Hartl FU, Bracher A, Hayer-Hartl M (2011) Structure and function of the AAA+ protein CbbX, a red-type rubisco activase. Nature 479:194–199

Niederhuber MJ, Lambert TJ, Yapp C, Silver PA, Polka JK (2017) Superresolution microscopy of the beta-carboxysome reveals a homogeneous matrix. Mol Biol Cell 28:2734–2745

Onizuka T, Endo S, Akiyama H, Kanai S, Hirano M, Yokota A, Tanaka S, Miyasaka H (2004) The *rbcX* gene product promotes the production and assembly of ribulose-1,5-bisphosphate carboxylase/oxygenase of *Synechococcus sp.* PCC7002 in *Escherichia coli*. Plant Cell Physiol 45:1390–1395

Pang A, Warren MJ, Pickersgill RW (2011) Structure of PduT, a trimeric bacterial microcompartment protein with a 4Fe-4S cluster-binding site. Acta Crystallogr D Biol Crystallogr 67:91–96

Pang A, Liang M, Prentice MB, Pickersgill RW (2012) Substrate channels revealed in the trimeric *Lactobacillus reuteri* bacterial microcompartment shell protein PduB. Acta Crystallogr D Biol Crystallogr 68:1642–1652

Park J, Chun S, Bobik TA, Houk KN, Yeates TO (2017) Molecular dynamics simulations of selective metabolite transport across the propanediol bacterial microcompartment shell. Journal Phys Chem B 121:8149–8154

Parsons JB, Dinesh SD, Deery E, Leech HK, Brindley AA, Heldt D, Frank S, Smales CM, Lunsdorf H, Rambach A, Gass MH, Bleloch A, Mcclean KJ, Munro AW, Rigby SEJ, Warren MJ, Prentice MB (2008) Biochemical and structural insights into bacterial organelle form and biogenesis. J Biol Chem 283:14366–14375

Parsons JB, Frank S, Bhella D, Liang M, Prentice MB, Mulvihill DP, warren MJ (2010) Synthesis of empty bacterial microcompartments, directed organelle protein incorporation, and evidence of filament-associated organelle movement. Mol Cell 38:305–315

Peña KL, Castel SE, de Araujo C, Espie GS, Kimber MS (2010) Structural basis of the oxidative activation of the carboxysomal γ-carbonic anhydrase, CcmM. Proc Natl Acad Sci U S A 107:2455–2460

Penrod JT, Roth JR (2006) Conserving a volatile metabolite: a role for carboxysome-like organelles in *Salmonella enterica*. J Bacteriol 188:2865–2874

Petit E, Latouf WG, Coppi MV, Warnick TA, Currie D, Romashko I, Deshpande S, Haas K, Alvelo-Maurosa JG, Wardman C, Schnell DJ, Leschine SB, Blanchard JL (2013) Involvement of a bacterial microcompartment in the metabolism of fucose and rhamnose by *Clostridium phytofermentans*. PLoS One 8:e54337

Pitts AC, Tuck LR, Faulds-Pain A, Lewis RJ, Marles-Wright J (2012) Structural insight into the *Clostridium difficile* ethanolamine utilisation microcompartment. PLoS One 7:e48360

Plegaria JS, Kerfeld CA (2018) Engineering nanoreactors using bacterial microcompartment architectures. Curr Opin Biotechnol 51:1–7

Price GD, Badger MR (1991) Evidence for the role of carboxysomes in the cyanobacterial CO_2-concentrating mechanism. Can J Bot 69:963–973

Rae BD, Long BM, Badger MR, Price GD (2012) Structural determinants of the outer shell of β-carboxysomes in *Synechococcus elongatus* PCC 7942: Roles for CcmK2, K3-K4, CcmO, and CcmL. PLoS One 7:e43871

Ravcheev DA, Moussu L, Smajic S, Thiele I (2019) Comparative genomic analysis reveals novel microcompartment-associated metabolic pathways in the human gut microbiome. Front Genet 10:636

Roberts EW, Cai F, Kerfeld CA, Cannon GC, Heinhorst S (2012) Isolation and characterization of the *Prochlorococcus* carboxysome reveal the presence of the novel shell protein CsoS1D. J Bacteriol 194:787–795

Ryan P, Forrester TJB, Wroblewski C, Kenney TMG, Kitova EN, Klassen JS, Kimber MS (2019) The small RbcS-like domains of the beta-carboxysome structural protein CcmM bind RubisCO at a site distinct from that binding the RbcS subunit. J Biol Chem 294:2593–2603

Sagermann M, Ohtaki A, Nikolakakis K (2009) Crystal structure of the EutL shell protein of the ethanolamine ammonia lyase microcompartment. Proc Natl Acad Sci U S A 106:8883–8887

Samborska B, Kimber MS (2012) A dodecameric CcmK2 structure suggests beta-carboxysomal shell facets have a double-layered organization. Structure 20:1353–1362

Savage DF, Afonso B, Chen AH, Silver PA (2010) Spatially ordered dynamics of the bacterial carbon fixation machinery. Science 327:1258–1261

Schindel HS, Karty JA, Mckinlay JB, Bauer CE (2019) Characterization of a glycyl radical enzyme bacterial microcompartment pathway in *Rhodobacter capsulatus*. J Bacteriol 201:e00343–e00318

Schmid MF, Paredes AM, Khant HA, Soyer F, Aldrich HC, Chiu W, Shively JM (2006) Structure of *Halothiobacillus neapolitanus* carboxysomes by cryo-electron tomography. J Mol Biol 364:526–535

Schumacher MA, Henderson M, Zhang H (2019) Structures of maintenance of carboxysome distribution Walker-box McdA and McdB adaptor homologs. Nucleic Acids Res 47:5950–5962

Seedorf H, Fricke WF, Veith B, Bruggemann H, Liesegang H, Strittmatter A, Miethke M, Buckel W, hinderberger J, Li F, Hagemeier C, Thauer RK, Gottschalk G (2008) The genome of *Clostridium kluyveri*, a strict anaerobe with unique metabolic features. Proc Natl Acad Sci U S A 105:2128–2133

Shively JM, Ball F, Brown DH, Saunders RE (1973) Functional organelles in prokaryotes: polyhedral inclusions (carboxysomes) of *Thiobacillus neapolitanus*. Science 182:584–586

Shively JM, Bradburne CE, Aldrich HC, Bobik TA, Mehlman JL, Jin S, Baker SH (1998) Sequence homologs of the carboxysomal polypeptide CsoS1 of the thiobacilli are present in cyanobacteria and enteric bacteria that form carboxysomes-polyhedral bodies. Can J Bot 76:906–916

Slininger Lee MF, Jakobson CM, Tullman-Ercek D (2017) Evidence for improved encapsulated pathway behavior in a bacterial microcompartment through shell protein engineering. ACS Synth Biol 6:1880–1891

Sommer M, Cai F, Melnicki M, Kerfeld CA (2017) β-Carboxysome bioinformatics: identification and evolution of new bacterial microcompartment protein gene classes and core locus constraints. J Exp Bot 68:3841–3855

Sommer M, Sutter M, Gupta S, Kirst H, Turmo A, Lechno-Yossef S, Burton RL, Saechao C, Sloan NB, Cheng X, Chan LG, Petzold CJ, Fuentes-Cabrera M, Ralston CY, Kerfeld CA (2019) Heterohexamers formed by CcmK3 and CcmK4 increase the complexity of beta carboxysome shells. Plant Physiol 179:156–167

Stojiljkovic I, Baumler AJ, Heffron F (1995) Ethanolamine utilization in *Salmonella typhimurium*: nucleotide sequence, protein expression, and mutational analysis of the *cchA cchB eutE eutJ eutG eutH* gene cluster. J Bacteriol 177:1357–1366

Sun Y, Casella S, Fang Y, Huang F, Faulkner M, Barrett S, Liu LN (2016) Light modulates the biosynthesis and organization of cyanobacterial carbon fixation machinery through photosynthetic electron flow. Plant Physiol 171:530–541

Sutter M, Roberts EW, Gonzalez RC, Bates C, Dawoud S, Landry K, Cannon GC, Heinhorst S, Kerfeld CA (2015) Structural characterization of a newly identified component of alpha-carboxysomes: the AAA+ domain protein CsoCbbQ. Sci Rep 5:16243

Sutter M, Greber B, Aussignargues C, Kerfeld CA (2017) Assembly principles and structure of a 6.5-MDa bacterial microcompartment shell. Science 356:1293–1297

Tabita FR, Hanson TE, Li H, Satagopan S, Singh J, Chan S (2007) Function, structure, and evolution of the RubisCO-like proteins and their RubisCO homologs. Microbiol Mol Biol Rev 71:576–599

Tabita FR, Satagopan S, Hanson TE, Kreel NE, Scott SS (2008) Distinct form I, II, III, and IV rubisco proteins from the three kingdoms of life provide clues about rubisco evolution and structure/function relationships. J Exp Bot 59:1515–1524

Takenoya M, Nikolakakis K, Sagermann M (2010) Crystallographic insights into the pore structures and mechanisms of the EutL and EutM shell proteins of the ethanolamine-utilizing microcompartment of *Escherichia coli*. J Bacteriol 192:6056–6063

Tanaka S, Kerfeld CA, Sawaya MR, Cai F, Heinhorst S, Cannon GC, Yeates TO (2008) Atomic-level models of the bacterial carboxysome shell. Science 319:1083–1086

Tanaka S, Sawaya MR, Phillips M, Yeates TO (2009) Insights from multiple structures of the shell proteins from the beta-carboxysome. Protein Sci 18:108–120

Tanaka S, Sawaya MR, Yeates TO (2010) Structure and mechanisms of a protein-based organelle in *Escherichia coli*. Science 327:81–84

Thompson MC, Crowley CS, Kopstein J, Bobik TA, Yeates TO (2014) Structure of a bacterial microcompartment shell protein bound to a cobalamin cofactor. Acta Crystallogr F Struct Biol Commun 70:1584–1590

Ting CS, Hsieh C, Sundararaman S, Mannella C, Marko M (2007) Cryo-electron tomography reveals the comparative three-dimensional architecture of *Prochlorococcus*, a globally important marine cyanobacterium. J Bacteriol 189:4485–4493

Tsai Y, Sawaya MR, Cannon GC, Cai F, Williams EB, Heinhorst S, Kerfeld CA, Yeates TO (2007) Structural analysis of CsoS1A and the protein shell of the *Halothiobacillus neapolitanus* carboxysome. PLoS Biol 5:e144

Tsai Y, Sawaya MR, Yeates TO (2009) Analysis of lattice-translocation disorder in the layered hexagonal structure of carboxysome shell protein CsoS1C. Acta Crystallogr D Biol Crystallogr 65:980–988

Tsai YC, Lapina MC, Bhushan S, Mueller-Cajar O (2015) Identification and characterization of multiple rubisco activases in chemoautotrophic bacteria. Nat Commun 6:8883

Urano N, Kataoka M, Ishige T, Kita S, Sakamoto K, Shimizu S (2011) Genetic analysis around aminoalcohol dehydrogenase gene of *Rhodococcus erythropolis* MAK154: a putative GntR transcription factor in transcriptional regulation. Appl Microbiol Biotechnol 89:739–746

Wagner HJ, Capitain CC, Richter K, Nessling M, Mampel J (2017) Engineering bacterial microcompartments with heterologous enzyme cargos. Eng Life Sci 17:36–46

Wang H, Yan X, Aigner H, Bracher A, Nguyen ND, Hee WY, Long BM, Price GD, Hartl FU, Hayer-Hartl M (2019) Rubisco condensate formation by CcmM in β-carboxysome biogenesis. Nature 566:131–135

Wheatley NM, Gidaniyan SD, Liu Y, Cascio D, Yeates TO (2013) Bacterial microcompartment shells of diverse functional types possess pentameric vertex proteins. Protein Sci 22:660–665

Wheatley NM, Sundberg CD, Gidaniyan SD, Cascio D, Yeates TO (2014) Structure and identification of a pterin dehydratase-like protein as a ribulose-bisphosphate carboxylase/oxygenase (RuBisCO) assembly factor in the alpha-carboxysome. J Biol Chem 289:7973–7981

Woodger FJ, Badger MR, Price GD (2005) Sensing of inorganic carbon limitation in *Synechococcus* PCC 7942 is correlated with the size of the internal inorganic carbon pool and involves oxygen. Plant Physiol 139:1959–1969

Yeates TO, Jorda J, Bobik TA (2013) The shells of BMC-type microcompartment organelles in bacteria. J Mol Microbiol Biotechnol 23:290–299

Zarzycki J, Sutter M, Cortina NS, Erb TJ, Kerfeld CA (2017) *In vitro* characterization and concerted function of three core enzymes of a glycyl radical enzyme - associated bacterial microcompartment. Sci Rep 7:42757

Zhang S, Laborde SM, Frankel LK, Bricker TM (2004) Four novel genes required for optimal photoautotrophic growth of the cyanobacterium *Synechocystis* sp. strain PCC 6803 identified by *in vitro* transposon mutagenesis. J Bacteriol 186:875–879

The Cyanophycin Granule Peptide from Cyanobacteria

Björn Watzer, Friederike Klemke, and Karl Forchhammer

Contents

1 Introduction 150
2 Cyanophycin: A Nitrogen-Rich Storage Compound 151
3 Structure of Cyanophycin 151
4 Physiological Function of Cyanophycin in the Context of Nitrogen Fixation 152
  4.1 Cyanophycin in Heterocyst-Forming Cyanobacteria 154
  4.2 Cyanophycin in Non-diazotrophic Cyanobacteria 157
5 Enzymes of Cyanophycin Metabolism 158
  5.1 Cyanophycin Synthetase (CphA) 158
  5.2 Cyanophycinase (CphB) 162
  5.3 Aspartyl-Arginine Dipeptidase 163
6 Genetic Organization and Expression of *cphA* and *cphB* 164
7 Biotechnological Production of Cyanophycin 164
  7.1 Cyanophycin Production Using Recombinant Production Hosts 165
  7.2 Dependence of CGP Production from Arginine Biosynthesis 167
  7.3 Overproduction of Cyanophycin by Genetic Engineering the P_{II} Signaling System 168
References 169

Abstract Cyanophycin (multi-L-arginyl-poly-L-aspartic acid) is a biopolymer of industrial interest, which naturally occurs in most cyanobacterial species and in a few heterotrophic bacteria. It is a non-ribosomally synthesized polyamide consisting of L-Arginine and L-Aspartate and serves as a biological nitrogen/carbon storage compound.

This chapter provides an overview about the natural occurrence of cyanophycin in cyanobacteria. Herein, we focus on the physiological function of cyanophycin in filamentous and unicellular cyanobacteria. Information about the enzymes involved

B. Watzer · K. Forchhammer (✉)
Interfakultäres Institut für Mikrobiologie und Infektionsmedizin, Lehrstuhl für Mikrobiologie/ Organismische Interaktionen, Eberhard Karls Universität Tübingen, Tübingen, Germany
e-mail: karl.forchhammer@uni-tuebingen.de

F. Klemke
Institut für Biologie, Strukturbiologie/Biochemie, Humboldt-Universität zu Berlin, Berlin, Germany

D. Jendrossek (ed.), *Bacterial Organelles and Organelle-like Inclusions*, Microbiology Monographs 34, https://doi.org/10.1007/978-3-030-60173-7_7

in cyanophycin metabolism and their regulations are also provided. Due to its unique properties, cyanophycin could be used in various industrial applications. This chapter also summarizes the current state of biotechnological cyanophycin production.

1 Introduction

Cyanobacteria are Gram-negative prokaryotes that perform oxygenic photosynthesis and fix carbon dioxide through the Calvin-Benson Cycle to produce reduced carbohydrates (Rippka et al. 1979). For the synthesis of cellular building blocks, required for cell growth, Cyanobacteria assimilate various inorganic and organic nitrogen sources, such as nitrate, ammonium, urea or a few other organic nitrogen compounds (e.g., amino acids, cyanate). The preferred nitrogen source is ammonium that is converted to glutamate via the coupled reactions of glutamine synthetase and glutamine-oxoglutarate aminotransferase (GS/GOGAT pathway) (Flores and Herrero pp.; Luque and Forchhammer 2008). According to their morphology, cyanobacteria are subdivided into five divisions (Rippka et al. 1979). Unicellular forms belong to the divisions I (order *Chroococcales*) and II (order *Pleurocapsales*). Cyanobacteria of division III (order *Oscillatoriales*) and IV (order *Nostocales*) form filaments of vegetative cells, with the latter being able to differentiate specialized cells. Division V cyanobacteria (order *Stigenomaticales*) are similar to division IV, but the filaments are branched. The capability to fix dinitrogen (N_2), called diazotrophy, is found in both unicellular and filamentous forms of cyanobacteria (Rippka et al. 1979; Fredriksson and Bergman 1997; Bandyopadhyay et al. 2013). In cyanobacteria of the order *Nostocales* and *Stigenomaticales*, nitrogen fixation takes place in specialized cells termed heterocysts (Herrero et al. 2016).

Cyanobacteria are widespread and well adapted to changing environmental conditions in terms of light conditions and nutrient supply. They colonize nearly all light-exposed habitats ranging from aquartical, including fresh- and seawater to terrestrial habitats (Whitton 2012). Some species can even tolerate extreme environments like hot springs, deserts or the arctic tundra (Miller and Castenholz 2000; Whitton 2012). Cyanobacteria are very interesting targets for biotechnological research because of their ability to synthesize reduced carbon compounds and various biopolymers, only by using water, carbon dioxide and light as an energy source. This chapter will focus on the biopolymer cyanophycin, a major nitrogen storage compound produced by many cyanobacteria.

2 Cyanophycin: A Nitrogen-Rich Storage Compound

Cyanophycin is a branched non-ribosomally synthesized polypeptide first found in cyanobacteria (Borzi 1887). Due to its composition of aspartate and arginine, cyanophycin has a high nitrogen to carbon ratio of 2:1, making it an excellent nitrogen storage substance (Simon and Weathers 1976). It serves as a dynamic nitrogen reservoir in most diazotrophic and non-diazotrophic cyanobacteria as well as in some heterotrophic bacteria (Ziegler et al. 2002; Füser and Steinbüchel 2007). Cyanophycin accumulates in the form of opaque and light-scattering cytoplasmic granules and was therefore mentioned cyanophycin granule peptide (CGP). Electron microscopic studies described CGP granules as electron dense, highly structured and membrane-less inclusions (Lang et al. 1972; Allen and Weathers 1980).

The polymer is synthesized by a single enzyme termed cyanophycin synthetase (CphA) (see below for details). Genes homologous to cyanobacterial *cphA* are found in a wide range of prokaryotes (Krehenbrink et al. 2002), indicating that cyanophycin has a broader biologicals significance, which is, however, mostly unexplored to date. In general, accumulation of the polymer increased under unbalanced growth conditions, e.g., in stationary growth phase or stress condition (high light, high CO_2 or nutrient stress) but only when nitrogen is available (Allen et al. 1980; Allen and Weathers 1980; Carr 1988; Herrero and Burnat 2014; Watzer and Forchhammer 2018b). These findings stand for a putative role of the polymer as a temporary nitrogen reserve molecule (Li et al. 2001).

CGP has been noticed for potential biotechnological use. Biodegradable polyamino acids like CGP could substitute synthetical polymers with similar properties in the fields of medical, food, feed and pharmaceutical industry (Frommeyer et al. 2016). Further, CGP can serve as a natural source of amino acids such as arginine and aspartate, and the poly-aspartate backbone of this polymer could also be used as a biodegradable substitute for polyacrylates (Obst and Steinbüchel 2004; Sallam et al. 2009a, b).

3 Structure of Cyanophycin

The nitrogen-rich polymer CGP consists of two amino acids, aspartate and arginine in equimolar amounts. The aspartic acid residues were linked by peptide bounds forming a poly-L-aspartic acid backbone. Arginine is linked by iso-peptide bond to the backbone, forming the branched polymer also called [multi-(L-arginyl-poly-Laspartic acid)] (Simon and Weathers 1976). One building unit of the polymer consists of the dipeptide β-Asp-Arg (Richter et al. 1999). Circular dichroism (CD) spectroscopy data suggest that a substantial fraction of CGP has β-pleated sheet structure (Simon et al. 1980). The individual CGP chains have no fixed length but show polydisperse molecular masses ranging from 25 to 100 kDa in cyanobacteria. This is in contrast to the native CGP producing heterotrophic bacteria

Acinetobacter sp. ADP1, which synthesizes CGP with a lower molecular weight ranging from 21 to 28 kDa (Elbahloul et al. 2005). Cyanophycins, produced by genetically modified microorganisms (GMOs) harboring cyanobacterial CGP synthesizing enzymes, have molecular masses in the range of 25–45 kDa (Frey et al. 2002; Steinle et al. 2009). Possible explanations for these differences could be that different ratios of CGP synthase to precursor molecules exist in GMOs or that CGP synthesis in a cyanobacterial background involves additional factors contributing the polymer length. These additional factors should also be absent in *Acinetobacter* sp. ADP1 (Watzer and Forchhammer 2018b). Cyanobacterial CGP is exclusively composed of aspartate and arginine. By contrast, CGP of recombinant microorganisms, further amino acids like lysin, citrulline and ornithine, have been detected (Steinle et al. 2009; Ziegler et al. 1998) (see below).

CGP granules were isolated for the first time in 1971 from *Anabaena cylindrical* by differential centrifugation (Simon 1971). Along with this study, CGP has shown its unique solubility behavior. CGP is insoluble under physiological conditions but soluble in either acidic, basic or highly ionic solutions. CGP can be solubilized by SDS but not by non-ionic detergents such as Triton X-100 (Wiefel and Steinbüchel 2014; Lang et al. 1972). Routinely, CGP is extracted by acidic treatment and subsequent precipitation at neutral pH (Frey et al. 2002).

4 Physiological Function of Cyanophycin in the Context of Nitrogen Fixation

CGP is produced by many different cyanobacteria, both diazotrophic and non-diazotrophic strains of unicellular or filamentous nature (Fig. 1). In diazotrophic strains, the formation of CGP is directly linked to the nitrogen-fixation process, whereas in non-diazotrophic strains, the biological significance remained elusive until recently (see below).

In principle, nature has evolved two solutions to the problem that oxygenic photosynthesis and nitrogen fixation are incompatible processes due to the extremely high oxygen sensitivity of the nitrogenase enzyme. Therefore, the two processes cannot take place at the same time at the same place. To solve this issue, the two processes may be separated on a temporal basis or may be confined to different compartments. Simple filamentous (Sect. 3) and unicellular nitrogen-fixing cyanobacteria temporarily separate nitrogen fixation and photosynthesis on a diurnal basis, as shown in the cyanobacterium *Cyanothece* sp. ATCC 51142; the cells perform photosynthesis during the day where they accumulate fixed carbon in the form of glycogen. This storage material is catabolized during the night to support nitrogen fixation with energy and carbon skeletons (Sherman et al. 1998). The nitrogen fixation products are immediately converted to CGP, which accumulates during the night. In the subsequent day period, when photosynthesis is active again, CGP is degraded to mobilize the fixed nitrogen for anabolic reactions to support cell

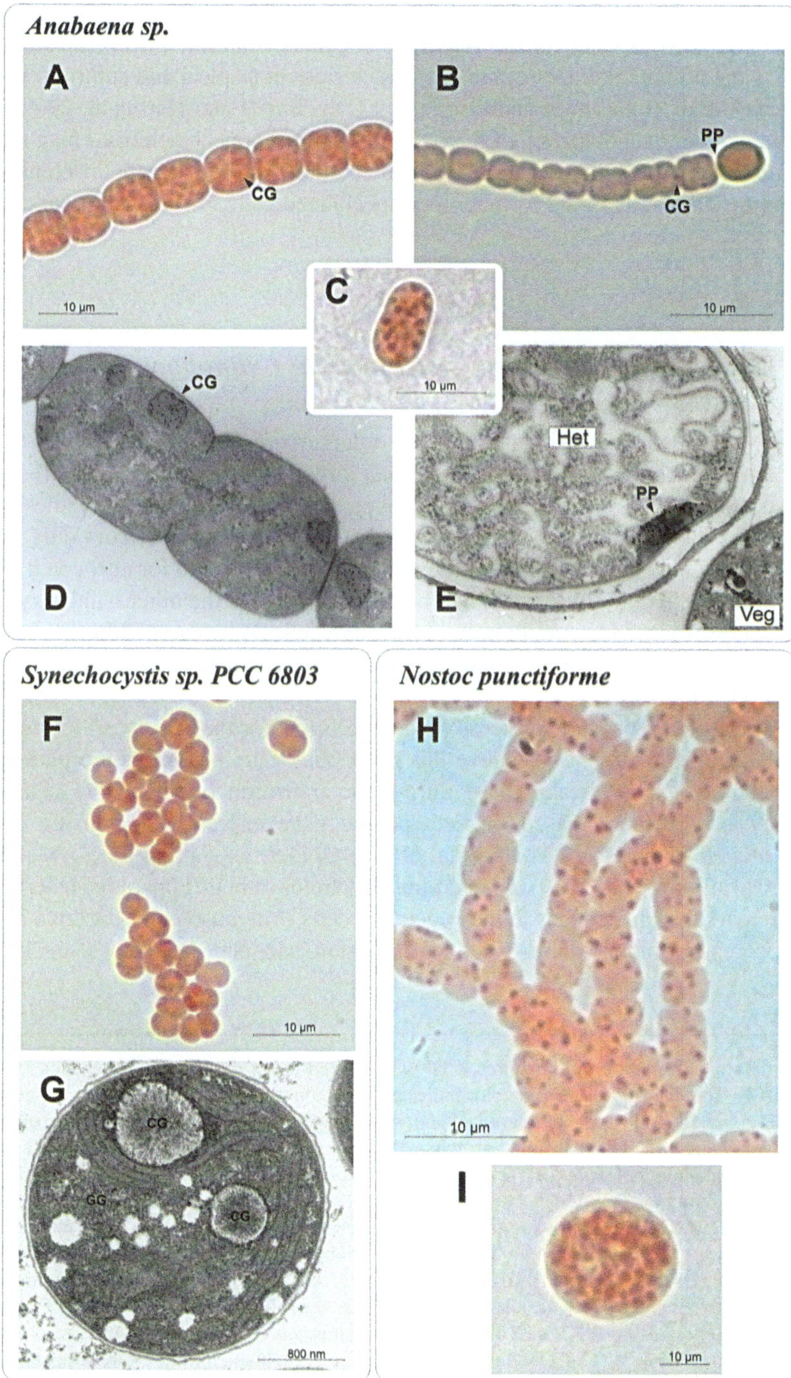

Fig. 1 Light and electron microscopic pictures of *Anabaena* sp. (**a–e**), *Synechocystis* sp. PCC 6803 (**f–g**) and *Nostoc punctiforme* ATCC 29133 (H-I) under CGP accumulating conditions. In light

growth (Sherman et al. 1998). A similar diurnal CPG accumulation during dark periods was also reported in the filamentous cyanobacterium *Trichodesmium* sp., which is the dominant diazotrophic cyanobacterium in tropical and subtropical seas and therefore of global importance in N and C cycling (Finzi-Hart et al. 2009).

Multicellular cyanobacteria of the order *Nostocales* are characterized by a spatial separation of photosynthesis and nitrogen fixation through a cell differentiation process which results in the formation of specialized cells for nitrogen fixation, the heterocysts.

4.1 Cyanophycin in Heterocyst-Forming Cyanobacteria

Heterocystous cyanobacteria comprise the orders *Nostocales* and *Stigenomaticales*. They are filamentous and true multicellular organisms in which two differently specialized cell types have a division of labor. Both processes, photosynthesis and nitrogen fixation, take place in different cells. Nitrogen fixation occurs only in the heterocysts, which are terminally differentiated cells specialized for nitrogen fixation (Kumar et al. 2010; Maldener and Muro-Pastor 2010). On the other hand, oxygenic photosynthesis takes place only in the vegetative cells. When no combined nitrogen sources are available, the multicellular organisms differentiate between 5 and 10% of cells into heterocysts.

The heterocysts provide a microoxic compartment for the activity of nitrogenase (Adams and Carr 1981). To achieve this functional state, numerous morphological and metabolic changes take place during the transition of vegetative cells into heterocysts, which are based on cell-specific gene expression occurring in the differentiated heterocysts (Wolk et al. 1994; Golden and Yoon 1998; Maldener et al. 2014; Flores et al. 2019). Concerning the photosynthetic apparatus, heterocysts show remarkably differences to the vegetative cells (Magnuson and Cardona 2016): The oxygen-producing photosystem II is absent and phycobilisomes are largely

---

Fig. 1 (continued) microscopic pictures, CGP was stained using the Sakaguchi reaction (Watzer et al. 2015). The intensity of the red color indicates the amount of arginine; therefore dark red dots are CGP granules. (**a**) Phosphate starved *Anabaena variabil*s ATT 29413 in the presence of nitrate as nitrogen source. (**b**) Diazothrophic growing filament of *Anabaena sp.* PCC7120 with terminal heterocyst containing polar plaques. (**c**) Mature akinete of *Anabaena variabilis* ATCC 29413. (**d–e**) Electron micrographs of *Anabaena variabilis* cells. Ultrastructural investigations were performed using electron microscopy following the immunocytochemical visualization of CGP with antisera raised against cyanophycin and a gold coupled anti-rabbit immunoglobulin G antibody (by Uwe Kahmann, Universität Bielefeld) (**d**) vegetative cells of *Anabaena variabilis,* magnification 28,500x; (**e**) heterocyst [Het] of *Anabaena variabilis,* magnification 50,000x. (**f–g**) Phosphate starved *Synechocystis* sp. PCC 6803 in light and transmission electron microscopy, respectively. (**h**) *Nostoc punctiforme* ATCC 29133 under phosphate starvation and nitrogen supplementation. (**i**) Mature akinete of *Nostoc punctiforme* ATCC 29133 (Watzer and Forchhammer 2018b). [CG] CGP granules, [PP] polar plaques, [GG] glycogen granules, [Het] heterocyst

degraded. Thylakoid membranes are re-organized and support cyclic electron transport via the photosystem I, to provide ATP (Wolk et al. 1994). In addition, respiration takes place in the thylakoid-derived honeycomb-membranes, where the molecular oxygen is reduced to water (Wolk 1996). In the microoxic environment of heterocysts, instead of CO_2 fixation by RubisCO, atmospheric N_2 is fixed by the key enzyme for nitrogen fixation, the nitrogenase. The nitrogenase encoded by *nif*HDK genes is cell specific and exclusively active in mature heterocysts (Golden et al. 1985; Fay 1992). The enzyme is very oxygen sensitive (Gallon 1981) and is irreversibly inactivated by oxygen (Hill et al. 1981; Golden et al. 1985; Fay 1992). It reduces molecular nitrogen to ammonium using reduction equivalents and ATP (Wolk et al. 1994; Kumar et al. 2010). The high energy demand for nitrogen fixation is partially provided from products of oxygenic photosynthesis performed in the neighboring vegetative cells of the filament. The division of labor between heterocysts and vegetative cells requires an efficient exchange of metabolites and signals along the filament (Wolk 1968; Wolk et al. 1974). Molecular diffusion of small molecules from cytoplasm to cytoplasm takes place through special cell-cell connections, so-called "septal junctions" (Maldener and Muro-Pastor 2010; Flores et al. 2016). The structure of the septal junctions has been resolved recently in more detail in *Anabaena sp.* PCC 7120 (Weiss et al. 2019). Each septal junction complex comprises a channel that spans the gap through the periplasmic space in the septal region and traverses the septal peptidoglycan though approx. 20 nm nanopores to connect two adjacent cells. On the cytoplasmic site of each entrance point, the channel is covered by a cap and plug structure that enables closure of the septal junctions (Weiss et al. 2019).

Mature heterocysts are characterized by a specialized cell wall composed of polysaccharide and glycolipid layers, which minimizes the diffusion of molecular oxygen into the specialized cell (Walsby 2007; Nicolaisen et al. 2009). A distinct morphological feature of heterocysts is so-called "polar plaques" or "polar nodes," which are specific structures containing CGP (Sherman et al. 2000) closely located to the neck and the adjacent vegetative cell. The polar nodes are easily recognized by light microscopy, and as such, they are characteristic morphological signatures of heterocysts (Ziegler et al. 2001) (Fig. 1). It was assumed that CGP located at these heterocystous structures is a key intermediate in the storage of fixed nitrogen products in heterocysts and acts as a dynamic nitrogen reservoir (Carr 1988; Sherman et al. 2000). CGP in the polar nodes could serve as a sink for fixed nitrogen in the heterocyst to avoid feedback inhibition of nitrogen assimilatory reactions from soluble products of nitrogen fixation (Burnat et al. 2014; Forchhammer and Watzer 2016).

Vegetative cells support the metabolism of heterocysts by supplying them with glutamate and reduced carbohydrates, e.g., alanine and the chemically inert disaccharide sucrose (Jüttner 1983; Curatti et al. 2002; Vargas et al. 2011) as a source for energy, reductant power, and the required carbon skeletons for assimilation of fixed nitrogen. Invertases degrade the disaccharides into glucose and fructose in heterocysts (Lopez-Igual et al. 2010; Vargas et al. 2011), which can be further catabolized via the oxidative pentose phosphate cycle to generate reducing power for nitrogen

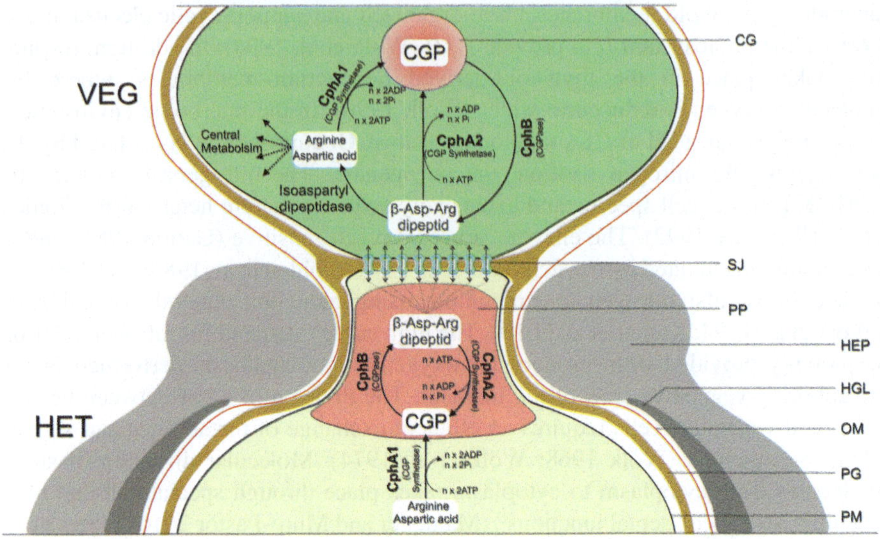

Fig. 2 Scheme of cyanophycin-related nitrogen transport between a heterocyst (HET) and a vegetative cell (VEG). Cyanophycin (CGP) is synthesized by CphA and accumulated in the polar plaques (PP). The degradation of CGP releases β-aspartyl-arginine dipeptides, which can be used to resynthesize CGP by CphA2 (see chapter "The Anammoxosome Organelle: The Power Plant of Anaerobic Ammonium-Oxidizing (Anammox) Bacteria") or transported to the adjacent vegetative cell by passing through the septal junctions (SJ). In the vegetative cells, the isoaspartyl dipeptidase releases monomeric arginine and aspartate, which flow into the central metabolism. Furthermore, monomeric amino acids or β-aspartyl-arginine dipeptides can be used by CphA1 or CphA2, respectively, to synthesize CGP, building a CGP granule (CG). *PM* Plasma membrane, *PG* Peptidoglycan, *OM* Outer membrane, *HGL* Heterocyst glycolipid layer, *HEP* Heterocyst envelope polysaccharide layer

fixation and respiration. Ammonium produced by nitrogenase activity is immediately assimilated by glutamine synthetase, whereas the GOGAT reaction takes place in adjacent vegetative cells (Martin-Figueroa et al. 2000), which requires an efficient shuffling of glutamate and glutamine between heterocysts and vegetative cells.

Vegetative cells are supplied by adjacent heterocysts with reduced nitrogen such as glutamine and the dipeptide β-aspartyl-arginine (Thomas et al. 1977; Burnat et al. 2014). Therefore, CGP in the outer region of the polar nodes is cleaved by the action of cyanophycinase into β-aspartyl-arginine dipeptides. CGP catabolic enzymes are present at significantly higher levels in vegetative cells than in heterocysts, indicating a sophisticated spatial organization of nitrogen metabolism (Burnat and Flores 2014; Burnat et al. 2014) (Fig. 2).

Surprisingly, under standard laboratory conditions, this mode of metabolic organization appears to be dispensable, since mutants of *Anabaena* sp. PCC 7120 or *Anabaena variabilis,* lacking the CGP synthetic genes, were little affected in diazotrophic growth under usual cultivation conditions (Ziegler et al. 2001). In this case, arginine and aspartate might be transferred directly to the vegetative cells, without an intermediate storage in CGP. However, mutants lacking the CGP synthetic genes were affected in diazotrophic growth under stressful high light conditions (Ziegler et al. 2001; Klemke et al. 2016), indicating that cyanopyhcin plays a role in coping with metabolic stress.

Moreover, diazotrophic growth is significantly decreased in strains, which are unable to degrade CGP (Burnat and Flores 2014; Burnat et al. 2014). These results identified β-aspartyl-arginine dipeptides as a nitrogen vehicle in heterocyst-forming cyanobacteria, next to glutamine, arginine, or aspartate alone. A benefit of β-aspartyl-arginine dipeptides as a nitrogen transport substance is avoiding the release of free arginine and aspartate back in the heterocyst. This indicates that CGP metabolism has evolved in heterocyst-forming cyanobacteria to increase the efficiency of nitrogen fixation (Watzer and Forchhammer 2018b; Flores et al. 2019a; Burnat et al. 2014, 2019).

Several of the heterocyst-forming cyanobacteria can differentiate another type of specialized cells, the akinetes. These are resting cells able to survive long periods of unfavorable conditions. During akinete development, the cells transiently accumulate CGP in addition to other storage compounds such as glycogen or lipid droplets (Perez et al. 2016) (Fig. 1). In mature akinetes, CGP disappears gradually, and when the akinetes germinate, CGP reappears transiently (Perez et al. 2018). Apparently, synthesis of CGP seems inessential, since *Anabaena variabilis* akinetes lacking CGP granules are not impaired in germination. This behavior agrees with early observations suggesting that CGP is not the direct nitrogen source for protein biosynthesis in germinating akinetes (Sutherland et al. 1985).

4.2 Cyanophycin in Non-diazotrophic Cyanobacteria

In non-diazotrophic cyanobacteria, CGP can only be detected under unbalanced growth conditions. In exponentially growing cells, its amount is usually less than 1% of the cell dry mass. However, when growth is retarded or arrested due to unfavorable conditions under nitrogen supply, (antibiotic treatment; sulfate, phosphate or potassium starvation), the cells may start to build up CGP, which can amount up to 18% of the cell dry mass (Allen et al. 1980; Watzer et al. 2015; Simon 1973). When *Synechocystis* cells recover from prolonged nitrogen starvation, by adding nitrate to the starved cells, they transiently accumulate CGP (Klotz et al. 2016; Allen and Hutchison 1980; Watzer and Forchhammer 2018a). This raised the question regarding the biological significance of transient CGP accumulation. This issue was recently solved by Watzer and Forchhammer (Watzer and Forchhammer 2018a). The transient accumulation of CGP becomes beneficial to the cells under conditions of fluctuating or limiting nitrogen supply. Whereas wild-type *Synechocystis* cells are able to fully recover from prolonged nitrogen chlorosis when they are provided with pulses of nitrate, recovery is severely delayed in a mutant deficient in CGP synthesis (*cphA* mutant, see below). This phenotype becomes even more prominent when cells are cultivated in 12 h day/night cycles. From these findings, it can be concluded that under natural conditions, characterized by fluctuating nutrient and light supply, cyanophycin synthesis allows intracellular sequestration of nitrogen assimilation products, so that under periods of nitrogen shortage, CGP degradation provides anabolic precursors to continue growth. Even under constant illumination and a constant but limited nitrogen supply, the wild type has a growth advantage over the *cphA* mutant. This suggests that CGP synthesis optimizes the nitrogen assimilation process.

The recent discovery of the ornithine–ammonia cycle (OAC) in cyanobacteria emphasizes the role of CGP as dynamic nitrogen storage (Zhang et al. 2018). The pathway starts with carbamoyl phosphate synthesis and ends with the conversion of arginine to ornithine, CO_2, and ammonia by a novel arginine dihydrolase (ArgZ). Disruption of the OAC-pathway by mutating *argZ* leads to impaired cell growth under fluctuating nitrogen supply. This demonstrates that the OAC-pathway allows rapid remobilization of nitrogen reserves under temporal starvation conditions. Since the *cphA*-deficient mutant showed similar growth defects under fluctuating nitrogen supply, it can be concluded that nitrogen, which is stored in CGP, can be remobilized by the OAC-pathway (Zhang et al. 2018).

5 Enzymes of Cyanophycin Metabolism

Enzymes of the CGP metabolism occur in a wide of range of phototrophic and chemolithoautotrophic bacteria (Füser and Steinbüchel 2007). In general, CGP is synthesized from aspartate and arginine by the cyanophycin synthetase (CphA1) in an ATP-depending elongation reaction. Intracellular CGP degradation is catalyzed by the cyanophycinase (CphB). The β-Asp-Arg dipeptides resulting from the cleavage of CGP are further hydrolyzed by isoaspartyl dipeptidase, releasing aspartate and arginine (Watzer and Forchhammer 2018b) (Fig. 3). The following paragraphs provide detailed information about the enzymes involved in CGP metabolism.

5.1 Cyanophycin Synthetase (CphA)

The non-ribosomal biosynthesis of cyanophycin takes place, as long as the cell has sufficient nitrogen and energy available by a single enzyme called cyanophycin synthetase (CphA) (Simon 1976). In the synthesis the amino acids aspartate and arginine condense in two steps with one ATP consumption per step (Simon 1976; Berg et al. 2000). Additionally, the reaction requires potassium ions, magnesium ions, cyanophycin primers (low-molecular-mass CGP), and a thiol reagent (such as β-mercaptoetanol or dithiothreitol) (Simon 1976). The requirement of low-molecular mass CGP in the reaction of CphA leads to the conclusion that CphA catalyzes the elongation of a CGP-filament but not its de novo synthesis.

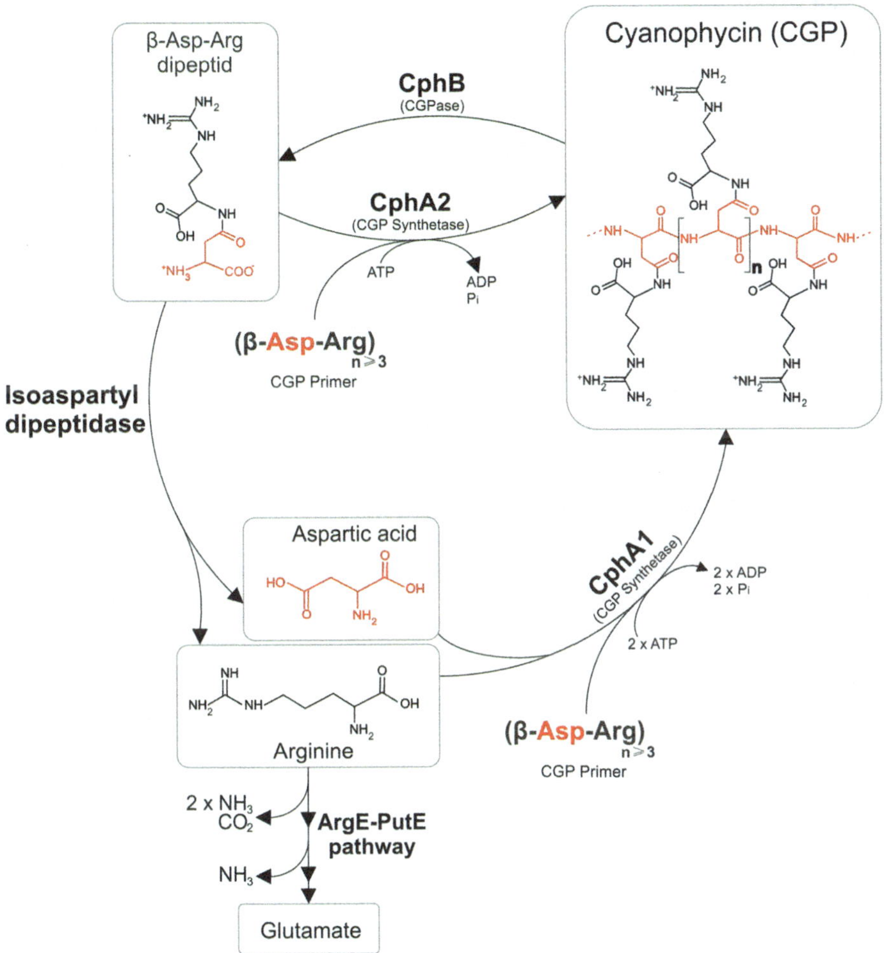

Fig. 3 Scheme of cyanophycin metabolism. Cyanophycin (CGP) is synthesized from aspartate and arginine by CGP synthetase (CphA1) in an ATP-depending elongation reaction using CGP primers, containing at least three Asp-Arg building blocks. The degradation of CGP by the cyanophycinase CphB releases β-aspartyl-arginine dipeptides, which can be further hydrolyzed by isoaspartyl dipeptidase to aspartate and arginine. In many nitrogen-fixing cyanobacteria, an additional CGP synthetase is present, termed CphA2. CphA2 can use β-aspartyl-arginine dipeptides to resynthesize CGP (Watzer and Forchhammer 2018b). The recently elucidated AgrE-PutA pathway allows rapid nitrogen mobilization by the catabolizes of arginine to glutamate, with concomitant release of ammonium (Burnat et al. 2019)

By using synthetic primers, Berg et al. could show that a single building unit (β Asp-Arg) does not serve as an efficient primer for the CphA1 elongation reaction. CphA1 needs at least an Asp-Arg trimer (β Asp-Arg)$_3$ as a primer for activity (Berg et al. 2000). Other peptides, like the peptidoglycan-pentapeptide or other cellular components, have been suggested to serve as an alternative priming substance,

Table 1 Overview of *cphA*-genes from different organisms

Gene names	Uniprot ID	Organism	References
cphA$_{6803}$; *slr2002*	P73833	*Synechocystis* sp. (strain PCC 6803/Kazusa)	Ziegler et al. (1998)
cphA$_{6308}$	P56947	*Geminocystis herdmanii* (strain PCC 6308) (*Synechocystis* sp. (strain PCC 6308))	Steinle et al. (2008)
cphA$_{MA19}$	Q8VTA5	*Synechococcus* sp. MA19	Hai et al. (1999)
cphA$_{7120}$; *all3879*	P58572	*Nostoc* sp. (strain PCC 7120/SAG 25.82/UTEX 2576)	Voss et al. (2004)
cphA$_{29413}$; *ava1814*	O86109	*Trichormus variabilis* (strain ATCC 29413/ PCC 7937) (*Anabaena variabilis*)	Ziegler et al. (1998)
cphA2$_{29413}$; *ava0335*	Q3MGC5	*Trichormus variabilis* (strain ATCC 29413/ PCC 7937) (*Anabaena variabilis*)	Klemke et al. (2016)
cphA2$_{7425}$; *cyan7425 4446*	B8HK06	*Cyanothece* sp. (strain PCC 7425/ATCC 29141)	Klemke et al. (2016)
cphA$_{NE1}$	A7YEP1	*Nostoc ellipsosporum* NE1	Hai et al. (2006)
cphA$_{TE}$; *tlr2170*	B9A0M2	*Thermosynechococcus elongatus* (strain BP-1)	Arai and Kino (2008)
cphA$_{ADP1}$; *ACIAD1279*	Q6FCQ7	*Acinetobacter baylyi* (strain ATCC 33305/ BD413/ADP1)	Krehenbrink et al. (2002)
cphA$_{DH}$; *dhaf0085*	B8FY49	*Desulfitobacterium hafniense* (strain DCB-2/ DSM 10664)	Ziegler et al. (2002)

which allows CphA1 activity in recombinant bacteria without ability to produce native CGP primers (Hai et al. 2002; Obst and Steinbüchel 2006). These alternative priming substances are also present in yeasts like *Saccharomyces cerevisiae* or *Pichia pastoris* but absent in higher plants (Steinle et al. 2008, 2010; Nausch et al. 2016). The primer-independent CphA of *Thermosynechococcus elongatus* BP-1 allows CGP production in higher plants (Arai and Kino 2008). Challenges and peculiarities of CGP production in plants have been reviewed by Nausch et al. (2016).

Up to date, several CphA1 enzymes from different bacteria, including cyanobacteria and heterotrophic bacteria, have been characterized (Table 1) (Ziegler et al. 1998; Arai and Kino 2008; Hai et al. 2002; Füser and Steinbüchel 2007; Aboulmagd et al. 2001a; Elbahloul et al. 2005; Du et al. 2019). The molecular masses of the characterized CphA1 enzymes range between 90 and 130 kDa. CphA1 from *Synechocystis* sp. PCC6803 and *Anabaena variabilis* PCC 7937 forms most likely a homodimer, while the primer-independent CphA1 from *Thermosynechoccus elongatus* forms a homotetramer (Ziegler et al. 1998; Aboulmagd et al. 2001a; Arai and Kino 2008). The primary structure of cyanobacterial CphA1 can be divided in two regions; each region contains an active site and an ATP binding site (Ziegler et al. 1998; Berg et al. 2000). Experiments revealed that arginine is probably bound at the C-terminal and aspartate at the N-terminal active site (Berg 2003).

The putative mechanism of CphA1 has been suggested by Berg et al. by measuring the step-wise incorporation of amino acids to the C-terminus of a CGP primer. First, the carboxylic acid group at the C-terminus of the polyaspartate backbone becomes activated by phosphorylation. Subsequently, one aspartate is bound at the C-terminus by its amino group, forming a peptide bound. The intermediate (-β-Asp-Arg)$_n$—Asp is transferred to the second active site of the N-terminal region of CphA1. The β-carboxyl group of aspartate becomes phosphorylated and is then linked to the α-aminogroup of arginine, forming an isopeptide-bound (Berg et al. 2000).

For CphA1 of *Synechocystis* sp. PCC 6308, the apparent K_M value in vitro for aspartate is 450 μM, for aginine 49 μM, for ATP 200 μM and for CGP as priming substance 35 μg/ml, indicating a higher affinity toward arginine compared to aspartate. During the in vitro reaction, CphA1 converts 1.3 ± 0.1 mol ATP per mol incorporated amino acid (Aboulmagd et al. 2001a).

Previous microscopical studies revealed different localizations of CphA depending on its activity in the non-diazortphic cyanobacterium *Synechocystis* sp. PCC 6803 (see below for more details) (Watzer and Forchhammer 2018a). Under balanced growth conditions, in which no CGP synthesis occurs, CphA is equally distributed in the cytoplasm. Immediately after the induction of CGP synthesis, CphA aggregates in foci that were randomly localized in the cell. These foci increase in size due to the cyanophycin synthesizing reaction of CphA and become visible as CGP granules. Doing this, CphA is exclusively attached on the granule surface. As CGP accumulates, the number of granules per cell continuously decreases (Watzer and Forchhammer 2018a). Electron micrographs could show that large CGP granule fuse when they collide, forming larger amorphous aggregates (Watzer et al. 2015). During CGP degradation, the localization of CphA changes from the granule surface back to the cytoplasm. This suggests that as CGP is degraded, CphA dissociates from the surface and converts to the inactive form, where it remains silent until CGP accumulation is triggered again.

In addition to the main enzyme for CGP synthesis CphA1, there is a second enzyme, CphA2 (Picossi et al. 2004). The gene for the enzyme CphA2 has been identified predominantly in diazotrophic cyanobacterial species that also possess the gene for CphA1 (Klemke et al. 2016). Therefore, in heterocystous cyanobacteria of Sect. 4, 71% and 82% of Sect. 5 contain a *cphA2* gene. CphA2 is a truncated enzyme version of CphA1, showing only a high homology in amino acid sequence to the N-terminal part of CphA1, but not to the C terminal part. The function of CphA2 is the addition of β-Asp-Arg dipeptide units to the growing CGP chain using one molecule of ATP (Klemke et al. 2016). CphA2 was found directly attached to the polymer like CphA1 in vegetative cells and heterocysts. By enzymatic activities of CphA2 and the degradation enzyme cyanophycinase (see below), a continuous pool of the dipeptide β-Asp-Arg in the filament is provided. If needed, CGP can be polymerized directly from β-Asp-Arg. A *cphA2* deletion mutant displays only a minor decrease in the overall CGP content. However, such a mutant showed defects under diazotrophic growth at high light conditions like a CphA1-deficient mutant. This observation implies that CGP hydrolysis and re-polymerization by CphA1 and CphA2 could run as a "futile cycle" in case of high photosynthetic activity (Fig. 3). Furthermore, CphA2 may be involved in regulating the cellular β-Asp-Arg pool and thereby restrict the degradation of the dipeptide by the isoaspartyl dipeptidase into the amino acids arginine and aspartate (Forchhammer and Watzer 2016).

5.2 *Cyanophycinase (CphB)*

CGP is resistant against hydrolytic cleavage by several arginases and proteases, probably due to its branched structure (Simon and Weathers 1976; Obst and Steinbüchel 2006). In 1999, Richter et al. reported a CGP hydrolyzing from the unicellular cyanobacterium *Synechocystis* sp. PCC 6803, called CphB (Richter et al. 1999). CphB is a 29.4 kDa, dimeric C-terminal exopeptidase, catalyzing the enzymatic cleavage of CGP to β-Asp-Arg dipeptides (Richter et al. 1999; Law et al. 2009). Based on sequence analysis, its sensitivity to serine protease inhibitors and site-directed mutagenesis, CphB can be classified as serine-type exopeptidase with the catalytic Ser at position 132 (Richter et al. 1999; Law et al. 2009). Structural modeling indicated that the substrate specificity of CphB for CGP is based on an extended conformation of the active site. The unique conformation of the active site requires β-linked aspartyl peptides for binding and catalysis, preventing CphB from non-specific cleavage of other polypeptides (Law et al. 2009; Watzer and Forchhammer 2018b).

In addition to the intracellular CphB, other versions of CGP hydrolyzing enzymes exist, catalyzing the extracellular cleavage of CGP. In 2002, Obst et al. isolated several Gram-negative, heterotrophic bacteria from different habitats, which were able to utilize CGP as a carbon, nitrogen, and energy source (Obst et al. 2002). One stain was identified as *Pseudomonas anguillisptica* strain BI. A novel CGPase was identified and called CphE in culture supernatant of this strain (Obst et al. 2002). CphE exhibits an amino acid sequence identity of 27–28% to cyanobacterial CphB. The primary degradation product of CGP digested with CphE are β-Asp-Arg dipeptides. Sensitivity to serine protease inhibitors has been reported, indicating that the catalytic mechanism of CphE is related to serein-type proteases (Obst et al. 2002). Today, extracellular CGPases have been reported in several bacteria ranging from Gram-positive to Gram-negative, from areobic to an anaerobic. This indicates that utilization of CGP as a carbon, nitrogen, and energy source seems to be a common principle in nature (Obst et al. 2002, 2004, 2005; Obst and Steinbüchel 2004; Sallam and Steinbüchel 2008; Watzer and Forchhammer 2018b).

5.3 Aspartyl-Arginine Dipeptidase

The last step of CGP catabolism is the cleavage of β-Asp-Arg dipeptides to monomeric amino acids. The ORF *sll0422* from *Synechocystis* sp. PCC 6803 and *all3922* from *Anabaena* sp. PCC 7120 are annotated on the basis of sequence similarity as "plant-type asparaginase" (PTA). In 2002, Hejazi et al. characterized several PTAs for substrate specificity, showing that *sll0422* and *all3922* are able to hydrolyze a wide range of isoaspartyl dipeptides including β-Asp-Arg.

The mature PTA of *Synechocystis* sp. PCC 6803 and *Anabaena* sp. PCC 7120 consists of two subunits, which are generated by autocleavage of the primary translation product (Hejazi et al. 2002). The native enzyme has a molecular mass of approximately 70 kDa, suggesting a subunit structure of $\alpha_2\beta_2$ (α derived from N-terminal part and β derived from the C-terminal part of the immature translation product).

A deletion mutant of gene *all3922* (coding for PTA in *Anabaena* sp. strain 7120) accumulated CGP and β-Asp-Arg and was strongly impaired in diazotrophic growth (Burnat et al. 2014). Further results showed that PTA was present in vegetative cells more than in heterocysts, although β-Asp-Arg was found in substantial amounts in heterocysts. It can be assumed that the dipeptide produced by cyanophycin degradation in heterocysts should be transferred to vegetative cells.

Furthermore, a compartmentalization was also found for arginine catabolizing enzymes. The arginine decarboxylase pathway that produces *sym*-homospermidine is also more present in vegetative cells than in nitrogen-fixing heterocysts (Burnat and Flores 2014). Distortion of the arginine decarboxylase pathway in *Anabaena* sp. PCC 7120 through the deletion of agmatinase *(speB)* led to the accumulation of high amounts of CGP and to impairment of diazotrophic growth (Burnat and Flores 2014). The arginine catabolic pathway of *Anabaena* sp. PCC 7120 has recently been elucidated by Burnat et al. 2019. This pathway consists of two genes, the arginine-guanidine removing enzyme *agrE* (*alr4995*) and a proline oxidase *putA* (*alr0540*). The AgrE-PutA pathway catabolizes arginine to produce proline and glutamate, with concomitant release of ammonium. Both genes, *agrE* and *putA*, were found to be expressed at higher levels in vegetative cells than in heterocysts, implying that arginine is catabolized mainly in the vegetative cells (Burnat et al. 2019; Flores et al. 2019a).

All these results suggest that CGP is degraded in the heterocysts to the dipeptide β-Asp-Arg and the dipeptide could serve as a mobile nitrogen reserve for the vegetative cells in addition to other molecules such as glutamine (Burnat et al. 2014, 2019; Flores et al. 2019a).

6 Genetic Organization and Expression of *cphA* and *cphB*

Genes involved in CGP metabolism are usually organized in clusters (Füser and Steinbüchel 2007). In the genome of *Synechocystis* sp. PCC 6803 *cphA* and *cphB* are localized side by side but are expressed independently (Mitschke et al. 2011). A hypothetical protein with unknown function (orf *slr2003)* is located downstream of *cphA* and both are transcribed in a polysistronic unit (Mitschke et al. 2011).

Anabaena sp. PCC 7120 contains two clusters, *cph1* and *cph2,* both containing cyanophycin biosynthetic genes with *cphA* and *cphB* (Picossi et al. 2004). The *cph1* cluster contains *cphB1* and *cphA1,* which are cotranscribed and form an operon. In addition, *cphA1* can be expressed from two independent promotors, of which one is constitutive and the other regulated by the global transcription factor NtcA (Picossi et al. 2004). The *chp1* operon was expressed under nitrate- and ammonia-supplemented growth, but the expression increased at diazotrophic growth conditions in heterocysts and vegetative cells. Cell-specific expression under diazotrophic conditions of the *cph1* gene cluster showed a global expression of *cphA1* and *cphB1* in vegetative cells and heterocysts, but an increased expression level was observed in heterocysts (Picossi et al. 2004). A higher enzyme activity of CphA1 and CphB1 was also observed in heterocysts than in vegetative cells (Carr 1988), which suggests a role of cyanophycin as a dynamic reservoir for fixed nitrogen (Li et al. 2001).

In the *cph2* cluster, *cphA2* and *cphB2* were found in opposite orientations and both were expressed monoistronically. Both genes were expressed under ammonia-, nitrate-, and N_2-supplemented growth, but the expression was higher in the absence of ammonia. Generally, the expression of *cph1* was higher compared to *cph2* (Picossi et al. 2004).

In addition to *cph1* and *cph2* clusters, a third cluster containing putative *cphA* and *cphB* genes was found in *Nostoc punctiforme* PCC 73102 and *Anabaena variabilis ATCC 29413* (Füser and Steinbüchel 2007).

7 Biotechnological Production of Cyanophycin

The natural compound cyanophycin has up to now no industrial application whereas chemical derivatives of CGP are of potential biotechnological interest. CGP can be converted via hydrolytic β-cleavage to poly(α-L-aspartic acid) (PAA) and free arginine. Arginine is a valuable product and PAA is biodegradable and has a high number of negatively charged carboxylic groups, making PAA a possible substituent for polyacrylates (Obst and Steinbüchel 2004; Sallam et al. 2009a, b; Khlystov et al. 2017). PAA can be employed as an anti-sealant or dispersing ingredient in many fields of applications, including washing detergents or suntan lotions. Furthermore, PAA has potential application areas as an additive in paper, paint, building, or oil industry (Sallam et al. 2009a, b; Obst and Steinbüchel 2004; Watzer and Forchhammer 2018b).

In addition to its use as polymer, CGP can also serve as a source for constituent dipeptides and amino acids in food, feed, and pharmaceutical industry. The amino acids arginine (semi-essential), aspartate (non-essential) and lysine (essential) derived from CGP have a broad spectrum of nutritional or therapeutic applications. Large-scale production of these amino acids, as mixtures or dipeptides, is established in the industry, with various commercial products already available on the market (Sallam and Steinbüchel 2010; Watzer and Forchhammer 2018b).

7.1 Cyanophycin Production Using Recombinant Production Hosts

Previous approaches to produce CGP mainly focused on heterotrophic bacteria, yeasts, or plants as production hosts. These recombinant production hosts were generated by transforming CGP synthetase genes, mostly from cyanobacterial origin. Therefore, biotechnologically established production strains like *E. coli*, *Cupriavidus necator* (formally known as *Ralstonia eutropha*), *Pseudomonas putida*, *Corynebacterium glutamicum*, or *Saccharomyces cerevisiae* were used for heterologous production of CGP (Aboulmagd et al. 2001b; Steinle et al. 2008; Frey et al. 2002; Watzer and Forchhammer 2018b; Du et al. 2019). In addition to production in microorganisms, production of CGP has also been attempted in several crop plants (reviewed by (Nausch et al. 2016).

In 1998, Ziegler et al. produced the first recombinant CGP by heterologous expression of *cphA* from *Synechocystis* sp. PCC 6803 in *E. coli*. CGP isolated from this recombinant producer strain showed some differences to those from cyanobacteria. In particular, a reduced polydispersity ranging from 20 to 40 kDa and an altered amino acid composition, containing low amounts of lysin next to arginine and aspartate, haven been reported (Ziegler et al. 1998). Based on this finding, several CGP producer strains have been generated in the following years to establish CGP for biotechnological industry.

E.coli DH1 harboring *cphA* from *Synechocystis* sp. PCC 6803 was used in the first large-scale CPG production study in a culture volume up to 500 liter, allowing CGP quantities in kilogram scale. During process optimization, the highest observed CGP amount was 24% (w/w) per cell dry mass (CDM). However, this optimal production rate required Terrific-Broth-Medium. In mineral salts medium, CGP production only occurred in the presence of additional casamino acids (Frey et al. 2002). An engineered CphA from *Nostoc ellipsosprum* could further increase the maximal CGP production up to 34.5% (w/w) of CDM, but also required complex growth media to yield such high amounts (Hai et al. 2006).

Cupriavidus necator and *Pseudomonas putida* are industry-established producer stains for several products and have therefore been considered as candidates for biotechnological CGP production (Aboulmagd et al. 2001b; Diniz et al. 2006). CPG synthase genes from *Synechocystis* sp. PCC 6308 or *Anabaena* sp. PCC 7120 have been used to enable CGP production in these species (Voss et al. 2004; Aboulmagd et al. 2001b). Metabolic engineering and process optimization studies showed that accumulation of CGP mainly depends on the origin of *cphA* gene, the accumulation of other storage compounds like polyhydroxyalkanoates (PHA), the availability of intermediates derived from the tricarboxylic acid cycle (TCA-cycle), and the addition of CGP precursor components like arginine to the medium (Voss et al. 2004; Diniz et al. 2006). Using this knowledge, it was possible to obtain CGP amounts up to 17.9% (w/w) per CDM at 30 liter scale using mineral salt medium (Diniz et al. 2006). CGP accumulation of *Cupriavidus necator* is also strongly affected by the *cphA* expression system. A stabilized multi-copy *cphA* expression system, using the 2-keto-3-deoxy-6-phospho-gluconate (KDPG) aldolase gene-dependent addiction system, allows cultivation without antibiotic selection. The multi-copy expression of *cphA* results in an increased CGP production between 26.9% and 40% (w/w) per CDM. The high amount of 40% (w/w) of CDM could be also observed in a 30- and 500-liter scale. Without the addition of amino acids to the medium, the amount of CGP was still between 26.9 and 27.7% (w/w) of CDM (Voss and Steinbüchel 2006).

The industry-established host *Saccharomyces cerevisiae* has also been considered for biotechnological CGP production (Steinle et al. 2008, 2009). By expressing *cphA* from *Synechocystis* sp. PCC 6803, it was possible to obtain CGP in amounts up to 6.9% (w/w) of CDM. In this case, two species of CGP could be observed: water-soluble and the typical water-insoluble CGP. Depending on the cultivation conditions, the ratio between water-soluble and insoluble CGP-species could be varied. Furthermore, the isolated polymer contained 2 mol% lysine, which can be increased up to 10 mol% by adding lysin to the medium (Steinle et al. 2008). During metabolic engineering studies, several arginine biosynthesis mutants of *Saccharomyces cerevisiae* have been characterized concerning their abilities to accumulate CGP. Next to arginine, aspartate, and lysine, further non-proteinogenic amino acids like citrulline and ornithine could be detected in isolated CGP derivates (Steinle et al. 2009). Depending on these results, customized CGP derivates with altered amino acid composition could also be produced using *Pseudomonas putida* ant the yeast *Pichia pastoris* as a production host (Wiefel et al. 2011; Frommeyer et al. 2016; Steinle et al. 2010). Customized CGP derivates are important sources for β-dipeptides, which can be used in several applications. The extracellular CGPase, CphE from *Pseudomonas alcaligens*, enables large-scale production of these customized β-dipeptides from CGP derivates (Sallam et al. 2009a, b; Sallam and Steinbüchel 2010).

The biotechnologically well-established bacterium *Corynebacterium glutamicum* has so far not been considered for CGP production due to low CGP quantities and difficulties in polymer extraction. Very recently, Wiefel et al. re-evaluated the capability of *Corynebacterium glutamicum* as a CGP producer strain, by using different expression vectors, *cphA*-genes, and cultivation conditions (Wiefel et al. 2019). It was possible to create strains which produce CGP up to 14% of the CDM, including water-soluble CGP containing more than 40 mol % of lysin as a side-chain. Additionally, a strain was generated which incorporated up to 6 mol % glutamic acid into the backbone as a substitution of aspartate. Interestingly, a previous study already reported a CGP-species from nitrogen-limited *Synechocystis* sp. PCC6308 which contained high amounts of glutamic acid (Merritt et al. 1994); however, the position in the polymer could not be resolved. The re-discovery of glutamic acid in the backbone of CGP in recombinant *Corynebacterium glutamicum* raises again the question of the biological significance of this peculiar CGP species in Cyanobacteria (Merritt et al. 1994; Wiefel et al. 2019).

7.2 Dependence of CGP Production from Arginine Biosynthesis

Generally, CGP accumulation is triggered by unfavorable conditions, such as entry into stationary phase of growth, limitation of macronutrients (with the exception of nitrogen starvation) or inhibition of translation by adding antibiotics like chloramphenicol (Allen et al. 1980; Simon 1973; Watzer and Forchhammer 2018b). All these conditions result in a reduced or arrested growth. In the exponential growth phase, amino acids like arginine and aspartate are mostly used for de novo protein biosynthesis. Under growth-limiting conditions, the protein biosynthesis is slowed down, resulting in an excess of free amino acids. This enrichment of monomeric amino acids allows CGP accumulation.

Acinetobacter calcoaceticus ADP1 accumulates only 3.5% (w/w) CGP of the cellular dry matter (CDM) when grown with ammonia as a nitrogen source and under phosphate starvation (Elbahloul et al. 2005). During process optimization it was shown that the CGP production of *A. calcoaceticus* ADP1 during growth with arginine as a sole carbon and nitrogen source increased drastically to 41.4% (w/w) of CDM. Surprisingly, a combined supply of arginine and aspartate had a much lower stimulating effect as arginine alone (Elbahloul et al. 2005). In the following metabolic engineering studies of *A. calcoaceticus* ADP1, several genes related to the arginine biosynthesis were modified to yield higher amounts of arginine. In consequence, significant higher amounts of CGP could be detected (Elbahloul and Steinbüchel 2006).

Another link between the arginine biosynthesis and CGP metabolism was observed during transposon mutagenesis studies in the filamentous cyanobacterium *Nostoc ellipsosorum*. Here, an arginine biosynthesis gene, *argL*, was interrupted by a transposon, leading to a partially impaired arginine biosynthesis. Without arginine supplementation, heterocyst failed to fix nitrogen, akinetes were unable to germinate and CGP granules did not appear. However, when arginine is provided, mutant could form CGP granules and was able to differentiate functional akinetes (Leganes et al. 1998).

These results point towards arginine as main bottleneck of CGP biosynthesis, while aspartate plays a minor role. High concentrations of arginine surpass the *Km* value of CphA, triggering CGP production. In agreement, the *Km* value for arginine is much lower compared to aspartate, indicating a higher sensitivity of CphA towards arginine in the unicellular cyanobacterium *Synechocystis* sp. PCC 6803 (Aboulmagd et al. 2001a). Whether *CphA* is subjected to additional activity control remains to be elucidated (Watzer and Forchhammer 2018a, b).

7.3 Overproduction of Cyanophycin by Genetic Engineering the P_{II} Signaling System

Since arginine synthesis is a limiting factor for CGP accumulation, overproduction of CGP requires enhanced arginine synthesis. Arginine synthesis is strictly controlled by the nutritional status of the cyanobacteria, mainly through tight regulation of the committed step of ornithine synthesis, the N-acetylglutamate kinase (NAGK) reaction. The NAGK enzyme is feedback controlled by arginine levels, and in addition, NAGK activity and arginine-feedback inhibition are modulated by the P_{II} signal transduction protein. The P_{II} signaling protein responds to the effector molecules 2-oxoglutarate (which reports the C/N metabolic balance) and to the adenylnucleotides ATP or ADP, which compete for the same binding site. Thereby, the P_{II} signaling protein integrates the energy status of the cells (ATP/ADP ratio) with the C/N metabolic balance, where high 2-OG levels indicates carbon-sufficient but nitrogen-limiting conditions (for review see (Forchhammer and Lüddecke 2016; Forcada-Nadal et al. 2018). P_{II} in the ATP-complexed state binds to NAGK, thereby enhancing its catalytic activity and tuning down arginine-feedback inhibition. This activation by P_{II} results in an increased metabolic flux into the arginine pathway (Maheswaran et al. 2004; Maheswaran et al. 2006; Watzer et al. 2015). A single amino acid replacement in the cyanobacterial P_{II} signaling protein (Ile86 to Asp86) generated a P_{II} variant (P_{II}-I86N) that constitutively binds to NAGK in vitro and enhances its activity (Fokina et al. 2010). Replacing the wild-type allele for P_{II} in *Synechocystis* by this variant resulted in a strain, which has constitutively high NAGK activity and ten-fold increased levels of arginine (BW86) (Watzer et al. 2015). Already under balanced growth conditions with nitrate as a nitrogen source, strain BW86 accumulated up to 15.6 ± 5.4% CGP relative to the cell dry mass (CDM), while CGP in the wild type was barely detectable. Phosphate or potassium starvation further increased the CGP content of strain BW86 up to 47.4 ± 2.3% or 57.3 ± 11.1% per CDM, respectively (Watzer et al. 2015). The cyanophycin, which is produced by strain BW86, showed a high polydispersity ranging from 25 to 100 kDa, similar to the polydispersity of wild-type cyanobacterial CGP (Simon 1971). This contrasted CGP from recombinant producer strains using heterologous expression systems such as heterotrophic bacteria, yeasts, or plants, whose recombinantly produced CGP shows polydispersity ranging from 25 to only 45 kDa (Frey et al. 2002; Steinle et al. 2008). The novel cyanobacterial producer stain for cyanophycin may provide a valuable resource for the synthesis of large quantities of natural long-chain cyanophycin with so-far unexplored properties.

References

Aboulmagd E, Sanio FBO, Steinbüchel A (2001a) Purification of *Synechocystis* sp strain PCC6308 cyanophycin synthetase and its characterization with respect to substrate and primer specificity. Appl Environ Microb 67(5):2176–2182. https://doi.org/10.1128/AEM.67.5.2176-2182.2001

Aboulmagd E, Voss I, Oppermann-Sanio FB, Steinbüchel A (2001b) Heterologous expression of cyanophycin synthetase and cyanophycin synthesis in the industrial relevant bacteria *Corynebacterium glutamicum* and *Ralstonia eutropha* and in *Pseudomonas putida*. Biomacromolecules 2(4):1338–1342. https://doi.org/10.1021/bm010075a

Adams DG, Carr NG (1981) The developmental biology of heterocyst and akinete formation in cyanobacteria. Crc Rev Microbiol 9(1):45–100. https://doi.org/10.3109/10408418109104486

Allen MM, Hutchison F (1980) Nitrogen limitation and recovery in the cyanobacterium *Aphanocapsa* sp. PCC 6308. Arch Microbiol 128(1):1–7. https://doi.org/10.1007/Bf00422297

Allen MM, Weathers PJ (1980) Structure and composition of cyanophycin granules in the cyanobacterium *Aphanocapsa* 6308. J Bacteriol 141(2):959–962

Allen MM, Hutchison F, Weathers PJ (1980) Cyanophycin granule polypeptide formation and degradation in the cyanobacterium *Aphanocapsa* 6308. J Bacteriol 141(2):687–693

Arai T, Kino K (2008) A cyanophycin synthetase from *Thermosynechococcus elongatus* BP-1 catalyzes primer-independent cyanophycin synthesis. Appl Microbiol Biotechnol 81(1):69–78. https://doi.org/10.1007/s00253-008-1623-y

Bandyopadhyay A, Elvitigala T, Liberton M, Pakrasi HB (2013) Variations in the rhythms of respiration and nitrogen fixation in members of the unicellular diazotrophic cyanobacterial genus *Cyanothece*. Plant Physiol 161(3):1334–1346. https://doi.org/10.1104/pp.112.208231

Berg H (2003) Untersuchungen zu Funktion und Struktur der Cyanophycin-Synthetase von *Anabaena variabilis* ATCC 29413. Dissertation Humboldt–Universität zu Berlin, Germany

Berg H, Ziegler K, Piotukh K, Baier K, Lockau W, Volkmer-Engert R (2000) Biosynthesis of the cyanobacterial reserve polymer multi-L-arginyl-poly-L-aspartic acid (cyanophycin) – Mechanism of the cyanophycin synthetase reaction studied with synthetic primers. Eur J Biochem 267 (17):5561–5570. https://doi.org/10.1046/j.1432-1327.2000.01622.x

Borzi A (1887) Le communicazioni intracellulari delle *Nostochinee*. Malpighia 1:28–74

Burnat M, Flores E (2014) Inactivation of agmatinase expressed in vegetative cells alters arginine catabolism and prevents diazotrophic growth in the heterocyst-forming cyanobacterium *Anabaena*. Microbiology 3(5):777–792. https://doi.org/10.1002/mbo3.207

Burnat M, Herrero A, Flores E (2014) Compartmentalized cyanophycin metabolism in the diazotrophic filaments of a heterocystforming cyanobacterium. Proc Natl Acad Sci U S A 111 (10):3823–3828. https://doi.org/10.1073/pnas.1318564111

Burnat M, Picossi S, Valladares A, Herrero A, Flores E (2019) Catabolic pathway of arginine in Anabaena involves a novel bifunctional enzyme that produces proline from arginine. Mol Microbiol 111(4):883–897. https://doi.org/10.1111/mmi.14203

Carr N (1988) Nitrogen reserves and dynamic reservoirs in cyanobacteria. In: Rogers LJ, Gallon JR (eds) Biochemistry of the algae and cyanobacteria. Annual proceedings of the phyto-chemical society of Europe. Clarendon Press, Oxford, UK, pp 13–21

Curatti L, Flores E, Salerno G (2002) Sucrose is involved in the diazotrophic metabolism of the heterocyst-forming cyanobacterium *Anabaena* sp. FEBS Lett 513(2–3):175–178. https://doi.org/10.1016/S0014-5793(02)02283-4

Diniz SC, Voss I, Steinbüchel A (2006) Optimization of cyanophycin production in recombinant strains of *Pseudomonas putida* and *Ralstonia eutropha* employing elementary mode analysis and statistical experimental design. Biotechnol Bioeng 93(4):698–717. https://doi.org/10.1002/bit.20760

Du J, Li L, Zhou S (2019) Microbial production of cyanophycin: from enzymes to biopolymers. Biotechnol Adv 37(7):107400. https://doi.org/10.1016/j.biotechadv.2019.05.006

Elbahloul Y, Steinbüchel A (2006) Engineering the genotype of *Acinetobacter* sp strain ADP1 to enhance biosynthesis of cyanophycin. Appl Environ Microbiol 72(2):1410–1419. https://doi.org/10.1128/Aem.72.2.1410-1419.2006

Elbahloul Y, Krehenbrink M, Reichelt R, Steinbüchel A (2005) Physiological conditions conducive to high cyanophycin content in biomass of *Acinetobacter calcoaceticus* strain ADP1. Appl Environ Microbiol 71(2):858–866. https://doi.org/10.1128/Aem.71.2.858-866.2005

Fay P (1992) Oxygen relations of nitrogen-fixation in cyanobacteria. Microbiol Rev 56(2):340–373

Finzi-Hart JA, Pett-Ridge J, Weber PK, Popa R, Fallon SJ, Gunderson T, Hutcheon ID, Nealson KH, Capone DG (2009) Fixation and fate of C and N in the cyanobacterium *Trichodesmium* using nanometer-scale secondary ion mass spectrometry. Proc Natl Acad Sci U S A 106 (24):9931–9931. https://doi.org/10.1073/pnas.0904281106

Flores E, Herrero A, Forchhammer K, Maldener I (2016) Septal junctions in filamentous heterocyst-forming cyanobacteria. Trends Microbiol 24(2):79–82. https://doi.org/10.1016/j.tim.2015.11.011

Flores E, Picossi S, Valladares A, Herrero A (2019) Transcriptional regulation of development in heterocyst-forming cyanobacteria. Biochim Biophys Acta Gene Regul Mech 1862(7):673–684. https://doi.org/10.1016/j.bbagrm.2018.04.006

Flores E, Arevalo S, Burnat M (2019a) Cyanophycin and arginine metabolism in cyanobacteria. Algal Res 42:101577. https://doi.org/10.1016/j.algal.2019.101577

Fokina O, Chellamuthu VR, Zeth K, Forchhammer K (2010) A novel signal transduction protein P_{II} variant from *Synechococcus elongatus* PCC 7942 indicates a two-step process for NAGK-P_{II} complex formation. J Mol Biol 399(3):410–421. https://doi.org/10.1016/j.jmb.2010.04.018

Forcada-Nadal A, Llacer JL, Contreras A, Marco-Marin C, Rubio V (2018) The P_{II}-NAGK-PipX-NtcA regulatory axis of cyanobacteria: a tale of changing partners, allosteric effectors and non-covalent interactions. Front Mol Biosci 5:91. https://doi.org/10.3389/fmolb.2018.00091

Forchhammer K, Lüddecke J (2016) Sensory properties of the P_{II} signalling protein family. FEBS J 283(3):425–437. https://doi.org/10.1111/febs.13584

Forchhammer K, Watzer B (2016) Closing a gap in cyanophycin metabolism. Microbiology 162:727–729. https://doi.org/10.1099/mic.0.000260

Fredriksson C, Bergman B (1997) Ultrastructural characterisation of cells specialised for nitrogen fixation in a non-heterocystous cyanobacterium, *Trichodesmium* sp. Protoplasma 197 (1–2):76–85. https://doi.org/10.1007/Bf01279886

Frey KM, Oppermann-Sanio FB, Schmidt H, Steinbüchel A (2002) Technical-scale production of cyanophycin with recombinant strains of *Escherichia coli*. Appl Environ Microbiol 68 (7):3377–3384. https://doi.org/10.1128/AEM.68.7.3377-3384.2002

Frommeyer M, Wiefel L, Steinbüchel A (2016) Features of the biotechnologically relevant polyamide family "cyanophycins"' and their biosynthesis in prokaryotes and eukaryotes. Crit Rev Biotechnol 36(1):153–164. https://doi.org/10.3109/07388551.2014.946467

Füser G, Steinbüchel A (2007) Analysis of genome sequences for genes of cyanophycin metabolism: identifying putative cyanophycin metabolizing prokaryotes. Macromol Biosci 7 (3):278–296. https://doi.org/10.1002/mabi.200600207

Gallon J (1981) The oxygen sensitivity of nitrogenase: a problem for biochemists and microorganisms. Trends Biochem Sci 6:19–23. https://doi.org/10.1016/0968-0004(81)90008-6

Golden JW, Yoon HS (1998) Heterocyst formation in *Anabaena*. Curr Opin Microbiol 1 (6):623–629. https://doi.org/10.1016/S1369-5274(98)80106-9

Golden JW, Robinson SJ, Haselkorn R (1985) Rearrangement of nitrogen-fixation genes during heterocyst differentiation in the cyanobacterium *Anabaena*. Nature 314(6010):419–423. https://doi.org/10.1038/314419a0

Hai T, Oppermann-Sanio FB, Steinbüchel A (1999) Purification and characterization of cyanophycin and cyanophycin synthetase from the thermophilic *Synechococcus* sp. MA19. FEMS Microbiol. Lett 181(2):229–236. https://doi.org/10.1111/j.1574-6968.1999.tb08849.x

Hai T, Oppermann-Sanio FB, Steinbüchel A (2002) Molecular characterization of a thermostable cyanophycin synthetase from the thermophilic cyanobacterium *Synechococcus* sp strain MA19 and *in vitro* synthesis of cyanophycin and related polyamides. Appl Environ Microbiol 68 (1):93–101. https://doi.org/10.1128/Aem.68.1.93-101.2002

Hai T, Frey KM, Steinbüchel A (2006) Engineered cyanophycin synthetase (CphA) from *Nostoc ellipsosporum* confers enhanced CphA activity and cyanophycin accumulation to *Escherichia coli*. Appl Environ Microbiol 72(12):7652–7660. https://doi.org/10.1128/Aem.01132-06

Hejazi M, Piotukh K, Mattow J, Deutzmann R, Volkmer-Engert R, Lockau W (2002) Isoaspartyl dipeptidase activity of plant-type asparaginases. Biochem J 364:129–136. https://doi.org/10.1042/bj3640129

Herrero A, Burnat M (2014) Cyanophycin: a cellular nitrogen reserve material. In: Flores E, Herrero A (eds) The cell biology of cyanobacteria. Caister Academic Press, Norfolk, pp 211–220

Herrero A, Stavans J, Flores E (2016) The multicellular nature of filamentous heterocyst-forming cyanobacteria. FEMS Microbiol Rev 40(6):831–854. https://doi.org/10.1093/femsre/fuw029

Hill S, Kennedy C, Kavanagh E, Goldberg RB, Hanau R (1981) Nitrogen-fixation gene (Nifl) involved in oxygen regulation of nitrogenase synthesis in *K Pneumoniae*. Nature 290 (5805):424–426. https://doi.org/10.1038/290424a0

Jüttner F (1983) C-14-labeled metabolites in heterocysts and vegetative cells of *Anabaena cylindrica* filaments and their presumptive function as transport vehicles of organic-carbon and nitrogen. J Bacteriol 155(2):628–633

Khlystov NA, Chan WY, Kunjapur AM, Shi WC, Prather KU, Olsen BD (2017) Material properties of the cyanobacterial reserve polymer multi-L- arginyl-poly-L-aspartate (cyanophycin). Polymer 109:238–245. https://doi.org/10.1016/j.polymer.2016.11.058

Klemke F, Nürnberg DJ, Ziegler K, Beyer G, Kahmann U, Lockau W, Volkmer T (2016) CphA2 is a novel type of cyanophycin synthetase in N-2-fixing cyanobacteria. Microbiology 162:526–536. https://doi.org/10.1099/mic.0.000241

Klotz A, Georg J, Budinska L, Watanabe S, Reimann V, Januszewski W, Sobotka R, Jendrossek D, Hess WR, Forchhammer K (2016) Awakening of a dormant cyanobacterium from nitrogen chlorosis reveals a genetically determined program. Curr Biol 26(21):2862–2872. https://doi.org/10.1016/j.cub.2016.08.054

Krehenbrink M, Oppermann-Sanio FB, Steinbüchel A (2002) Evaluation of non-cyanobacterial genome sequences for occurrence of genes encoding proteins homologous to cyanophycin synthetase and cloning of an active cyanophycin synthetase from *Acinetobacter* sp strain DSM 587. Arch Microbiol 177(5):371–380. https://doi.org/10.1007/s00203-001-0396-9

Kumar K, Mella-Herrera RA, Golden JW (2010) Cyanobacterial heterocysts. Csh Perspect Biol 2 (4):a000315. https://doi.org/10.1101/cshperspect.a000315

Lang NJ, Simon RD, Wolk CP (1972) Correspondence of cyanophycin granules with structured granules in *Anabaena cylindrica*. Arch Mikrobiol 83(4):313–320. https://doi.org/10.1007/Bf00425243

Law AM, Lai SW, Tavares J, Kimber MS (2009) The structural basis of beta-peptide-specific cleavage by the serine protease cyanophycinase. J Mol Biol 392(2):393–404. https://doi.org/10.1016/j.jmb.2009.07.001

Leganes F, Fernandez-Pinas F, Wolk CP (1998) A transposition-induced mutant of *Nostoc ellipsosporum* implicates an arginine-biosynthetic gene in the formation of cyanophycin granules and of functional heterocysts and akinetes. Microbiology 144:1799–1805. https://doi.org/10.1099/00221287-144-7-1799

Li H, Sherman DM, Bao SL, Sherman LA (2001) Pattern of cyanophycin accumulation in nitrogen-fixing and non-nitrogen-fixing cyanobacteria. Arch Microbiol 176(1–2):9–18. https://doi.org/10.1007/s002030100281

Lopez-Igual R, Flores E, Herrero A (2010) Inactivation of a heterocyst-specific invertase indicates a principal role of sucrose catabolism in heterocysts of *Anabaena* sp. J Bacteriol 192 (20):5526–5533. https://doi.org/10.1128/Jb.00776-10

Luque I, Forchhammer K (2008) Nitrogen assimilation and C/N balance sensing. In: Herrero A, Flores E (eds) The cyanobacteria: molecular biology, genomics and evolution. Caister Academic Press, Norfolk, pp 335–382

Magnuson A, Cardona T (2016) Thylakoid membrane function in heterocysts. BBA-Bioenergetics 1857(3):309–319. https://doi.org/10.1016/j.bbabio.2015.10.016

Maheswaran M, Urbanke C, Forchhammer K (2004) Complex formation and catalytic activation by the P_{II} signaling protein of N-acetyl-L-glutamate kinase from *Synechococcus elongatus* strain PCC 7942. J Biol Chem 279(53):55202–55210. https://doi.org/10.1074/jbc.M410971200

Maheswaran M, Ziegler K, Lockau W, Hagemann M, Forchhammer K (2006) P_{II}-regulated arginine synthesis controls accumulation of cyanophycin in *Synechocystis* sp strain PCC 6803. J Bacteriol 188(7):2730–2734. https://doi.org/10.1128/Jb.188.7.2730-2734.2006

Maldener I, Muro-Pastor AM (2010) Cyanobacterial heterocysts. In: eLS. Wiley, Chichester. https://doi.org/10.1002/9780470015902.a0000306.pub2

Maldener I, Summers ML, Sukenik A (2014) Cellular differentiation in filamentous cyanobacteria. In: Flores E, Herrero A (eds) The cell biology of cyanobacteria. Caister Academic Press, Norfolk, pp 263–291

Martin-Figueroa E, Navarro F, Florencio FJ (2000) The GS-GOGAT pathway is not operative in the heterocysts. Cloning and expression of *glsF* gene from the cyanobacterium *Anabaena* sp PCC 7120. FEBS Lett 476(3):282–286. https://doi.org/10.1016/S0014-5793(00)01722-1

Merritt MV, Sid SS, Mesh L, Allen MM (1994) Variations in the amino acid composition of cyanophycin in the cyanobacterium *Synechocystis* sp. PCC 6308 as a function of growth conditions. Arch Microbiol 162(3):158–166

Miller SR, Castenholz RW (2000) Evolution of thermotolerance in hot spring cyanobacteria of the genus *Synechococcus*. Appl Environ Microbiol 66(10):4222–4229. https://doi.org/10.1128/aem.66.10.4222-4229.2000

Mitschke J, Georg J, Scholz I, Sharma CM, Dienst D, Bantscheff J, Voss B, Steglich C, Wilde A, Vogel J, Hess WR (2011) An experimentally anchored map of transcriptional start sites in the model cyanobacterium *Synechocystis* sp PCC 6803. Proc Natl Acad Sci U S A 108 (5):2124–2129. https://doi.org/10.1073/pnas.1015154108

Nausch H, Huckauf J, Broer I (2016) Peculiarities and impacts of expression of bacterial cyanophycin synthetases in plants. Appl Microbiol Biotechnol 100(4):1559–1565. https://doi.org/10.1007/s00253-015-7212-y

Nicolaisen K, Mariscal V, Bredemeier R, Pernil R, Moslavac S, Lopez-Igual R, Maldener I, Herrero A, Schleiff E, Flores E (2009) The outer membrane of a heterocyst-forming cyanobacterium is a permeability barrier for uptake of metabolites that are exchanged between cells. Mol Microbiol 74(1):58–70. https://doi.org/10.1111/j.1365-2958.2009.06850.x

Obst M, Steinbüchel A (2004) Microbial degradation of poly(amino acid)s. Biomacromolecules 5 (4):1166–1176. https://doi.org/10.1021/bm049949u

Obst M, Steinbüchel A (2006) Cyanophycin—an ideal bacterial nitrogen storage material with unique chemical properties. In: Shively JM (ed) Inclusions in prokaryotes. Springer, Berlin, Heidelberg, pp 167–193. https://doi.org/10.1007/3-540-33774-1_7

Obst M, Oppermann-Sanio FB, Luftmann H, Steinbüchel A (2002) Isolation of cyanophycin-degrading bacteria, cloning and characterization of an extracellular cyanophycinase gene (*cphE*) from *Pseudomonas anguilliseptica* strain BI. The *cphE* gene from *P anguilliseptica* BI encodes a cyanophycin hydrolyzing enzyme. J Biol Chem 277(28):25096–25105. https://doi.org/10.1074/jbc.M112267200

Obst M, Sallam A, Luftmann H, Steinbüchel A (2004) Isolation and characterization of gram-positive cyanophycin-degrading bacteria – kinetic studies on cyanophycin depolymerase activity in aerobic bacteria. Biomacromolecules 5(1):153–161. https://doi.org/10.1021/bm034281p

Obst M, Krug A, Luftmann H, Steinbüchel A (2005) Degradation of cyanophycin by *Sedimentibacter hongkongensis* strain KI and *Citrobacter amalonaticus* strain G isolated from an anaerobic bacterial consortium. Appl Environ Microbiol 71(7):3642–3652. https://doi.org/10.1128/Aem.71.7.3642-3652.2005

Perez R, Forchhammer K, Salerno G, Maldener I (2016) Clear differences in metabolic and morphological adaptations of akinetes of two *Nostocales* living in different habitats. Microbiology 162:214–223. https://doi.org/10.1099/mic.0.000230

Perez R, Wörmer L, Sass P, Maldener I (2018) A highly asynchronous developmental program triggered during germination of dormant akinetes of filamentous diazotrophic cyanobacteria. FEMS Microbiol Ecol 94(1):fix131. https://doi.org/10.1093/femsec/fix131

Picossi S, Valladares A, Flores E, Herrero A (2004) Nitrogen-regulated genes for the metabolism of cyanophycin, a bacterial nitrogen reserve polymer – Expression and mutational analysis of two cyanophycin synthetase and cyanophycinase gene clusters in the heterocyst-forming cyanobacterium *Anabaena* sp PCC 7120. J Biol Chem 279(12):11582–11592. https://doi.org/10.1074/jbc.M311518200

Richter R, Hejazi M, Kraft R, Ziegler K, Lockau W (1999) Cyanophycinase, a peptidase degrading the cyanobacterial reserve material multi-L-arginyl-poly-L-aspartic acid (cyanophycin) – Molecular cloning of the gene of *Synechocystis* sp PCC 6803, expression in *Escherichia coli*, and biochemical characterization of the purified enzyme. Eur J Biochem 263(1):163–169. https://doi.org/10.1046/j.1432-1327.1999.00479.x

Rippka R, Deruelles J, Waterbury JB, Herdman M, Stanier RY (1979) Generic assignments, strain histories and properties of pure cultures of cyanobacteria. J Gen Microbiol 111(Mar):1–61. https://doi.org/10.1099/00221287-111-1-1

Sallam A, Steinbüchel A (2008) Anaerobic and aerobic degradation of cyanophycin by the denitrifying bacterium *Pseudomonas alcaligenes* strain DIP1 and role of three other coisolates in a mixed bacterial consortium. Appl Environ Microbiol 74(11):3434–3443. https://doi.org/10.1128/Aem.02575-07

Sallam A, Steinbüchel A (2010) Dipeptides in nutrition and therapy: cyanophycin-derived dipeptides as natural alternatives and their biotechnological production. Appl Microbiol Biotechnol 87(3):815–828. https://doi.org/10.1007/s00253-010-2641-0

Sallam A, Kast A, Przybilla S, Meiswinkel T, Steinbüchel A (2009a) Biotechnological process for production of beta-dipeptides from cyanophycin on a technical scale and its optimization. Appl Environ Microbiol 75(1):29–38. https://doi.org/10.1128/AEM.01344-08

Sallam A, Steinle A, Steinbüchel A (2009b) Cyanophycin: biosynthesis and applications. In: Rehm BHA (ed) Microbial production of biopolymers and polymer precursors. Caister Academic Press, Norfolk, pp 79–99

Sherman LA, Meunier P, Colon-Lopez MS (1998) Diurnal rhythms in metabolism: a day in the life of a unicellular, diazotrophic cyanobacterium. Photosynth Res 58(1):25–42. https://doi.org/10.1023/A:1006137605802

Sherman DM, Tucker D, Sherman LA (2000) Heterocyst development and localization of cyanophycin in N2-fixing cultures of *Anabaena* sp PCC 7120 (cyanobacteria). J Phycol 36 (5):932–941. https://doi.org/10.1046/j.1529-8817.2000.99132.x

Simon RD (1971) Cyanophycin granules from the blue-green alga *Anabaena cylindrica*: a reserve material consisting of copolymers of aspartic acid and arginine. Proc Natl Acad Sci U S A 68 (2):265–267. https://doi.org/10.1073/pnas.68.2.265

Simon RD (1973) The effect of chloramphenicol on the production of cyanophycin granule polypeptide in the blue green alga *Anabaena cylindrica*. Arch Mikrobiol 92(2):115–122. https://doi.org/10.1007/bf00425009

Simon RD (1976) The biosynthesis of multi-L-arginyl-poly(L-aspartic acid) in the filamentous cyanobacterium *Anabaena cylindrica*. Biochim Biophys Acta 422(2):407–418. https://doi.org/10.1016/0005-2744(76)90151-0

Simon RD, Weathers P (1976) Determination of the structure of the novel polypeptide containing aspartic acid and arginine which is found in cyanobacteria. Biochim Biophys Acta Protein Struct 420(1):165–176. https://doi.org/10.1016/0005-2795(76)90355-x

Simon RD, Lawry NH, McLendon GL (1980) Structural characterization of the cyanophycin granule polypeptide of *Anabaena cylindrica* by circular dichroism and Raman spectroscopy. Biochim Biophys Acta 626(2):277–281. https://doi.org/10.1016/0005-2795(80)90121-x

Steinle A, Oppermann-Sanio FB, Reichelt R, Steinbüchel A (2008) Synthesis and accumulation of cyanophycin in transgenic strains of *Saccharomyces cerevisiae*. Appl Environ Microbiol 74 (11):3410–3418. https://doi.org/10.1128/AEM.00366-08

Steinle A, Bergander K, Steinbüchel A (2009) Metabolic engineering of *Saccharomyces cerevisiae* for production of novel cyanophycins with an extended range of constituent amino acids. Appl Environ Microbiol 75(11):3437–3446. https://doi.org/10.1128/Aem.00383-09

Steinle A, Witthoff S, Krause JP, Steinbüchel A (2010) Establishment of cyanophycin biosynthesis in *Pichia pastoris* and optimization by use of engineered cyanophycin synthetases. Appl Environ Microbiol 76(4):1062–1070. https://doi.org/10.1128/AEM.01659-09

Sutherland JM, Reaston J, Stewart WDP, Herdman M (1985) Akinetes of the cyanobacterium *Nostoc* PCC 7524 – Macromolecular and biochemical-changes during syndronous germination. J Gen Microbiol 131:2855–2863. https://doi.org/10.1099/00221287-131-11-2855

Thomas J, Meeks J, Wolk CP, Shaffer P, Austin S (1977) Formation of glutamine from [13n] ammonia, [13n] dinitrogen, and [14C] glutamate by heterocysts isolated from *Anabaena cylindrica*. J Bacteriol 129(3):1545–1555

Vargas WA, Nishi CN, Giarrocco LE, Salerno GL (2011) Differential roles of alkaline/neutral invertases in *Nostoc* sp. PCC 7120: Inv-B isoform is essential for diazotrophic growth. Planta 233(1):153–162. https://doi.org/10.1007/s00425-010-1288-5

Voss I, Steinbüchel A (2006) Application of a KDPG-aldolase gene-dependent addiction system for enhanced production of cyanophycin in *Ralstonia eutropha* strain H16. Metab Eng 8(1):66–78. https://doi.org/10.1016/j.ymben.2005.09.003

Voss I, Diniz SC, Aboulmagd E, Steinbüchel A (2004) Identification of the *Anabaena* sp. strain PCC 7120 cyanophycin synthetase as suitable enzyme for production of cyanophycin in gram-negative bacteria like *Pseudomonas putida* and *Ralstonia eutropha*. Biomacromolecules 5 (4):1588–1595. https://doi.org/10.1021/bm049861g

Walsby AE (2007) Cyanobacterial heterocysts: terminal pores proposed as sites of gas exchange. Trends Microbiol 15(8):340–349. https://doi.org/10.1016/j.tim.2007.06.007

Watzer B, Forchhammer K (2018a) Cyanophycin synthesis optimizes nitrogen utilization in the unicellular cyanobacterium *Synechocystis* sp strain PCC 6803. Appl Environ Microbiol 84(20): e01298-18. https://doi.org/10.1128/AEM.01298-18

Watzer B, Forchhammer K (2018b) Cyanophycin: a nitrogen-rich reserve polymer. In: Tiwari A (ed) Cyanobacteria. IntechOpen, London, 85–107, https://doi.org/10.5772/intechopen.77049

Watzer B, Engelbrecht A, Hauf W, Stahl M, Maldener I, Forchhammer K (2015) Metabolic pathway engineering using the central signal processor P_{II}. Microb Cell Factories 14:192. https://doi.org/10.1186/s12934-015-0384-4

Weiss GL, Kieninger AK, Maldener I, Forchhammer K, Pilhofer M (2019) Structure and function of a bacterial gap junction analog. Cell 178(2):374–384. https://doi.org/10.1016/j.cell.2019.05.055

Whitton BA (2012) Ecology of cyanobacteria II: their diversity in space and time. Springer, New York

Wiefel L, Steinbüchel A (2014) Solubility behavior of cyanophycin depending on lysine content. Appl Environ Microbiol 80(3):1091–1096. https://doi.org/10.1128/Aem.03159-13

Wiefel L, Bröker A, Steinbüchel A (2011) Synthesis of a citrulline-rich cyanophycin by use of *Pseudomonas putida* ATCC 4359. Appl Microbiol Biotechnol 90(5):1755–1762. https://doi.org/10.1007/s00253-011-3224-4

Wiefel L, Wohlers K, Steinbüchel A (2019) Re-evaluation of cyanophycin synthesis in Corynebacterium glutamicum and incorporation of glutamic acid and lysine into the polymer. Appl Microbiol Biotechnol 103(10):4033–4043. https://doi.org/10.1007/s00253-019-09780-5

Wolk CP (1968) Movement of carbon from vegetative cells to heterocysts in *Anabaena cylindrica*. J Bacteriol 96(6):2138–2143

Wolk CP (1996) Heterocyst formation. Annu Rev Genet 30(1):59–78. https://doi.org/10.1146/annurev.genet.30.1.59

Wolk CP, Austin SM, Bortins J, Galonsky A (1974) Autoradiographic localization of N-13 after fixation of N-13-labeled nitrogen gas by a heterocyst-forming blue-green-alga. J Cell Biol 61 (2):440–453. https://doi.org/10.1083/jcb.61.2.440

Wolk CP, Ernst A, Elhai J (1994) Heterocyst metabolism and development. In: Bryant DA (ed) The molecular biology of cyanobacteria. Advances in photosynthesis, vol 1. Springer, Dordrecht, pp 769–823

Zhang H, Liu Y, Nie X, Liu L, Hua Q, Zhao GP, Yang C (2018) The cyanobacterial ornithine-ammonia cycle involves an arginine dihydrolase. Nat Chem Biol 14(6):575–581. https://doi.org/10.1038/s41589-018-0038-z

Ziegler K, Diener A, Herpin C, Richter R, Deutzmann R, Lockau W (1998) Molecular characterization of cyanophycin synthetase, the enzyme catalyzing the biosynthesis of the cyanobacterial reserve material multi-L-arginyl-poly-L-aspartate (cyanophycin). Eur J Biochem 254 (1):154–159. https://doi.org/10.1046/j.1432-1327.1998.2540154.x

Ziegler K, Stephan DP, Pistorius EK, Ruppel HG, Lockau W (2001) A mutant of the cyanobacterium *Anabaena variabilis* ATCC 29413 lacking cyanophycin synthetase: growth properties and ultrastructural aspects. FEMS Microbiol Lett 196(1):13–18. https://doi.org/10.1016/S0378-1097(01)00027-1

Ziegler K, Deutzmann R, Lockau W (2002) Cyanophycin synthetase-like enzymes of non-cyanobacterial eubacteria: characterization of the polymer produced by a recombinant synthetase of *Desulfitobacterium hafniense*. Zeitschrift Für Naturforschung Section C-a J Biosci 57(5–6):522–529. https://doi.org/10.1515/znc-2002-5-621

Storage Polysaccharides in Prokaryotes: Glycogen, Granulose, and Starch-Like Granules

Matthieu Colpaert, Malika Chabi, Ugo Cenci, and Christophe Colleoni

Contents

1 Introduction 178
2 Glycogen Particles, Starch-Like Granules, and Granulose in Prokaryotes 179
3 Three Biosynthesis Glycogen Pathways in Prokaryotes 183
3.1 Nucleotide-Sugar-Based Glycogen Biosynthetic Pathway in Bacteria: GlgC Pathway 183
3.2 Alternative Glycogen Biosynthesis Pathways in Prokaryotes: The GlgE Pathway . 190
3.3 Other Glycogen Biosynthesis Pathways in Prokaryotes: The MalQ Path 191
4 Glycogen Catabolic Paths in Prokaryotes 192
4.1 Glycogen Phosphorylase 192
4.2 Glycogen Debranching Enzyme 194
5 Starch Granules in Unicellular Diazotrophic Cyanobacteria: Convergent Evolution of Glucan Trimming Mechanism in the Starch Biosynthetic Pathway of *Cyanobacterium* sp. CLg1 195
5.1 Why Do Unicellular Nitrogen-Fixing Cyanobacteria Synthesize Starch-Like Granules Rather Than Glycogen? 197
6 Granulose: Another Example of Transition from Soluble to Insoluble Storage Polysaccharide? 199
6.1 *Firmicutes* Phylum: *Clostridium* sp. 199
6.2 *Candidatus Methylacidiphilum fumariolicum* 200
7 Storage Polysaccharide and Fitness in Prokaryotes 200
7.1 The Accumulation of Abnormal Glycogen Affects the Aging of Prokaryotes 202
8 Conclusion and Perspectives 203
References 204

Abstract Homopolymers of D-glucose represent the most successful and abundant polysaccharides found in nature. In this chapter, we will focus on α-glucan polysaccharides in particular glycogen and its derivatives (i.e., granulose, starch) that define probably one of the oldest forms of carbon storage among prokaryotes. They are made of linear chains of glucosyl units joined by α-1,4 glycosidic bonds and

M. Colpaert · M. Chabi · U. Cenci · C. Colleoni (✉)
Univ. Lille, CNRS, UMR8576-UGSF-Unité de Glycobiologie Structurale et Fonctionnelle, Lille, France
e-mail: christophe.colleoni@univ-lille1.fr

D. Jendrossek (ed.), *Bacterial Organelles and Organelle-like Inclusions*, Microbiology Monographs 34, https://doi.org/10.1007/978-3-030-60173-7_8

hooked to each other by α-1,6 glycosidic bonds, referred to as branching points. The glucan chains are organized in such manner that α-glucan polysaccharides appear mostly in the form of tiny hydrosoluble or insoluble water particles in the cytosol of bacteria. Because "Nothing in biology makes sense except in the light of evolution" to quote Dobzhansky, this chapter aims to emphasize the importance of structure-function relationship that determines the physicochemical and biochemical properties of storage polysaccharides in microorganisms.

1 Introduction

Energy homeostasis is a critical issue for any living organism. Prior to the emergence of energy-carbon-based storage compounds, several reports speculate that polyphosphate granules were probably the first form of energy storage compound that evolved in the prebiotic history of life (Achbergerová and Nahálka 2011; Albi and Serrano 2016; Piast and Wieczorek 2017).

In the prokaryotic world, energy-carbon or energy storage compounds have evolved in various forms such as polysaccharides, lipids, poly-β-hydroxybutyrate (PHB)/polyhydroxyalkanoate (PHA) and polyphosphate. Despite their distinct physicochemical properties, energy-carbon compounds share specific features such as a high molecular weight in order to reduce internal osmotic pressure within cells, and they are produced when there is an excess of energy or are degraded for the maintenance of the cells in times of starvation. For this reason, trehalose and sucrose might not be considered sensu stricto as energy-carbon storage compounds in prokaryotes but more likely as transitory carbohydrate compounds produced in response to environmental stresses (e.g., cold, salt).

In the present book chapter, we will confine our attention to glycogen, granulose, and starch-like granules, polysaccharides composed of chains of α-1,4 linked D-glucosyl units that are interconnected through α-1,6 linkages or branching points. We will exclude consideration of other polysaccharides such as dextrans, fructans, PHB/PHA, and polyphosphate. Nevertheless, energy storage functions of PHB/PHA and polyphosphate are detailed in recent reviews (Jendrossek and Pfeiffer 2014; Albi and Serrano 2016). Glycogen particles are ubiquitous in the tree life that encompasses the *Archaea*, *Bacteria*, and *Eukaryota* domains. We have to keep in mind that most of our major understanding of glycogen metabolism pathways essentially relies on the studies in *Escherichia coli* and yeast *Saccharomyces cerevisiae*, which are the usual representatives of prokaryotes and heterotroph eukaryotes, respectively (for review see Wilson et al. 2010). Nevertheless early electron microscopy observations point out that some prokaryotes substitute hydrosoluble glycogen particles by abnormal glycogen particles such as starch-like granules described in a small group of cyanobacteria or granuloses synthesized in *Clostridium* species. Fortunately, the improvement of high-throughput genome sequencing brings some insight in storage polysaccharide metabolism network. Thus, the plethora of bacterial genomes outlines the diversity of carbon storage pathways as well as the fact that some bacterial species have lost the ability to synthesize carbohydrate storage polysaccharides. The aim of this chapter is to give an overview of basic and alternative glycogen metabolism pathways in prokaryotes and how storage polysaccharide impacts cell physiology.

2 Glycogen Particles, Starch-Like Granules, and Granulose in Prokaryotes

Although glycogen, starch, starch-like granules, and granulose are composed of glucan chains made of glucose residues linked in α-1,4 position and hooked together by α-1,6 linkages, the organization of glucan chains inside polysaccharides gives rise to distinct properties as resumed in Table 1. The glycogen synthesis occurs in the cytosol of prokaryotes and eukaryotic organisms as tiny hydrosoluble particles of 40–60 nm. Detailed structural analyses confirm that glycogen particles from eukaryotes and prokaryotes are very similar in terms of glucan chain content (chain length distribution is monomodal with a major frequency of glucan chain of 6–7 units glucose) as well as high percentage of branching points (9–12% that correspond to an average of two α-1,6-linkages per glucan chains as determined in Melendez-Hevia et al. 1993). Interestingly, taking into account these parameters, mathematical modeling predicts a size of 42 nm in granulose granule diameter and explains this self-limitation due to homogenous distribution of branching points. This particular organization leads quickly to an exponential increase in the number of non-reducing ends at the surface of polysaccharide (Fig. 1a) and then will limit the access to the catalytic sites of biosynthetic enzymes and per se the size of glycogen particles (Meléndez-Hevia et al., 1993; Meléndez et al. 1999). Interestingly, further mathematical modeling infers that glycogen particles can be considered as fractal objects, made by iteration of a single motif: one glucan chain supporting two branches. Thanks to this particular glucan chain organization, a single catabolic enzyme can release 19,000 glucose molecules in 20 s (Meléndez et al. 1997). Hence, glycogen particles are dynamic polysaccharides, which can meet the immediate need of eukaryotic and prokaryotic cells and can be used as an intermediate buffer or osmotically inert sink for carbon (Table 1).

On the contrary, starch-like granules and granulose are not soluble in water, and granule sizes are above 42 nm. We know from the structural characterization of starch-like granules produced by two unicellular diazotrophic cyanobacteria *Cyanobacterium* sp. CLg1 and *Cyanothece* ATCC51142 that those granules are similar to starch granules found in plants. Those semi-crystalline nonaqueous polysaccharides are composed of two polysaccharides: amylopectin and amylose. The former, the major fraction, is a branched polysaccharide containing 5% of branching points, while amylose, a dispensable fraction of starch, is mostly composed of slightly

Table 1 Properties of storage polysaccharides

Organisms	Polysaccharides	Size	Solubility in water	Semi-crystallinity	Polysaccharide fraction	% α-1,6	DP	MW (×10^6)	References
Bacteria	Glycogen	30–60 nm	Yes	No	100%	9–12%	6–7	2–13	Park et al. (2011)
Archaea									Meléndez et al. (1997)
Eukaryota									
Archaeplastida	Starch/starch-like	1–100µm	No	Yes	≈75% amylopectin	5–6%	11–12	10–100	Vamadevan and Bertoft (2014)
Cyanobacteria					≈25% amylose	<1%	900–3300		Nakamura et al. (2005)
Clostridium sp.	Granulose	210–270 nm	No	N.D.	N.D.	N.D	N.D	N.D	Tracy et al. (2008)
Methylacidiphilum fumariolicum	Abnormal glycogen	100–200 nm							Khadem et al. (2010)

Glycogen, starch/starch-like, granulose, and abnormal glycogen are composed of α-1,4 linked glucose residues that are interconnected through α-1,6 linkages. Glycogen is the main storage form of carbon storage in living organisms. Starch and starch-like granules have been reported in Archaeplastida lineages (plants/algae, red algae, and glaucophytes) and in a small group of unicellular nitrogen-fixing cyanobacteria. To date, granulose and abnormal glycogen have been observed in *Clostridium* sp. and *Candidatus Methylacidiphilum fumariolicum*. Despite sharing the primary structure, those homopolymers of glucose harbor various physicochemical properties in terms of size, solubility in water, and crystallinity. Starch and starch-like granules contain two polysaccharides: amylopectin fraction and branched polysaccharide that represents up to 75% (w/w) of starch granules and amylose fraction, mostly composed of long linear glucan chains. The organization of glucan chains inside polysaccharides and the frequency of branching point (α-1,6), the maximum abundance of glucan chain (degree of polymerization (DP) represents the number of glucose residues that composes glucan chain), explain the physicochemical properties of polysaccharides

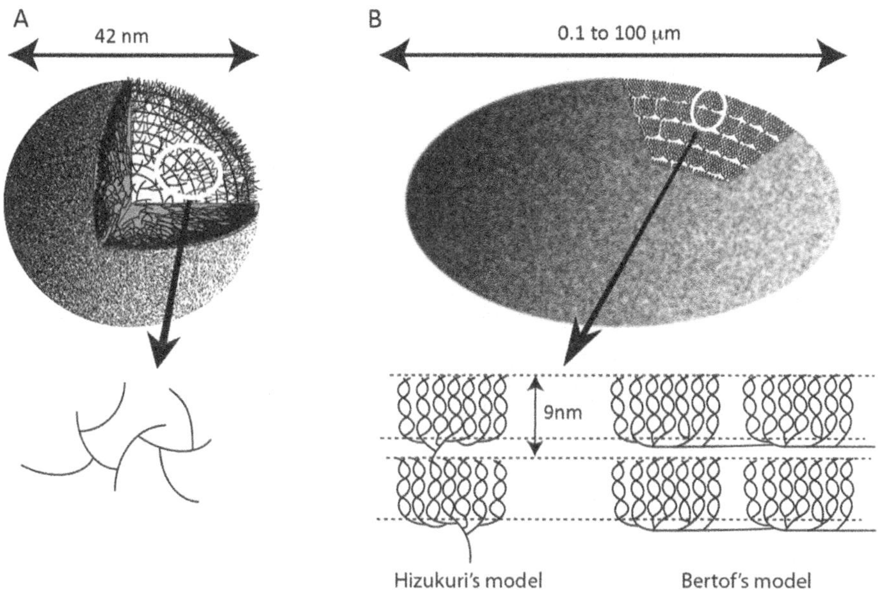

Fig. 1 Structural comparison between glycogen and amylopectin. Both polysaccharides are composed of glucan chains made of α-1,4 glucose residues and branched in α-1,6 position. (**a**) The uniform distribution of branches leads to an exponential increase that self-limits glycogen particle size to 42 nm. Mathematical modeling suggests that glycogen particles are fractal objects made of repeating patterns of glucan chains (black lines) harboring two glucan chains (intersection lines). (**b**) The size of semi-crystalline starch granule is variable depending on the source. Clusters of amylopectin are generated through the asymmetric distribution of branches, localized in the amorphous lamellae, while the intertwined glucan chains define crystalline lamellae. The sum of one amorphous and crystalline lamella is constant to 9 nm independently of amylopectin clusters examined. In the model proposed by Hizukuri (1986), a long glucan chain interconnects two clusters, whereas in the alternate model of Bertoft (2004), the clusters are anchored to a backbone consisting of a long glucan chain

branched (<1%) linear glucan chains. First proposed by Hizukuri (1986) and then modified by Bertoft (2004), these models propose a specific localization of branching points allowing the formation of double helices of glucan (Fig. 1b). Thus, this particular "cluster" organization allows more efficient storage of glucose residues without the size limitations imposed on glycogen particles. However, the downside is that synthesis and degradation of starch are more complex and less dynamic. To our knowledge, structural characterization of granulose has been limited to the measurement of absorbency of iodine-granulose complex, which appears similar to iodine-amylopectin complex (see below). Data of wide-angle X-ray diffraction analysis should discriminate between the semi-crystallinity and amorphous nature of granulose.

Table 2 Distribution and number of isoforms involved in the storage polysaccharide pathway in cyanobacteria

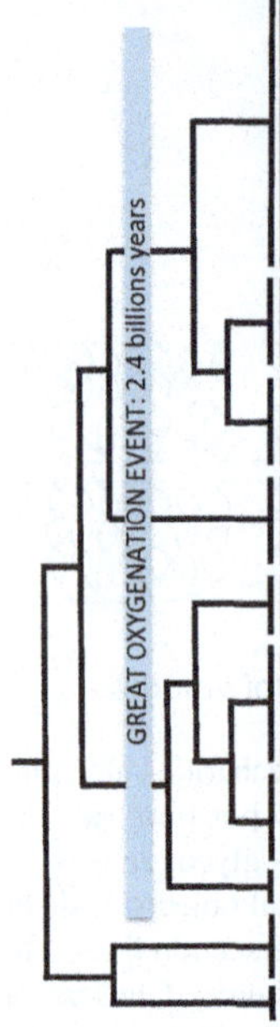

	Diazotroph	Glg C	GlgA1	GlgA2	GBSS	GlgB-GH13	GlgX/TreX-GH13_11	Amp-GH13_20	amylase-GH13	TreY-GH13_26	TreZ-GH13_10	DBE-GH133	GlgP/MalP-GT35	MalQ-GH77	BE-GH57	Amp-GH57	GH-GH57	Susy
Cyanobacterium sp. CLg1★	Y	1	1	1	1	3	2	1	2	-	-	1	1	1	1	1	-	-
C. watsonii WH8501★	Y	1	1	1	1	3	1	1	2	-	-	1	1	1	1	1	-	-
Cyanobacterium MIBC10216★	Y	1	1	1	-	3	-	1	1	-	-	-	2	1	1	1	-	-
Cyanothece CCY0110★	Y	2	1	1	-	3	1	1	3	-	1	-	1	1	1	1	1	-
Cyanothece ATCC51142★	Y	2	1	1	-	3	1	1	4	-	-	1	3	1	1	1	1	-
Cyanothece PCC8801★	Y	1	1	1	-	2	3	1	1	1	1	1	2	1	1	1	-	-
Cyanothece PCC7424	Y*	1	1	1	-	2	2	1	1	1	1	1	4	1	1	1	1	1
Synechococcus PCC7002	N	1	1	1	-	1	-	-	1	-	-	-	2	1	1	1	-	-
M. aeruginosa NIES843	N	1	1	1	-	1	1	-	1?	-	-	1	3	1	1	1	1	1
Synechocystis PCC6803	N	1	1	1	-	1	2	1	1	-	-	-	2	1	1	1	-	-
Oscillatoria sp. PCC6506	Y	1	1	-	-	1	1	1	-	1	1	1	2	1	1	-	1	-
A.platensis NIES39	Y	1	1	1	-	1	2	1	1	1	1	1	2	3	1	-	-	1
T.erythraeum IMS101	Y	1	1	-	-	1	3	1	1	-	-	1	2	1	1	-	-	-
Nostoc PCC7120	Y	1	1	1	-	1	-	1	1	1	-	1	2	1	1	1	1	2
Anabaena ATCC29413	Y	1	1	1	-	1	-	1	-	1	1	1	2	1	1	1	1	2
N.punctiforme PCC73102	Y	1	1	-	-	1	-	1	2?	1	1	1	1	1	1	1	1	1
A.marina MBIC11017	N	1	1	1	-	1	-	1	2	-	-	1	2	1	1	1	1	1
T.elongatus BP1	N	1	1	-	-	1	-	-	-	1	-	-	3	1	1	1	1	1
Cyanothece PCC7425	Y	1	1	1	-	1	-	1	-	-	-	1	4	1	1	1	1	2
S.elongatus PCC7942	N	1	1	-	-	1	1	-	-	-	-	1	1	1	1	-	-	-
S.elongatus PCC6301	N	1	1	-	-	1	1	-	-	-	-	1	1	1	1	-	-	-
Synechococcus CC9605	N	1	1	-	-	1	2	1	-	-	-	-	1	1	1	-	-	-
Synechococcus CC9311	N	1	1	-	-	1	1	1	-	-	-	-	1	1	1	-	-	-
Synechococcus CC9902	N	1	1	-	-	1	1	1	-	-	-	-	1	1	1	-	-	-
Synechococcus RCC307	N	1	1	-	-	1	1	1	-	-	-	-	1	1	1	-	-	-
P.maritimus MIT 9303	N	1	1	-	-	1	1	1	-	-	-	-	1	1	1	-	-	-
P.maritimus MIT9313	N	1	1	-	-	1	1	1	-	-	-	-	1	1	1	-	-	-
P.maritimus MIT 9312	N	1	1	-	-	1	1	-	-	-	-	-	1	1	1	-	-	-
Leptolyngbya sp. PCC7376	Y	1	1	1	-	1	-	1	1	-	-	-	1	1	1	-	1	-
Synechococcus JA-3-3Ab	Y	1	1	1	-	1	-	-	1	1	1	1	2	1	1	1	1	-
G.violaceus PCC7421	N	1	1	-	-	1	-	1	1?	-	-	1	4	1	1	1	-	1
G. lithophora alchichica D10	N	1	1	-	-	1	1	-	-	-	-	1	3	1	1	1	1	-
E.coli K12 MG1655	-	1	1	-	-	1	1	-	2	-	-	-	2	1	-	-	-	-

Black lines depict the evolutionary relationship between cyanobacterial species in respect to the Great Oxygenation Event (GOE). Representatives of each cyanobacteria clade were selected based on the availability of genome sequences. The number of enzyme isoforms was determined using Blast searches on NCBI, CyanoBase, and CAZy. Cyanobacteria strains labeled with a star synthesize starch-like granules. The ability of nitrogen fixation in aerobic conditions is indicated by Y (Yes) or N (No) and by Y* (Yes) in anoxic conditions. Key: ADP-glucose pyrophosphorylase (GlgC); glycogen synthase isoforms (GlgA1, GlgA2); granule-bound starch synthase (GBSS); branching enzyme GH13 family (GlgB-GH13); debranching enzyme GH13 subfamily 11 (GlgX/TreX-GH13_11); amylopullulanase-GH13 subfamily 20 (Amp-GH13_20); amylase-GH13; maltooligosyltrehalose synthase GH13 subfamily 26 (TreY-GH13_26); maltooligosyltrehalose hydrolase GH13 subfamily 10 (TreZ-GH13_10); debranching enzyme or amylo-1,6 glucosidase (DBE-GH133); glycogen phosphorylase/maltodextrin phosphorylase (GlgP/MalP-GT35); α-1,4 glucanotransferase (MalQ-GH77); putative branching enzyme GH57 family (BE-GH57); putative amylopullulanase GH57 family (Amp-GH57); putative glycosyl hydrolase family GH57 (GH-GH57). *Escherichia coli* is used as reference for the type of isoforms found in cyanobacteria (gray columns)

3 Three Biosynthesis Glycogen Pathways in Prokaryotes

Over the past decades, our understanding of glycogen biosynthetic pathway in prokaryotes relies mostly on functional approaches using *Escherichia coli*, as a model. At the present, most of the activities involving carbohydrate metabolism pathways and their regulation are well understood (for reviews see Wilson et al. 2010; Preiss 2014). In prokaryotes, the most predominant glycogen biosynthetic pathway (GlgC) is based on the use of nucleotide-sugars as activated substrates for the polymerization. Nevertheless, both high-throughput genome sequencing and comparative genome analyses have evidenced two alternative glycogen biosynthetic pathways in bacteria known as the GlgE pathway and the MalQ pathway, respectively (Fig. 2).

3.1 Nucleotide-Sugar-Based Glycogen Biosynthetic Pathway in Bacteria: GlgC Pathway

In the GlgC path, the biosynthesis of glycogen involves a suite of three enzymatic reactions for catalyzing: (1) the nucleotide-sugar synthesis mediated by ADP-glucose pyrophosphorylase (GlgC), (2) the formation of α-1,4 linkage catalyzed by glycogen synthase (GlgA), and (3) the formation of α-1,6 linkages mediated by glycogen branching enzyme (GlgB). Glycogen metabolizing genes are commonly organized in operon such as the *glgBXCAP* well conserved in Enterobacteriales and the sister lineage Pasteurales (Almagro et al. 2015). Both *glgX* and *glgP* genes encode for glycogen catabolic enzymes (see Glycogen Catabolic Paths in Prokaryotes).

Nucleotide-Sugar Synthesis: UDP-/ADP-Glucose Pyrophosphorylase (EC 2.7.7.27)

In 1957, Leloir and Cardini were the first to establish the importance of nucleotide-sugar in the biosynthesis of storage polysaccharides rather than glucose-1-phosphate which had been earlier thought to define the polymerization substrate (Leloir and Cardini 1957). In prokaryotes, genetic evidence of ADP-glucose pyrophosphorylase (GlgC) as a pivotal enzyme for glycogen biosynthesis in *E. coli* was published in 1968 (Damotte et al. 1968). Following this study, correlation between GlgC activity and glycogen biosynthesis was confirmed in many other prokaryotes as well as in cyanobacteria (Iglesias et al. 1991). In regard to its role in the glycogen pathway, ADP-glucose pyrophosphorylase is an allosteric enzyme negatively or positively regulated by fructose-1,6-bisphosphate or AMP, respectively. In cyanobacteria, the ADP-glucose pyrophosphorylase is subjected to redox regulation as well as to the ratio of 3-phosphoglycerate/orthophosphate that reflects the activity of

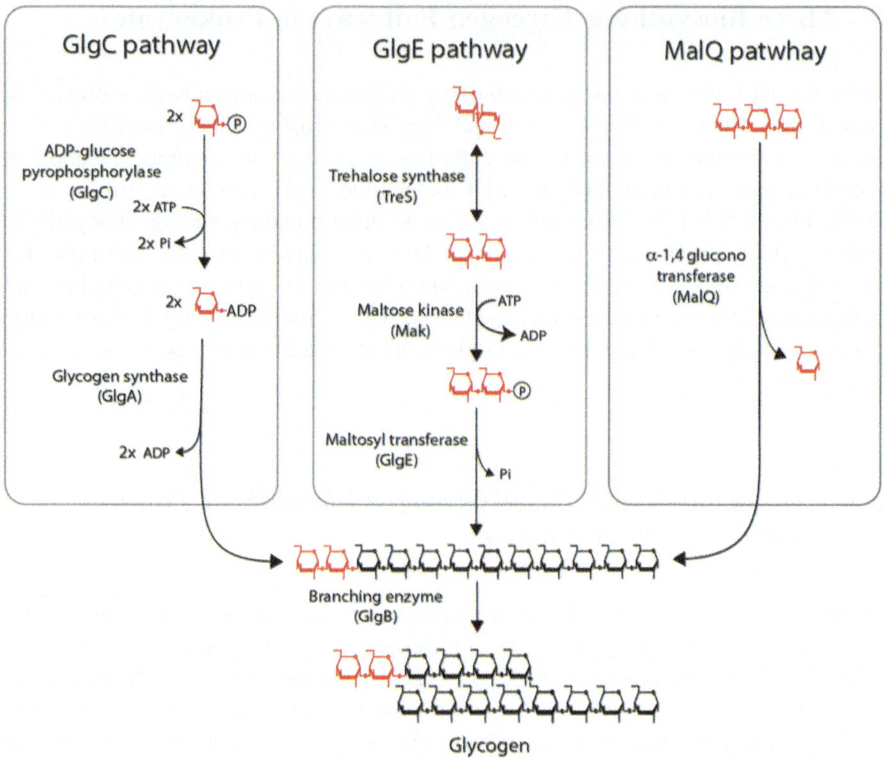

Fig. 2 Glycogen anabolism pathways in prokaryotes. The most predominant glycogen biosynthesis pathway, GlgC pathway, involves an ADP-glucose pyrophosphorylase (GlgC) and glycogen synthase (GlgA) for synthesizing the glucan chains. The former synthesizes the nucleotide-sugar, ADP-glucose, from ATP and glucose-1-phosphate (G1P). Glycogen synthase then transfers the glucose moiety of ADP-glucose onto the non-reducing end of growing glucan chain. In the GlgE pathway, non-reducing disaccharide, trehalose, is converted reversibly into maltose by trehalose synthase (TreS) followed by a phosphorylation onto the reducing end of maltose. The latter reaction processed by maltose kinase (Mak) found in some bacterial species fused with TreS activity. Maltose-1-phosphate (M1P) is incorporated onto the non-reducing end of glucan chain thanks to maltosyl transferase (GlgE). MalQ pathway, described in few species, relies onto pre-existing short glucan chains. α-1,4 Glucanotransferase (MalQ) activity disproportionates short glucan chains. Then, in both GlgC-GlgE pathways and malQ pathway, an a-1,6 linkage is produced when the neo-glucan chain fits the catalytic site of branching enzyme (GlgB). The incorporation of two glucose moieties (in red) requires 2, 1, and 0 ATP for GlgC, GlgE, and MalQ pathways, respectively

Calvin-Benson cycle (Ballicora et al. 2003; Díaz-Troya et al. 2014). It should be stressed out that some bacteria species like *Prevotella bryantii* B14, a common bacterium found in the rumen, rely apparently on UDP-glucose pyrophosphorylase activity for glycogen synthesis instead of ADP-glucose pyrophosphorylase. Accordingly, glycogen anabolism in this species depends on UDP-glucose-dependent glycogen synthase (Lou et al. 1997). Besides NDP-glucose pyrophosphorylase activity, another possible but yet to be confirmed source of nucleotide-sugar was revealed by the study of sucrose metabolism in the filamentous nitrogen-fixing cyanobacterium *Anabaena* ATCC29213. To recall, sucrose synthesis is induced in response to osmotic stress in cyanobacteria and then catabolized either by invertase or by sucrose synthase (SuSy) activities. In plants, several studies suggested that SuSy activity was involved in the cell wall synthesis by catalyzing the release of UDP-glucose and fructose from UDP and sucrose. Unexpectedly, mutants of SuSy of *Anabaena* displayed a more complex phenotype. Indeed, SuSy null mutants harbored a decreased glycogen content, despite twofold increases of ADP-glucose pyrophosphorylase activity and are impaired in the nitrogen fixation process (Porchia et al. 1999). Further biochemical characterizations of SuSy in both *Anabaena* ATCC29413 and *Thermosynechococcus elongatus* have revealed that SuSy catalyzed preferentially the synthesis of ADP-glucose and fructose by cleaving sucrose in the presence of ADP (Porchia et al. 1999; Figueroa et al. 2013). These astonishing results contrast to the dedicated role of SuSy of plants, which is to supply in UDP-glucose for cell wall synthesis. Subsequent studies based on metabolic network modeling in *Anabaena* sp. PCC7120 may suggest that ADP-glucose pyrophosphorylase activity alone is insufficient to fulfill the need in ADP-glucose for glycogen synthesis (Cumino et al. 2007). Unexpectedly, these results suggest that sucrose in *Anabaena* sp. PCC7120 possesses an additional function besides being an osmotic protectant. It was proposed that sucrose among other metabolites might be a carrier metabolite linking the vegetative cells and the heterocyst cells. This would be similar to some plants where sucrose is a carrier metabolite that links source organs (leaves) and sink organs (tubers). Indeed, in the filamentous cyanobacterium *Anabaena*, energy costly nitrogen fixation pathway is catalyzed by an oxygen-sensitive nitrogenase activity localized in a specialized cell termed the heterocyst. In turn, to maintain an anoxic environment and high ATP level, both reductive pentose phosphate pathway and photosynthesis II activity (oxygen evolution) are shut down in this dedicated cell. In this context, the heterocyst cell may be considered as a heterotrophic cell, comparable to sink tissue cells in plants that heavily rely on neighboring photosynthetic cells for ATP and sucrose intakes.

Formation of α-1,4 Linkages: Glycogen/Starch Synthase

Glycogen/starch synthase activities transfer the glucose moiety of nucleotide-sugar onto the non-reducing end of α-glucan chains. Structural analysis of the enzymes suggests that two families exist named according to Carbohydrate Active EnZyme classification (CAZy, www.cazy.org): glycosyl transferase 3 (GT3) and glycosyl transferase 5 (GT5) (Coutinho et al. 2003). The former is exclusively found in glycogen metabolism of fungi and animal cells, while the GT5 family is widespread among Archaeplastida (plants/green algae, red alga, and glaucophytes) and prokaryotes. With the exception of *Dictyostelium discoideum*, social amoeba possesses an eukaryotic (GT3) as well as a bacterial (GT5) glycogen synthase (Ball et al. 2015). Because of the general absence of ADP-glucose in glycogen-storing eukaryotes, most of glycogen synthases GT3 characterized in eukaryotes are UDP-glucose and oligosaccharide primer dependent (Ball et al. 2011). In prokaryotes, the specificity of glycogen synthase GT5 is not so stringent. For instance, the characterization of glycogen synthase of *Prevotella bryantii* shows clearly an elongation activity using UDP-glucose (Lou et al. 1997). More recently, Gehre and collaborators have unraveled a surprising adaptation of GlgA activity of *Chlamydia trachomatis* (Gehre et al. 2016). In etiologic agents of human and animal diseases, *Chlamydiales* have a biphasic development that requires a eukaryotic host to complete their life cycle. Inside the eukaryotic cell, they are located inside a specific inclusion vesicle, which is associated with a large accumulation of glycogen in the lumen of vesicle inclusion. Through genetic and biochemical approaches, Gehre and collaborators have evidenced that luminal glycogen pool is mostly de novo synthesized due to an evolution of GlgA activity to polymerize UDP-glucose as well as the recruitment of host UDP-glucose transporter on the inclusion membrane (Gehre et al. 2016).

While most prokaryotes possess a *glgA* gene, cyanobacteria genomes encode two isoforms named *glgA1* and *glgA2* (Ball et al. 2015). Surprisingly, both *Cyanobacterium* sp. CLg1 and *Crocosphaera watsonii* encode a third *glgA* gene, which is phylogenetically related to granule-bound starch synthase (GBSS) of plants. In plants or green algae, GBSS activities have been shown to be responsible for synthesizing the amylose fraction in starch (Delrue et al. 1992). Mass spectrometry analysis onto protein attached to starch granules of *Cyanobacterium* sp. CLg1 and incubation experiments of purified starch granules with ^{14}C-ADP-glucose have shown that GBSS-like cyanobacteria has maintained similar enzymatic properties of plants, i.e., the synthesis of long linear glucan chains giving rise to the amylose fraction of starch-like granules (Deschamps et al. 2008). The functions of GlgA1 and GlgA2 activities in glycogen biosynthesis were investigated mainly in *Synechocystis* PCC6803 (Gründel et al. 2012; Yoo et al. 2014). Single knockout mutants were unaffected in growth and glycogen biosynthesis suggesting an overlap in function of these two enzymes. However, a detailed biochemical analysis of glycogen synthases suggested that GlgA2 is responsible for long glucan chain synthesis, while the distributive activity of GlgA1 is responsible for short glucan chain (Yoo et al. 2014). More recently, the characterization of a mutant of the starch-accumulating *Cyanobacterium* sp. CLg1 strain impaired in GlgA2 activity had comforted the idea of the processive nature of GlgA2 (Kadouche et al. 2016). Indeed, the absence of GlgA2 activity resulted specifically in the disappearance of starch granules without affecting glycogen biosynthesis (Kadouche et al. 2016). The simplest explanation for this phenotype is to propose that GlgA2 activity is mandatory for the synthesis of long glucan chains indispensable for the establishment of the cluster organization of amylopectin. In comparison to the widespread distribution of the GlgA1 isoform among prokaryotes, GlgA2 is restricted to a small number of bacterial species including *Chlamydiales*. More surprisingly, GlgA2 is very similar to the genes of soluble starch synthases III and IV (SSIII and SSIV) found in Archaeplastida (Kadouche et al. 2016). This suggests that the SSIII/SSIV genes of plants have been inherited from a prokaryote and were possibly maintained in the Archaeplastida because of its remarkable ability to synthesize long glucan chains.

Formation of α-1,6 Linkages: Glycogen/Starch Branching Enzyme

Branching enzyme (GlgB)—α-1,4-glucan: α-1,4-D-glucan 6-α glucosyl transferase—catalyzes the transglucosylation reaction by cleaving α-1,4 linkages and by transferring the generated reducing ends to C6 hydroxyls. Branching enzyme plays a critical role in determining the branching pattern of both glycogen and starch (Sawada et al. 2014). However, no clear correlation has been found between amino acid sequence of GlgB and the molecular properties of the branched glucan produced. In prokaryotes, branching enzyme activities are distributed in two CAZy families: glycosyl hydrolase 13 (GH13) and 57 (GH57) families (Suzuki and Suzuki 2016). Family GH13, also known as the α-amylase family, is the largest sequence-based family of glycoside hydrolases characterized with $(\alpha/\beta)_8$ barrel domain. Based on phylogenetic tree of family GH13, 35 subfamilies emerge containing various enzyme activities and substrate specificities acting on α-glycosidic bonds (Stam et al. 2006). GH13 branching enzymes of eukaryotic and prokaryotic cells are distributed into two subfamilies: GH13_8 and GH13_9 (Stam et al. 2006). In 2000, Binderup and his collaborators have shown that the N-terminal end of the *E. coli* branching enzyme influences the chain transfer pattern (Binderup et al. 2000). The deletion of 112 amino acids at the N-terminus led to a truncated form of *E. coli* branching enzymes, which preferentially transfers longer chains having a degree of polymerization (DP) of more than 20 residues of glucose than the unmodified branching enzyme (DP<14). Subsequent studies confirmed the role of the N-terminus of glycogen branching enzymes in *E. coli* as well as for other GlgB proteins of other species (Devillers et al. 2003; Palomo et al. 2009; Jo et al. 2015; Wang et al. 2015). Branching enzymes belonging to the GH57 family, first described in thermophilic bacteria *Dictyoglomus thermophilum* and *Pyrococcus furiosus*, are now described in cyanobacteria and other prokaryote species (Fukusumi et al. 1988; Laderman et al. 1993; Suzuki and Suzuki 2016). Like GH13 family, the GH57 family contains several carbohydrate-active enzyme activities: amylopullulanase, α-galactosidase, α-1,4 glucanotransferase, α-amylase, branching enzyme, and uncharacterized activities sharing a catalytic $(\beta/\alpha)_7$ barrel fold and five conserved domains (Zona et al. 2004; Murakami et al. 2006; Santos et al. 2011). In cyanobacteria, the distribution of GH57 family enzymes among species is variable, and their exact catalytic functions have yet to be defined (Colleoni and Suzuki 2012). Interestingly, putative GH57 branching enzyme sequences are well conserved among cyanobacteria including the reduced genomes of the *Synechococcus*/*Prochlorococcus* genus as well as in the early diverging cyanobacteria species such as *Gloeobacter violaceus*, *Synechococcus* JA-3-3Ab, and *Gloeomargarita lithophora*. So far the role of this GH57 family in the physiology of cyanobacteria has not yet been explored. Because the GH57 family is mainly composed of thermostable enzymes, it is tempting to suggest this family of carbohydrate-active enzymes plays an important role in thermophile strains (e.g., *Thermosynechococcus elongatus* BP1 and *Synechococcus* JA3-3Ab) or under thermal stress conditions that require a set of stable enzymes.

The fact that most of glycogen-accumulating prokaryotes possess at least one GH13 branching enzyme activity or one GH57 branching enzyme (Suzuki and Suzuki 2016) emphasizes the mandatory role of branching enzymes in storage polysaccharide biosynthesis. Cyanobacteria species possess both GH13 and GH57 families, and for some cyanobacterial species, additional GH13 branching enzyme isoforms are observed. It is worth noting that starch-accumulating cyanobacteria harbor in total 3–4 candidate branching enzyme activities. Structural and biochemical characterization of GH13 branching enzymes of *Cyanobacterium* sp. NBRC 102756 and *Cyanothece* ATCC51142 has revealed different branching patterns that might be correlated with the ability of those strains to synthesize starch-like granules (Hayashi et al. 2015, 2017; Suzuki et al. 2015).

De Novo Biosynthesis of Glucan Chains

Following the discovery of glycogen synthase by Leloir and Cardini in 1957, the main question during many years concerned about the ability of priming glucan synthesis from nucleotide-sugar. Apparently, glycogen synthase GT3 and GT5 can be distinguished by their ability to prime de novo synthesis of a glucan. In eukaryotic cells, the initiation of glycogen synthesis appears to require two activities: (1) an UDP-glucose-specific glycosyl transferase of CAZy family GT8 (CAZy classification), which displays self-glycosylation properties (i.e., glycogenin) and is able to synthesize a short glucan made of 8 up to 36 glucose residues using UDP-glucose as nucleotide-sugar (Albrecht et al. 2004), and (2) a glycogen synthase GT3 which then elongates the glycogenin-dependent primer in order to further elongate a glucan which thereby becomes accessible for branching enzyme. The study of yeast null mutants affected in both glycogenin isoforms, Glg1p and Glg2p, supported the mandatory nature of glycogenin-controlled glucan priming for synthesis of glycogen in fungi (Torija et al. 2005). However, it should be stressed that glycogen synthesis occurs spontaneously in 2–3% of glycogenin-null mutant of yeast depending on growth conditions (i.e., nitrogen limitation) and it might reach up to 98% if the amount of UDP-glucose is upraised in the cytosol (Torija et al. 2005). More recently, the ability of GT3-glycogen synthase to initiate glycogen synthesis in the absence of glycogenin has been experimented in animals (Testoni et al. 2017). Against all odds, not only glycogenin-deficient mice (Gyg knockout) accumulate glycogen, but also an excess of large glycogen particles is observed in both homozygous and heterozygous strains. Interestingly, these results enlighten, first, that the lack of glycogenin proteins does not prevent glycogen synthesis and, second, that glycogenin might possess an unexpected function in the glycogen biosynthesis pathway such as a regulator of glycogen synthase activity through protein-protein interactions (Testoni et al. 2017).

In prokaryotes, preliminary studies reported by Krisman and his colleagues suggested that the initiation of glycogen synthesis in *E. coli* depends as in eukaryotic cells on glucoprotein formation by a glycogenin-like protein and, therefore, is similar to the eukaryotic path (Barengo et al. 1975; Barengo and Krisman 1978). However, detailed enzymatic analyses suggested that both functions, primer initiation and elongation activity, might be probably under sole control of the *E. coli* glycogen synthase (Kawaguchi et al. 1978; Holmes and Preiss 1979). The discovery of a 7.9 kDa protein (60 amino acids) named GlgS (S stands for Stimulate) in 1997 questioned the existence of a glycogenin-like function in *E. coli* (Beglova et al. 1997). Historically, the GlgS mutant of *E. coli* was first identified as glycogen-less phenotype through random transposon insertion mutagenesis and iodine screening of cell patches (Hengge-Aronis and Fischer 1992). *glgS* gene does not belong to the *glgBXCAP* operon; its overexpression and its upregulation by sigmaS during entry into stationary phase suggested a function in the glycogen metabolism pathway. Therefore, the authors proposed that GlgS might be dedicated to glycogen initiation. However, in 2003, the expression of recombinant glycogen synthase of *Agrobacterium tumefaciens* in glycogen-less background of *E. coli* confirmed the ability of prokaryote glycogen synthase to de novo synthesize a glucan (Ugalde et al. 2003). This result was further confirmed by the restoration of starch synthesis in the double mutant SSIII/SSIV of *A. thaliana* by the transgenic expression of the *A. tumefaciens* glycogen synthase (Crumpton-Taylor et al. 2013).

Then, both transcriptomic and molecular analyses have definitely invalidated a direct function of GlgS in glycogen metabolism of *E. coli* (Rahimpour et al. 2013). It has been shown that GlgS renamed ScoR (Surface composition Regulator) controls the cell wall composition on the surface (flagella, adhesins, exopolysaccharides). Thus, in absence of ScoR or GlgS, mutant strains displayed a higher number of flagella, and synthesis of exopolysaccharides was increased at the extent of glycogen synthesis. This pleiotropic effect disappeared when GlgS is overexpressed. The authors proposed to explain glycogen-less phenotype of ScorR/GlgS mutants that both abnormal cell mobility and exopolysaccharides synthesis drained all carbon sources normally stored at the onset of stationary phase in the form of glycogen.

Although the ability of prokaryote glycogen synthase to synthesize de novo a glucan is well established, it appears that the recombinant GlgA2 isoform of *Cyanobacterium* sp. CLg1 produced in a glycogen-less background of *E. coli* is not capable of initiating a glucan synthesis by itself. The expression of GlgA2 or GlgA1 in the single *glgA* mutant background restores glycogen synthesis in *E. coli* only when cells are grown in the presence of maltose as carbon source and not in the presence of glycerol or mannitol. In addition, electrophoretic separation of enzyme followed by in situ activity assays (i.e., zymograms) indicated that the priming reaction inside a native polyacrylamide gel depends on the presence of a glucosylated protein, named X-factor, that can be supplied by *Cyanobacterium* sp. CLg1 crude extracts (Kadouche et al. 2016). Further investigation will be necessary to evidence the exact nature of this X-factor.

3.2 *Alternative Glycogen Biosynthesis Pathways in Prokaryotes: The GlgE Pathway*

In contrast to the nucleotide-sugar-based glycogen synthesis pathway, this alternative glycogen pathway, named the GlgE pathway, relies on a maltosyl transferase activity (GH13 subfamily 3 according to CAZy classification) that transfers maltosyl groups from maltose-1-phosphate onto the non-reducing end of glucan chains. The evidence supporting the physiological significance of this alternative pathway in prokaryotes arose from the study of *Mycobacterium tuberculosis*. Mycobacteria are pathogenic bacteria surrounded by a capsule, composed of 80% of glucan made of glucose residue linked in α-1,4 and α-1,6 positions (Lemassu and Daffé 1994). To determine whether the mycobacterial capsule contributes significantly in the virulence and depends on the putative orthologous *glgC*, *glgA*, and *glgB* genes that have been identified in the genome of *M. tuberculosis*, a series of knockout mutants were generated either in ADP-glucose pyrophosphorylase activity (*ΔglgC*) or glycogen synthase (*ΔglgA*) (Sambou et al. 2008). Surprisingly, both deletions of GlgC and GlgA genes led only to a reduction but not to the expected wipeout of both glycogen and extracellular polysaccharide formation, thus emphasizing the existence of an alternative route for glycogen synthesis. In 2016, a better understanding and clarification of glycogen metabolism in *M. tuberculosis* was achieved when Koliwer-Brandl and collaborators included null *glgE* mutants in their studies in various genetic backgrounds (Koliwer-Brandl et al. 2016). Historically, this GlgE activity was first thought to consist of a glucanase activity releasing maltose-1-phosphate (M1P) from glycogen (Belanger and Hatfull 1999). Unexpectedly, the inactivation of GlgE activity led to both an accumulation of M1P and a decrease of glycogen content suggesting a function of GlgE in the biosynthesis rather than degradation. In order to determine the source of M1P, further investigations established that GlgA is surprisingly not involved in the elongation process of glucan chain like normal glycogen synthase. On the contrary, GlgA catalyzes preferentially the formation of M1P from ADP-glucose and glucose-1-phosphate. Finally, combination of mutations points out two independent routes, GlgA-GlgC and TreS-Mak (trehalose synthase and maltose kinase) activities, that converge on the synthesis of M1P. Despite the name, trehalose synthase activity converts reversibly trehalose (alpha-D-glucose-1,1-alpha-D-glucose) into maltose. The non-reducing disaccharide trehalose functions as an important intracellular osmoprotectant in a wide range of bacterial species, fungi, and amoeba. In the GlgE pathway, maltose produced from trehalose by the TreS activity is then phosphorylated into M1P by maltose kinase activity (Mak or Pep2). Interestingly, M1P synthesis is favored by the formation of an heterocomplex between TreS and Mak (Roy et al. 2013) or by the appearance of TreS-Mak fused protein in some bacterial species (Fraga et al. 2015). M1P is then incorporated onto the non-reducing end of glucan chains via GlgE (Kalscheuer et al. 2010). The pivotal role of GlgE in the glycogen synthesis pathway of mycobacteria and the high cell toxicity of M1P explain the growing interest in the understanding of its regulation (Leiba et al. 2013) and its 3D structure (Syson et al. 2014) as a potential target for new drugs (Leiba et al. 2013). At the present, comparative genomic analysis suggests that the GlgE pathway is not restricted to mycobacteria species but is found in 14 % of sequenced genomes from diverse bacteria such as in *Pseudomonas*, *Xanthomonas*, and *Burkholderia* species (Chandra et al. 2011). In addition, both GlgE pathway and GlgC pathway can coexist in the same bacterium cell such as the Gram-positive *Corynebacterium glutamicum*. We have little information about the coordination of these pathways in the glycogen synthesis of such bacteria. Recently, Clermont and his collaborators have noticed that glycogen synthesis in the double knockout *glgC* and *glgA* of *C. glutamicum* depended on carbon source supplied in the medium culture. Thus, no glycogen synthesis has been assayed in the presence of glucose, while normal amounts of glycogen in comparison to wild-type strain were inferred in the presence of maltose (Clermont et al. 2015). At first glance, it is tempting to suggest that *C. glutamicum* synthesizes glycogen via either GlgE pathway or the GlgC pathway depending on the carbon source available. However, since maltose is used as a precursor to the GlgE pathway as well as in the MalQ pathway (see below), further investigations are required to evaluate the contribution of these pathways in the glycogen biosynthesis of *C. glutamicum*.

3.3 Other Glycogen Biosynthesis Pathways in Prokaryotes: The MalQ Path

As third glycogen biosynthetic path in prokaryotes, the MalQ pathway relies on the ability of MalQ—an α-1,4 glucanotransferase (GH77) or amylomaltase—to disproportionate short glucans into longer glucans (Monod and Torriani 1950). When long enough to fit the catalytic site of branching enzyme, intramolecular or intermolecular transfer generates branched glucan. In the *E. coli* model, the uptake of extracellular maltose induces the transcription of several genes of the maltose metabolism. Among them both maltodextrin glucanotransferase (MalQ) and maltodextrin phosphorylase (MalP) contribute to the catabolism of maltodextrin. In response to maltose or malto-oligosaccharides (MOS: glucan chains made of 3–12 glucosyl residues), the *malPQ* operon and other genes (such as *malZ* and *malS* encoding α-glucosidase and periplasmic α-amylase, respectively) are transcriptionally activated through the regulator MalT. In contrast to glycogen phosphorylase (GlgP), MalP activity is poorly active on branched polysaccharide while it releases glucose-1-phosphate from the non-reducing end of MOS with a minimum of four residues of glucose. In the absence of MalP, *E. coli* accumulates long glucan chains responsible for the dark-blue iodine staining of cell patches (Schwartz 1967). Interestingly, the lack of MalP activity leads to abnormal cell morphology ("snake" morphology) in an *E. coli* mutant (Schwartz 1967) as well as in a mutant of *Corynebacterium glutamicum* (Seibold et al. 2009).

MalQ catalyzes the transfer of a maltosyl group from maltotriose onto the non-reducing end of maltose or onto short glucan chains made of 3–7 glucosyl units (i.e., malto-oligosaccharide (MOS)). Therefore, MalQ generates larger malto-oligosaccharides (DP>4) suitable for MalP activity. MalQ mutants of *E. coli* do not grow in the presence of maltose but can grow in the presence of MOS with a higher DP. This growth appears dependent on the MalP activity since double mutants *ΔmalPQ* are capable of growing in maltose or MOS as carbon source. By a combination of mutation involving maltose catabolizing enzymes and glycogen synthase activity, Park and his collaborators have shown that normal glycogen synthesis occurs in a *glgA* mutant background (glycogen-less strain) when MalP activity is missing (Park et al. 2011). Interestingly recent survey of glycogen gene content in sequenced genomes reveals that the artificial situation described by Park and his collaborators might arise in some bacterial species belonging to β-proteobacteria. Thus, glycogen synthesis in both *Ralstonia eutropha* H16 and *Bordetella parapertussis* 1282 relies on the *glgBXmalQ* operon, while in *Burkholderia* sp. 383 the *treZ* gene is inserted between *glgX* and *malQ* genes (Almagro et al. 2015). Besides the lack of GlgC and GlgA activities, open questions remain concerning the nature of glycogen metabolism regulation in those species.

4 Glycogen Catabolic Paths in Prokaryotes

In enterobacteria, the glycogen catabolism pathway is pretty well understood. Glycogen phosphorylase (GlgP) and glycogen debranching enzyme (direct debranching; GlgX) work in synergy with α-1,4 glucanotransferase (MalQ) and maltodextrin phosphorylase (MalP) to essentially convert glycogen to glucose-1-phosphate. Glycogen phosphorylase catabolizes the first catabolic step by releasing G-1-P from the non-reducing ends of glucan chains. This reaction stops around four residues of glucose before a branch point. Then short-branched glucans are specifically trimmed by the GlgX activity (Dauvillée et al. 2005; Suzuki et al. 2007). Short glucans released in the cytosol are further metabolized by MalP and MalQ activities. The latter disproportionates short glucans (DP<4) in longer glucans accessible to the catalytic site of MalP (Fig. 3).

4.1 Glycogen Phosphorylase

Thus, in the model organism *E. coli*, two phosphorylase activities are involved in the glycogen catabolism pathway; GlgP and MalP show clear substrate preference for glycogen and for malto-oligosaccharides, respectively. Interestingly, both GlgP and GlgX enzymes are produced during glycogen biosynthesis. In order to avoid a futile cycle, ADP-glucose acts as a competitive inhibitor with respect to G-1-P for the GlgP activity in most prokaryotes during the biosynthesis of glycogen (Chen and

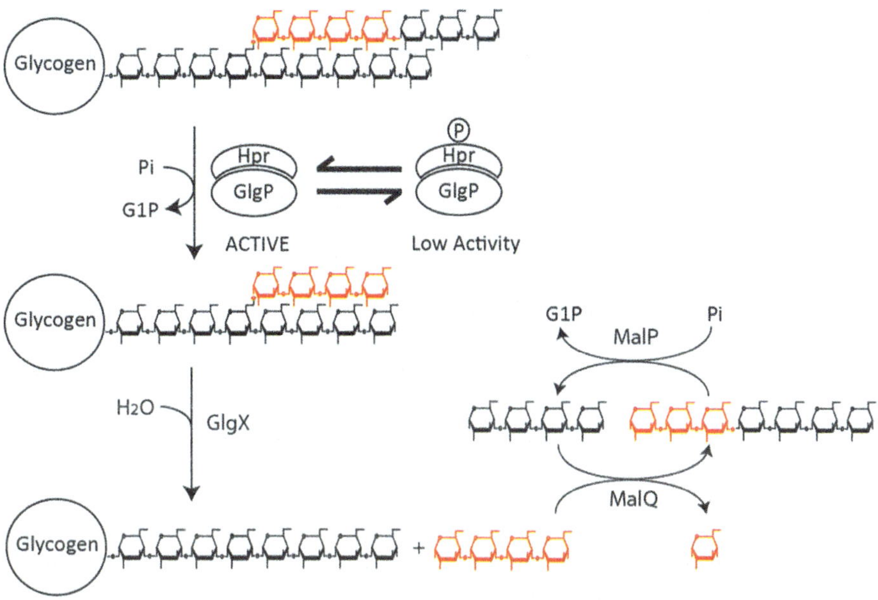

Fig. 3 Glycogen catabolism pathway. Glycogen phosphorylase (GlgP) and histidine carrier protein (Hpr) form a stable complex. Hpr proteins favor the formation of dimer, tetramer, and octamer of GlgP (not shown here for the sake of simplicity). Both phosphorylated Hpr proteins and competitive inhibitor, i.e., ADP-glucose, reduce drastically the GlgP activity during glycogen biosynthesis. The degradation of glycogen is initiated when Hpr proteins are dephosphorylated. In the presence of phosphate inorganic (Pi), GlgP activities release glucose-1-phosphate (G1P) from the non-reducing of glucan chains of glycogen. The enzymatic reactions stop four units of glucose residues before reaching an α-1,6 linkage or branching point. Debranching enzyme activities (GlgX) recognize and trim specifically short-branched maltotetraosyl residues. The new linear glucan chains produced can be then further digested by GlgP activity. Maltotetraose molecules are metabolized in glucose and G1P via the synergic action of α-1,4 glucanotransferase (MalQ) and maltodextrin phosphorylase (MalP)

Segel 1968; Takata et al. 1998). Subsequent studies on GlgP have revealed that the latter forms a tight interaction with the histidine phosphocarrier protein (Hpr) (Seok et al. 1997). The latter is a component of the phosphoenolpyruvate:sugar phosphotransferase system (PTS). This PTS system is involved in the transport and the phosphorylation of many sugars (Deutscher et al. 2006). According to the phosphorylation state of Hpr, the affinity of Hpr will change for GlgP. Thus, the phosphorylated form of Hpr (P-Hpr) displays a fourfold higher affinity for GlgP than the unphosphorylated form. However, only the Hpr form allosterically activates GlgP by 2.5-fold. Interestingly, because the concentration of Hpr is much higher than that of GlgP, this indicates that GlgP is always complexed with either P-Hpr or Hpr (Mattoo and Waygood 1983). The phosphorylation state of Hpr varies with the physiological state of the cell. Thus, at the onset of stationary phase, which coincides with glycogen operon activation and glycogen synthesis, most of Hpr is found to be phosphorylated form. Both the predominant form of P-Hpr and the increase in

ADP-glucose level will prevent the dimerization and activity of GlgP, respectively, and therefore glycogen degradation (Chen and Segel 1968; Seok et al. 2001).

In contrast to *E. coli*, the set of genes involved in the catabolic pathway varies greatly in cyanobacteria species (Colleoni and Suzuki 2012). So far, the reasons for this multiplicity of degradation pathways are unclear. The functions of phosphorylase activities were investigated in *Synechocystis* PCC8803 (Fu and Xu 2006). The characterization of knockout mutants in the two phosphorylase genes sll1367 and sll1356 revealed that GlgP (sll1367) is clearly involved in glycogen catabolism, while GlgP (sll1356) is essential for growth at high temperature. Interestingly, despite the increase of GlgP-sll1356 activity in the null GlgP-sll1367 mutant, glycogen content decreased only slightly due to mobilization at night (Fu and Xu 2006). More recently, the functions of glycogen phosphorylase isoforms were investigated in the resuscitation or awakening process, which consists in the ability of non-nitrogen-fixing cyanobacterium to exit from the dormant stage to the vegetative stage (Doello et al. 2018). In nitrogen limitation condition, light-harvesting pigments are degraded, and glycogen synthesis is induced until photosynthesis and metabolic activities are strongly reduced. Hence, this survival cell response termed "chlorosis" is maintained until the nitrogen source is supplied. The resuscitation process requires the mobilization of glycogen as energy-carbon source until the photosynthetic apparatus becomes functional. Characterization of single or double mutants of glycogen phosphorylase isoforms in *Synechocystis* PCC6803 reveals a minor role of GlgP-sll1356, while GlgP-sll1367 null mutants were impaired in the resuscitation process and revealed as a major contribution of this enzyme to glycogen catabolism during resuscitation (Doello et al. 2018). Further investigations are required to understand the function of the GlgP-sll1356 isoform. One reasonable explanation might be that sll1356 gene encodes a maltodextrin phosphorylase, which as described previously is more active on maltodextrin than glycogen. Hence, as with GlgP mutants in *E. coli* showing a defect in glycogen breakdown, the putative-MalP (sll1356) isoform cannot compensate for the loss of function of GlgP (Alonso-Casajús et al. 2006).

4.2 Glycogen Debranching Enzyme

Surprisingly, several cyanobacterial genomes, if complete, do not seem to encode any classical GlgX-GH13-type debranching enzyme (Colleoni and Suzuki 2012). However, such genomes appear to contain amylopullulanase (GH13-20 and/or GH57) or amylo-1,6 glucosidase (GH133) genes that could exert the corresponding functions. For instance, the early diverging *Synechococcus* sp. JA-3-3Ab strain contains two candidate debranching enzymes activities: the putative amylopullulanase-GH57 and amylo-1,6-glucosidase-GH133. Little is known, however, about their roles in the glycogen catabolism pathway. Recently, an amylopullulanase type of GH13 enzyme has been characterized from the filamentous cyanobacterium *Nostoc punctiforme* (Choi et al. 2009). In contrast to the GlgX activity, this enzyme displays hydrolysis activity toward both α-1,6 and α-1,4 linkages when incubated with soluble starch or amylopectin. Incubation experiments have shown that amylopullanase-GH13 hydrolyzes first α-1,6 linkages of long-branched glucans (4>DP>10) and chews up long glucan chains up to 8 residues of glucose from the reducing end. Because *N. punctiforme* does not contain any GlgX or TreX GH13 activity but harbors both maltooligosyltrehalose synthase (TreY) and maltooligosyltrehalose hydrolase (TreZ), it is quite possible that the amylopullulanase-GH13 substitutes for the missing TreX activity, which is required to produce the suitable malto-oligosaccharide for TreY activity. Amylo-1,6-glucosidase (GH133) shows similarity to the amino acid sequence of the C-terminal domain of indirect debranching enzymes of animals and fungi. By analogy with this eukaryotic enzyme, the function of this activity may possibly reside in debranching single α-1,6 glucose residues branched on linear glucan chains. So far, no biochemical characterization has ever been carried out on these cyanobacterial amylo-1,6 glucosidases. However, the characterization of the null GlgX mutant of *Synechococcus* sp. PCC7942 brought some insights on the putative function of the amylo-1,6 glucosidase activity (Suzuki et al. 2007).

As described above, characterizations of *glgX* mutants in both *E. coli* and *Synechococcus* PCC7942 confirm its function in the catabolism pathway. Nevertheless, characterization of *glgX*-deficient mutants of the starch-accumulating *Cyanobacterium* sp. CLg1 suggests its involvement in the crystallization process of storage polysaccharides.

5 Starch Granules in Unicellular Diazotrophic Cyanobacteria: Convergent Evolution of Glucan Trimming Mechanism in the Starch Biosynthetic Pathway of *Cyanobacterium* sp. CLg1

Until the end of the twentieth century, the mechanisms underlying the crystallization of starch were poorly understood. The characterization of mutants substituting starch by glycogen biosynthesis in green algae and different plant species surprisingly revealed a defect in the isoamylase type of debranching enzyme. In order to explain this unexpected function for hydrolytic enzymes in the amylopectin synthesis, a glucan trimming model was proposed in 1996 (Mouille et al. 1996). This model suggests that an isoamylase type debranching enzyme trims a highly branched precursor named pre-amylopectin and enables thereby the formation of closely packed double helices of glucan that precipitate out of the solution. It implies that this isoamylase preferentially trims those loosely branched glucans that prevent the clustering of double helices of glucan. In the absence of this enzyme, a highly branched hydrosoluble polysaccharide accumulates in the deficient mutants. Until 1994, starch accumulation had been described solely in lower photosynthetic eukaryotes, i.e., algae. The observation of thin sections of nitrogen-fixing cyanobacterium *Cyanothece* ATCC51142 through day-night cycles has revealed that round-shaped materials accumulate between the cyanobacterial thylakoid membranes during the day and disappear at night (Schneegurt et al. 1994). Those large bodies were partially characterized and described as "abnormal" glycogen particles and not as starch-like material (Schneegurt et al. 1997). The authors of the study did not realize at the time that they were dealing in fact with starch-like structures. Later on, a survey of storage polysaccharides in different species of cyanobacteria reported the presence of solid granules in others cyanobacterial species (Nakamura et al. 2005). Subsequently, detailed characterization indicated that such carbohydrate granules, including those of *Cyanothece* ATCC51142, are composed of high-molecular weight polysaccharides similar to amylopectin (Suzuki et al. 2013). Around the same time, *Cyanobacterium* sp. CLg1, a new strain isolated in the tropical North Atlantic Ocean phylogenetically related to *Crocosphaera watsonii*, was shown to accumulate starch granules made of both amylopectin and amylose fractions (Falcon et al. 2002; Deschamps et al. 2008). Interestingly, as in plants, the presence of amylose was correlated with the identification of a polypeptide showing a high similarity in amino acid sequence to GBSS (granule-bound starch synthase) of plants (Deschamps et al. 2008). To the present day, less than a dozen of starch-accumulating cyanobacteria have been identified. They are all unicellular capable of nitrogen fixation and belong to the Chroococcales order (Colleoni and Suzuki 2012). Deciphering the storage polysaccharide metabolism pathways is pretty challenging since those strains are refractory to all available transformation protocols. Nevertheless in order to tackle this question, an UV mutagenesis was carried out on wild-type *Cyanobacterium* sp. CLg1. Based on the iodine staining of cell patches, more than a hundred mutants were identified and subsequently categorized according to the ratio of soluble to insoluble polysaccharide. Among them, a dozen mutants harboring an increase in the hydrosoluble glycogen-like fraction and the disappearance of starch granules were impaired in one of two of the debranching enzyme isoforms GlgX2 (Cenci et al. 2013). This striking result suggests that such as in green alga and plants, crystallization process of amylopectin relies on the presence of a debranching enzyme activity. However, detailed phylogeny analysis built with debranching enzyme sequences of Archaeplastida[1] and Bacteria indicates that the eukaryotic sequences do not have a cyanobacterial origin but are derived from obligate intracellular pathogens and symbionts of the order *Chlamydiales*. It should be stressed out that crystallization process is not clearly associated with GlgX2-like sequence among starch-accumulating cyanobacteria. For instance, the genome of starch-accumulating *Cyanothece* ATCC51142 contains a single *glgX* gene encoding a GH13-debranching enzyme (cce_3465: GH13 family according to CAZy classification) that is closely related to GlgX1 isoform and not to GlgX2 isoform of *Cyanobacterium* sp. CLg1. Interestingly the role of GlgX-GH13 of *Cyanothece*

[1]Archaeplastida phylum encompasses three photosynthetic eukaryotes lineages: plants/algae, red algae, and glaucophytes. They share a common ancestor that emerged 1.5 billion years ago when a heterotrophic eukaryote engulfs a cyanobacterium.

sp. ATCC51142 has been recently clarified in starch metabolism pathway. Indeed, knockout mutant of GlgX (cce_3465) harbors a starch granule excess phenotype similar to *E. coli* or *Synechococcus* sp. PCC7942 mutant strains impaired in the debranching enzyme activity. Hence, these data strongly suggest a function catabolic pathway rather in crystallization process (Liberton et al. 2019). If GlgX is involved in the catabolism of α-1,6 linkages, this might suggest that amylo-1,6 glucosidase (cee_3194 gene) belonging to GH133 family according to CAZy classification might have evolved to cleave off branched glucan.

5.1 *Why Do Unicellular Nitrogen-Fixing Cyanobacteria Synthesize Starch-Like Granules Rather Than Glycogen?*

Cyanobacteria represent one of the oldest phyla of prokaryotes on earth (Summons et al. 1999). They are unique microorganisms capable of performing oxygenic photosynthesis and for some of them to reduce molecular nitrogen through nitrogenase activity. Nitrogenase activity loses covalent bonds between two nitrogen (and two carbon or one carbon and one nitrogen) atoms in an ATP-dependent manner and requires anoxic condition. Because of the lack of substrate specificity, it is hypothesized that the function of nitrogenase activity might have evolved over time: first, by reducing the effects of toxic cyanide compounds and then by reducing atmospheric nitrogen in ammonium. Over billion years of evolution, oxygenic photosynthesis activities of cyanobacteria have changed earth's atmosphere composition, thereby generating the Great Oxygenation Event (GOE) 2.4 billion years ago (Kopp et al. 2005). The transition from a reductive to an oxidative environment was a trigger for the diversification of cyanobacteria lineages and the appearance of new traits (e.g., size and cell morphology) (Sánchez-Baracaldo et al. 2014). A remarkable adaptation was achieved in nitrogen-fixing cyanobacteria or diazotrophic cyanobacteria. Indeed, the reduction of dinitrogen to ammonium is catalyzed by a protein complex called nitrogenase that displays an extreme sensitivity to oxygen. As oxygen level rose, diazotrophic cyanobacteria developed different strategies to protect the nitrogenase activity: (1) by inhabiting anoxic biotopes, (2) by confining nitrogenase activity in specialized cells called heterocysts, and (3) by separating temporally oxygenic photosynthesis and nitrogen fixation. In diazotrophic filamentous cyanobacteria, nitrogenase activity is located in heterocysts harboring a thick cell wall and lacking a photosystem II, which is responsible for photosynthetic oxygen evolution. Neighbor cells performing normal oxygenic photosynthesis activity supply the large amount of energy required to fuel the nitrogenase complex. Hence, physical separation of two exclusive biological processes, i.e., photosynthesis and nitrogen fixation, is possible through the evolution of dedicated cell. Consistent with this view, it was believed that nitrogen fixation could not occur in unicellular cyanobacteria. The discovery of two unicellular cyanobacteria species, *Gloeothece* sp. and *Cyanothece* sp., capable of fixing dinitrogen challenged this paradigm and, for the first time, provided evidence that in microorganisms primary metabolic processes are regulated by the circadian clock regulation (Wyatt and Silvey 1969; Singh 1973). In contrast to filamentous diazotrophic cyanobacteria that perform nitrogen fixation during the day, unicellular diazotrophic cyanobacteria carry out this activity exclusively at night. Unicellular cyanobacteria thus have developed a temporal separation of those two incompatible processes, which take place for some of them exclusively in micro-aerobic and for a small group of cyanobacteria also in aerobic conditions. In the latter case, unicellular nitrogen-fixing cyanobacteria exhibit high rates of dark respiration, which both provide the energy required for nitrogen fixation and while lowering the oxygen levels through respiratory consumption eventually lead to microaerophilic or anoxic conditions (Compaoré and Stal 2010). In order to achieve the required high rates of dark respiration, unicellular diazotrophic cyanobacteria are speculated to have evolved a more efficient storage polysaccharide allowing an increase in carbon storage in the light phase. Recently, the measurement of nitrogen fixation in six *Cyanothece* species under various growth and incubations conditions strengthened this assumption. In this report, starch-accumulating *Cyanothece* spp. exhibit a higher rate of nitrogen fixation than glycogen-accumulating *Cyanothece* species in aerobic growth conditions (Bandyopadhyay et al. 2013). This indirect proof reinforces the idea that the GOE, 2.4 billion years ago, was probably a driving force for the transition from glycogen to semi-crystalline storage polysaccharides. It should be stressed out that distribution of GlgX/isoamylase type debranching enzyme sequences does not necessarily correlate with the accumulation of starch in cyanobacteria. This suggests that the transition from glycogen to starch occurred in these cyanobacteria by either recruiting another debranching enzyme or calling for a different, yet to be described, mechanism of granule aggregation. Another striking observation from these bioinformatic genome-mining approaches consists of the absence of candidate debranching enzyme sequences of all known types in some starch-accumulating organisms. Genome analyses of starch-accumulating *Cyanobacterium* MIBC10216 and red algal-derived organisms did not reveal any gene encoding a GH13-type debranching enzyme. In such a case, the crystallization of amylopectin may indeed follow a different mechanism. A preliminary answer to this question has been addressed through the characterization of debranching enzyme and chloroplastic amylase null mutants in *Arabidopsis*. In the absence of chloroplastic amylase, DBE null mutants switch from accumulation of glycogen-like polysaccharide to synthesis of starch-like polysaccharides (Streb et al. 2008). Overall, those studies suggest that the combination of different cyanobacterial branching enzymes alone possibly offers a specific branching pattern that promotes a cluster organization of amylopectin. In turn, this could explain why usually three to four branching enzymes are usually found in these genomes.

6 Granulose: Another Example of Transition from Soluble to Insoluble Storage Polysaccharide?

Or, in other words, is there another selection pressure, which led to transition from soluble to insoluble storage polysaccharide? Few cases of abnormal glycogen/carbohydrate granules are described in the literature. It should be stressed that in the examples described below, the carbohydrate granules were not characterized structurally. Thus, this absence of information forbids us to distinguish between insoluble glycogen particles (absence of crystallinity) and true starch-like (semi-crystalline) granules. The evidence that they are actually carbohydrate granules is based on glucose assay, electronic microscope observations (size, electron density), and density (sedimentation at low speed). Therefore, further characterization of storage polysaccharides in these microorganisms should definitively improve our understanding of alternative aggregation mechanisms.

6.1 Firmicutes *Phylum:* Clostridium *sp.*

Clostridium species belong to the *Firmicutes* phylum and define Gram-positive anaerobic endospore-forming bacteria mainly used to produce acid and organic solvents (Dash et al. 2016). Back in 1950, Hopson and his collaborators reported in *C. butyricum* the accumulation of amylopectin-like polysaccharides that exert a strong interaction with iodine. The wavelength of the maximum absorbency of the iodine-polysaccharide complex (λmax) was reported to average 545 nm, which is close to the λmax value of amylopectin (550–560 nm) (Bergère et al. 1975). This material named "granulose," was found in most *Clostridium* species (Reysenbach et al. 1986). Based on electron microscopy observations, granulose sizes vary from 210 to 270 nm which is by far larger than normal glycogen particles (Tracy et al. 2008). Granulose biosynthesis occurs prior to the end of the exponential growth phase when the shift from acid to solvent production takes place while its degradation coincides with the endospore formation (Johnstone and Holland 1977). Characterization of mutants in *C. pasteurianum* and *C. acetobutylicum* ATCC824 impaired in granulose synthase harbor granulose-less phenotypes and are incapable of initiating sporulation suggesting that granuloses act as major carbon and energy source (Robson et al. 1974; Ehsaan et al. 2016). A survey of *Clostridium* genomes has evidenced the presence of one branching isoform belonging either to the GH13 family or the GH57 family (Suzuki and Suzuki 2016) and one glycogen synthase isoform named "granulose synthase." Interestingly, the latter is co-purified with native granulose (Robson et al. 1974). At the present day, the genetic determinants responsible for the synthesis of granulose in *Firmicutes* are still unknown.

6.2 Candidatus Methylacidiphilum fumariolicum

The second example concerns *Ca. M. fumariolicum*, which belongs to the *Verrucomicrobia* phylum (Pol et al. 2007). *Planctomycetes*, *Verrucomicrobia*, and *Chlamydiae* form the PVC superphylum, which to date includes two additional phyla *Lentisphaerae* and *Poribacteria*. *Ca. M. fumariolicum* is a methanotrophic bacterium that uses methane as both carbon and energy source. Interestingly, *Ca. M. fumariolicum* is able to fix nitrogen and carbon dioxide through the nitrogenase complex and the Calvin-Benson cycle, respectively (Khadem et al. 2010, 2011). In the absence of the nitrogen source (ammonium or dinitrogen gas), *Ca. M. fumariolicum* accumulates both electron-dense bodies of 100–200 nm with an elliptical or circular shape (that have been identified as abnormal glycogen particles) and soluble polysaccharide. The physiological function of the abnormal glycogen particles was further investigated by growth experiments which consisted of depletion of methane and/or nitrogen source (Khadem et al. 2012a). Overall, these experiments suggest that in absence of the major carbon dioxide and energy source (i.e., methane), the carbohydrate granules are the main source of carbon for maintaining the survival of cells. Although little is known about the fine structure of abnormal glycogen, the glycogen metabolism pathway relies on set of enzymes that is similar to that of *E. coli* (Khadem et al. 2012b).

7 Storage Polysaccharide and Fitness in Prokaryotes

A large number of studies on enterobacteria indicates an interconnection of glycogen metabolism with many cellular processes, including nitrogen, iron and magnesium metabolisms, stress responses, and RNA metabolisms. I invite readers to read the excellent review on these aspects (Wilson et al. 2010). Similar interconnections probably occur in cyanobacteria species. Furthermore, both photosynthetic activity and carbon fixation add a layer of complexity in terms of regulation and cross talks between metabolic networks. At the present, there is compelling evidence that glycogen is critical for abiotic stress responses (e.g., salinity, temperature) and for survival under day-night growth conditions (Miao et al. 2003; Suzuki et al. 2010; Gründel et al. 2012). Why do glycogen-less cyanobacteria grow normally in continuous light and not under day-night conditions? The answer starts with the characterization of *rpaA*-null mutant in *Synechococcus*. RpaA is an important transcriptional factor interconnecting circadian clock and different metabolic pathways. It activates the transcription of key genes belonging to the glycogen catabolism pathway (glycogen phosphorylase, *glgP*), glycolysis (glyceraldehyde-3-phosphate dehydrogenase, *gap*; fructose-1,6-bisphosphatase, *fbp*), and the oxidative pentose phosphate pathway (glucose-6-phosphate dehydrogenase, *zwf*; 6-phosphogluconate dehydrogenase, *gnd*). Like glycogen-less mutants, *rpaA*-null mutants grow at the same rate as the wild-type strain in continuous light, but they are not viable under day-night cycle. Furthermore, despite normal GlgC and GlgA activities, the amount of glycogen is strongly reduced in *rpaA* mutants (Puszynska and O'Shea 2017). Two independent research groups investigated the *rpaA* mutant phenotype in *Synechococcus*. First, Diamond and colleagues posit that abnormal accumulation of reactive oxidative species (ROS) produced at night in *rpaA* mutants might explain day-night lethality. Their hypothesis is sustained by the fact that by modulating light intensity per se by modifying ROS levels or/and by reducing NADPH,H+ consuming pathways (knockout of valine, leucine, and isoleucine biosynthetic pathways), the *rpaA* mutant is enabled to grow under day-night cycle (Diamond et al. 2017). Thus, they propose that glycogen degradation and the OPP pathway fuel the cell in reducing power (NADPH,H+) at night which prevent the accumulation of ROS. Further investigations, conducted by Puszynska and O'Shea (2017), suggest that the inability of *rpaA* mutant to maintain the adenylate energetic charge at night might explain the lethality. In order to validate their hypothesis, the genes for the glucose transporter (GalP) as well as for GlgP, Gap, and Zwf activities were expressed in *rpaA* mutant. The authors correlate the phenotype rescue with a restoration of adenylate energetic charge in the *rpaA* mutant. Unfortunately, no information is available concerning the ROS levels in the rescued *rpaA* mutant. Nevertheless, both articles from Diamond et al. (2017) and Puszynska and O'Shea (2017) emphasize the critical role of carbon stores at night. Like in enterobacteria, under macro-element (i.e., N, P) limitation, non-nitrogen-fixing cyanobacteria accumulate a large amount of glycogen between thylakoid membranes. Nitrogen starvation triggers the degradation of photosynthetic pigments in many photosynthetic organisms. This mechanism named the "chlorosis response" induces dormant stage where only basal metabolic and photosynthesis activities occur as long as the nitrogen source is not available. Interestingly, only 48 h is required for a dormant cyanobacterium to be fully metabolically active and perform photosynthesis activity again when nitrogen is supplied (Klotz et al. 2016). In addition, there is now compelling evidence that both chlorosis response and survival at the dormant stage depend explicitly on glycogen metabolism and not on alternative carbon source such as poly-β-hydroxybutyrate stores (Damrow et al. 2016). More recently, Doello and his collaborators investigated the importance of glycogen pool during this resuscitation or awakening process. This study sheds light on the importance of glycogen catabolism in the few hours of awakening when the photosynthetic apparatus is not active yet to supply the cells in ATP (Doello et al. 2018). Glycogen is degraded when the glycogen phosphorylase isoform, GlgP-slr1367, is activated. The glycogen phosphorylase isoform is synthesized during the biosynthesis of glycogen. So far, the mechanisms that control glycogen phosphorylase (GlgP-slr1367) activation or inhibition during degradation and synthesis, respectively, are still unknown. In addition, this work outlines that the main glycolytic pathway routes consist of both oxidative pentose phosphate pathway and Entner-Doudoroff (ED) pathways, recently discovered in cyanobacteria, and not as expected, the Embden-Meyerhof-Parnas (EMP) pathway that yields twice more ATP than ED pathway (Chen et al. 2016). Despite the absence of alternative sources of ATP (i.e., photosynthesis activity) during the awakening process, the thermodynamic constraints of the EMP pathway might favor the use of the ED pathway (Flamholz et al. 2013).

7.1 *The Accumulation of Abnormal Glycogen Affects the Aging of Prokaryotes*

Another intriguing aspect concerns the interconnection between glycogen metabolism and prokaryotic aging or senescence process. The latter is defined by an accumulation of cellular damage within the mother cell, which leads to a decrease in reproductive rates and increase of mortality with age. First described in eukaryotes including yeast, bacteria were considered immortal and free of aging because it was supposed that the cell division gives rise to two identical cells. However, early in 2000, two studies reveal that aging processes occur in bacteria as well (Ackermann et al. 2003; Stewart et al. 2005). Asymmetric partitioning of cellular components in rod-shaped bacteria such as *E. coli* and *Caulobacter crescentus* enables mother and daughter cells to be distinguished. Thus, by tracking cell divisions of rod-shaped bacteria and more precisely by following the polar cells, it appears that bacteria are affected by aging like most cells. More recently, thanks to microfluidic system devices, Boehm and his collaborators have pointed out that the replicative lifespan of *E. coli* is genetically controlled and is linked to the glycogen metabolism pathway (Boehm et al. 2016). In this study, a mutation in the carbon storage regulator gene (*csrA*) led to a drastic reduction of cell division numbers from 150 to 5 for a mother's cell. Null *csrA* mutants in *E. coli* caused an increase both in gluconeogenesis and glycogen biosynthesis and an inhibition of glycolysis as well (Sabnis et al. 1995). That this *csrA* phenotype is rescued in glycogen-less strain mutants deleted in the glycogen synthase gene (*glgA*) further emphasizes the importance of glycogen in the aging process. How can we explain that an excess of glycogen accumulation is responsible for a decrease of replicative lifespan in *E. coli*?

Based on previous reports on effect of partitioning of protein aggregates or inclusion bodies in the aging process in *E. coli* cells (Lindner et al. 2008), the authors localized glycogen particles upon cell division. Interestingly, they observed through fluorescent probe GlgA-GFP in both wild-type and *csrA* mutant that most of glycogen particles occur at the pole of the mother cell, while new released cell is free of glycogen. In the case of *csrA* mutant, the glycogen accumulation gradually increases from the pole of the mother cell more rapidly after each cell division than wild-type cell. Interestingly GglA-GFP fluorescence signal was observed across the entire cell usually after the fifth cell division; *csrA* mutant seems then deprived of the genome, which would explain why these cells then stop growing and dividing (Boehm et al. 2016).

8 Conclusion and Perspectives

The survival of free-living cells depends on their ability to maintain their energy status when faced with fluctuating environmental conditions. This critical issue requires the synthesis of energy-carbon storage compounds such as α-polysaccharides, for fueling the cell in the absence of exogenous energy. Glycogen particles represent the most primitive and abundant form of carbon storage in the living world. Described in species of all domains of life, it seems clear that glycogen particles are optimized to meet the specific needs of the cell and that evolution has spawned various biochemical pathways achieving its synthesis. Hence, mathematical modeling and biological evidence outline the tight relationship between structure and function of glycogen particles. In addition, a large body of evidence suggests that any structural alteration of glycogen leads to deleterious effects for glycogen-accumulating cells. In some cases, however, the glycogen particle was not designed to fulfill the cell requirement. Thus, transition state occurs from tiny hydrosoluble polysaccharide, glycogen to insoluble water polysaccharide starch-like. First described in Archaeplastida lineages (plants/green algae, red algae, glaucophytes), unicellular nitrogen-fixing cyanobacteria synthesize amylopectin/starch granules as well. Although there is no direct proof, it is tempting to suggest that both the ATP cost and anoxia required by nitrogen fixation pathway at night were the main driving force to substitute glycogen toward a better storage polysaccharide like starch. We can hypothesize that the same selection pressure led the nitrogen-fixing *Candidatus Methylacidiphilum fumariolicum* to synthesize carbohydrate granules rather than glycogen particles. Interestingly, the energy cost of endospore formation during solvent production in *Clostridium* might be, in that case, the driving force to substitute glycogen by granulose. Although little is known about the degradation pathway of carbohydrate granules produced in both organisms, those reflect again a compromise between storage capacity and availability of carbon where solid granules, starch and granulose, are optimized to store a maximum of glucose moieties in a limited space.

In addition, we must bear in mind that occurrence of storage polysaccharides is usually associated with the presence of other polymeric materials, such as polyphosphate granules, lipids, and poly-β-hydroxybutyrate, that can be used as energy-carbon store or as energy store exclusively (Rao et al. 2009; Achbergerová and Nahálka 2011; Achbergerová and Nahálka 2011). Thus, it is worth noting that some organisms are completely devoid of storage polysaccharide. In 2002, Henrissat and his associates pointed out the loss of carbohydrate-active enzymes concerns human or mammalian pathogenic bacteria (Henrissat et al. 2002). This observation can be extended to symbiotic bacteria either with bugs (Degnan et al. 2009; Nikoh et al. 2011) or with eukaryotes (Newton et al. 2007).

Acknowledgment I would like to thank Prof. Steven Ball for his critical reading and helpful comments.

References

Achbergerová L, Nahálka J (2011) Polyphosphate—an ancient energy source and active metabolic regulator. Microb Cell Fact 10:1–14

Ackermann M, Stearns SC, Jenal U (2003) Senescence in a bacterium with asymmetric division. Science. 300:1920–1920

Albi T, Serrano A (2016) Inorganic polyphosphate in the microbial world. Emerging roles for a multifaceted biopolymer. World J Microbiol Biotechnol 32:1–12

Albrecht T, Haebel S, Koch A, Krause U, Eckermann N, Steup M (2004) Yeast glycogenin (Glg2p) produced in *Escherichia coli* is simultaneously glucosylated at two vicinal tyrosine residues but results in a reduced bacterial glycogen accumulation. Eur J Biochem 271:3978–3989

Almagro G, Viale AM, Montero M, Rahimpour M, Muñoz FJ, Baroja-Fernández E, Bahaji A, Zúñiga M, González-Candelas F, Pozueta-Romero J (2015) Comparative genomic and phylogenetic analyses of gammaproteobacterial glg genes traced the origin of the *Escherichia coli* glycogen glgBXCAP operon to the last common ancestor of the sister orders Enterobacteriales and Pasteurellales. PLoS One 10:1–30

Alonso-Casajús N, Dauvillée D, Viale AM, Muñoz FJ, Baroja-Fernández E, Morán-Zorzano MT, Eydallin G, Ball S, Pozueta-Romero J (2006) Glycogen phosphorylase, the product of the glgP gene, catalyzes glycogen breakdown by removing glucose units from the nonreducing ends in Escherichia coli. J Bacteriol 188:5266–5272

Ball S, Colleoni C, Cenci U, Raj JN, Tirtiaux C (2011) The evolution of glycogen and starch metabolism in eukaryotes gives molecular clues to understand the establishment of plastid endosymbiosis. J Exp Bot 62:1775–1801

Ball S, Colleoni C, Arias MC (2015) The transition from glycogen to starch metabolism in cyanobacteria and eukaryotes. In Starch—metabolism and structure. Springer, Tokyo, pp 93–158

Ballicora MA, Iglesias AA, Preiss J (2003) ADP-glucose pyrophosphorylase, a regulatory enzyme for bacterial glycogen synthesis. Microbiol Mol Biol Rev 67:213–225

Bandyopadhyay A, Elvitigala T, Liberton M, Pakrasi HB (2013) Variations in the rhythms of respiration and nitrogen fixation in members of the unicellular diazotrophic cyanobacterial genus cyanothece. Plant Physiol 161:1334–1346

Barengo R, Krisman CR (1978) Initiation of glycogen biosynthesis in *Escherichia coli* studies of the properties of the enzymes involved. BBA Gen Subj 540:190–196

Barengo R, Flawiá M, Krisman CR (1975) The initiation of glycogen biosynthesis in *Escherichia coli*. FEBS Lett 53:274–278

Beglova N, Fischer D, Hengge-Aronis R, Gehring K (1997) 1H, 15N and 13C NMR assignments, secondary structure and overall topology of the *Escherichia coli* GlgS protein. Eur J Biochem 246:301–310

Belanger AE, Hatfull GF (1999) Exponential-phase glycogen recycling is essential for growth of Mycobacterium smegmatis. J Bacteriol 181:6670–6678

Bergère J, Rousseau M, Mercier C (1975) Ann Microbiol (Paris). Ann Microbiol 126:295–314

Bertoft E (2004) On the nature of categories of chains in amylopectin and their connection to the super helix model. Carbohydr Polym 57:211–224

Binderup K, Mikkelsen R, Preiss J (2000) Limited proteolysis of branching enzyme from Escherichia coli. Arch Biochem Biophys 377:366–371

Boehm A, Arnoldini M, Bergmiller T, Röösli T, Bigosch C, Ackermann M (2016) Genetic manipulation of glycogen allocation affects replicative lifespan in E. coli. PLoS Genet 12:1–17

Cenci U, Chabi M, Ducatez M, Tirtiaux C, Nirmal-Raj J, Utsumi Y, Kobayashi D, Sasaki S, Suzuki E, Nakamura Y et al (2013) Convergent Evolution of Polysaccharide Debranching Defines a Common Mechanism for Starch Accumulation in Cyanobacteria and Plants. Plant Cell 25:3961–3975

Chandra G, Chater KF, Bornemann S (2011) Unexpected and widespread connections between bacterial glycogen and trehalose metabolism. Microbiology 157:1565–1572

Chen GS, Segel IH (1968) Purification and properties of glycogen phosphorylase from *Escherichia coli*. Arch Biochem Biophys 127:175–186

Chen X, Schreiber K, Appel J, Makowka A, Fähnrich B, Roettger M, Hajirezaei MR, Sönnichsen FD, Schönheit P, Martin WF et al (2016) The Entner–Doudoroff pathway is an overlooked glycolytic route in cyanobacteria and plants. Proc Natl Acad Sci USA 113:5441–5446

Choi J-H, Lee H, Kim Y-W, Park J-T, Woo E-J, Kim M-J, Lee B-H, Park K-H (2009) Characterization of a novel debranching enzyme from Nostoc punctiforme possessing a high specificity for long branched chains. Biochem Biophys Res Commun 378:224–229

Clermont L, Macha A, Müller LM, Derya SM, von Zaluskowski P, Eck A, Eikmanns BJ, Seibold GM (2015) The α-glucan phosphorylase MalP of *Corynebacterium glutamicum* is subject to transcriptional regulation and competitive inhibition by ADP-glucose. J Bacteriol 197:1394–1407

Colleoni C, Suzuki E (2012) Storage polysaccharide metabolism in cyanobacteria. In: Tetlow IJ (ed) Essential reviews in experimental biology: starch: origins, structure and metabolism. The Society for Experimental Biology, London, pp 217–253

Compaoré J, Stal LJ (2010) Oxygen and the light-dark cycle of nitrogenase activity in two unicellular cyanobacteria. Environ Microbiol 12:54–62

Coutinho PM, Deleury E, Davies GJ, Henrissat B (2003) An evolving hierarchical family classification for glycosyltransferases. J Mol Biol 328:307–317

Crumpton-Taylor M, Pike M, Lu KJ, Hylton CM, Feil R, Eicke S, Lunn JE, Zeeman SC, Smith AM (2013) Starch synthase 4 is essential for coordination of starch granule formation with chloroplast division during Arabidopsis leaf expansion. New Phytol 200:1064–1075

Cumino AC, Marcozzi C, Barreiro R, Salerno GL (2007) Carbon cycling in Anabaena sp. PCC 7120. Sucrose synthesis in the heterocysts and possible role in nitrogen fixation. Plant Physiol 143:1385–1397

Damotte M, Cattanéo J, Sigal N, Puig J (1968) Mutants of *Escherichia coli* K12 altered in their ability to store glycogen. Biochem Biophys Res Commun 32:916–920

Damrow R, Maldener I, Zilliges Y (2016) The multiple functions of common microbial carbon polymers, glycogen and PHB, during stress responses in the non-diazotrophic cyanobacterium synechocystis sp. PCC 6803. Front Microbiol 7. https://doi.org/10.3389/fmicb.2016.00966

Dash S, Ng CY, Maranas CD (2016) Metabolic modeling of clostridia: current developments and applications. FEMS Microbiol Lett. 363:1–10

Dauvillée D, Kinderf IS, Li Z, Kosar-Hashemi B, Samuel MS, Rampling L, Ball S, Morell MK (2005) Role of the *Escherichia coli* glgX gene in glycogen metabolism. J Bacteriol 187:1465–1473

Degnan PH, Yu Y, Sisneros N, Wing RA, Moran NA (2009) *Hamiltonella defensa*, genome evolution of protective bacterial endosymbiont from pathogenic ancestors. Proc Natl Acad Sci USA 106:9063–9068

Deschamps P, Colleoni C, Nakamura Y, Suzuki E, Putaux JL, Buléon A, Haebel S, Ritte G, Steup M, Falcón LI et al (2008) Metabolic symbiosis and the birth of the plant kingdom. Mol Biol Evol 25:536–548

Deutscher J, Francke C, Postma PW (2006) How phosphotransferase system-related protein phosphorylation regulates carbohydrate metabolism in bacteria. Microbiol Mol Biol Rev 70:939–1031

Devillers CH, Piper ME, Ballicora MA, Preiss J (2003) Characterization of the branching patterns of glycogen branching enzyme truncated on the N-terminus. Arch Biochem Biophys 418:34–38

Delrue B, Fontaine T, Routier F, Decq A, Wieruszeski JM, Van Den Koornhuyse N, Maddelein ML, Fournet B, Ball S (1992) Waxy Chlamydomonas reinhardtii: monocellular algal mutants defective in amylose biosynthesis and granule-bound starch synthase activity accumulate a structurally modified amylopectin. J Bacteriol 174:3612–3620

Diamond S, Rubin BE, Shultzaberger RK, Chen Y, Barber CD, Golden SS (2017) Redox crisis underlies conditional light–dark lethality in cyanobacterial mutants that lack the circadian regulator, RpaA. Proc Natl Acad Sci USA 114:E580–E589

Díaz-Troya S, López-Maury L, Sánchez-Riego AM, Roldán M, Florencio FJ (2014) Redox regulation of glycogen biosynthesis in the cyanobacterium synechocystis sp. PCC 6803: analysis of the AGP and glycogen synthases. Mol Plant 7:87–100
Doello S, Klotz A, Makowka A, Gutekunst K, Forchhammer K (2018) A specific glycogen mobilization strategy enables rapid awakening of dormant cyanobacteria from chlorosis. Plant Physiol 177:594–603
Ehsaan M, Kuit W, Zhang Y, Cartman ST, Heap JT, Winzer K, Minton NP (2016) Mutant generation by allelic exchange and genome resequencing of the biobutanol organism Clostridium acetobutylicum ATCC 824. Biotechnol Biofuels. 9:4–24
Falcon LI, Cipriano F, Chistoserdov AY, Carpenter EJ (2002) Diversity of diazotrophic unicellular cyanobacteria in the tropical North Atlantic Ocean. Appl Environ Microbiol 68:5760–5764
Figueroa CM, Asención Diez MD, Kuhn ML, McEwen S, Salerno GL, Iglesias AA, Ballicora MA (2013) The unique nucleotide specificity of the sucrose synthase from Thermosynechococcus elongatus. FEBS Lett 587:165–169
Flamholz A, Noor E, Bar-Even A, Liebermeister W, Milo R (2013) Glycolytic strategy as a tradeoff between energy yield and protein cost. Proc Natl Acad Sci USA 110:10039–10044
Fraga J, Maranha A, Mendes V, Pereira PJB, Empadinhas N, Macedo-Ribeiro S (2015) Structure of mycobacterial maltokinase, the missing link in the essential GlgE-pathway. Sci Rep 5:8026
Fu J, Xu X (2006) The functional divergence of two glgP homologues in Synechocystis sp. PCC 6803. FEMS Microbiol Lett 260:201–209
Fukusumi S, Kamizoto A, Horinouchi S, Teruhiko B (1988) Cloning and nucleotide sequence of a heat-stable amylase gene from an anaerobic thermophile, Dictyoglomus thermophilum. Eur J Biochem 174:15–21
Gehre L, Gorgette O, Perrinet S, Prevost M, Ducatez M, Giebel AM, Nelson DE, Ball SG (2016) Sequestration of host metabolism by an intracellular pathogen. Elife. 5:1–22
Gründel M, Scheunemann R, Lockau W, Zilliges Y (2012) Impaired glycogen synthesis causes metabolic overflow reactions and affects stress responses in the cyanobacterium Synechocystis sp. PCC 6803. Microbiol (United Kingdom) 158:3032–3043
Hayashi M, Suzuki R, Colleoni C, Ball SG, Fujita N, Suzuki E (2015) Crystallization and crystallographic analysis of branching enzymes from Cyanothece sp. ATCC 51142. Acta Crystallogr Sect Struct Biol Commun. 71:1109–1113
Hayashi M, Suzuki R, Colleoni C, Ball SG, Fujita N, Suzuki E (2017) Bound substrate in the structure of cyanobacterial branching enzyme supports a new mechanistic model. J Biol Chem. 292:5465–5475
Hengge-Aronis R, Fischer D (1992) Identification and molecular analysis of glgS, a novel growth-phase-regulated and rpoS-dependent gene involved in glycogen synthesis in Escherichia coli. Mol Microbiol 6:1877–1886
Henrissat B, Deleury E, Coutinho PM (2002) Glycogen metabolism loss: a common marker of parasitic behaviour in bacteria? Trends Genet 18:437–440
Hizukuri S (1986) Polymodal distribution of the chain lengths of amylopectin and its significance. Carbohydr Res 147:342–347
Holmes E, Preiss J (1979) Characterization of Escherichia coli B glycogen synthase enzymatic reactions and products. Arch Biochem Biophys 196:436–448
Iglesias A, Kakefuda G, Preiss J (1991) Regulatory and structural properties of the cyanobacterial ADPglucose pyrophosphorylases. Plant Physiol 97:1187–1195
Jendrossek D, Pfeiffer D (2014) New insights in the formation of polyhydroxyalkanoate granules (carbonosomes) and novel functions of poly(3-hydroxybutyrate). Environ Microbiol 16:2357–2373
Jo HJ, Park S, Jeong HG, Kim JW, Park JT (2015) *Vibrio vulnificus* glycogen branching enzyme preferentially transfers very short chains: N1 domain determines the chain length transferred. FEBS Lett 589:1089–1094
Johnstone K, Holland KT (1977) Ultrastructural changes during sporulation of *Clostridium bifermentans*. J Gen Microbiol 100:217–220

Kadouche D, Ducatez M, Cenci U, Tirtiaux C, Suzuki E, Nakamura Y, Putaux J-L, Terrasson AD, Diaz-Troya S, Florencio FJ et al (2016) Characterization of function of the GlgA2 glycogen/starch synthase in *Cyanobacterium* sp. Clg1 highlights convergent evolution of glycogen metabolism into starch granule aggregation. Plant Physiol 171:1879–1892

Kalscheuer R, Syson K, Veeraraghavan U, Weinrick B, Biermann KE, Liu Z, Sacchettini JC, Besra G, Bornemann S, Jacobs WR (2010) Self-poisoning of *Mycobacterium tuberculosis* by targeting GlgE in an α-glucan pathway. Nat Chem Biol 6:376–384

Kawaguchi K, Fox J, Holmes E, Boyer C, Preiss J (1978) De Novo synthesis of *Escherichia coli* glycogen is due to primer associated with glycogen synthase and activation by branching enzyme. Arch Biochem Biophys 190:385–397

Khadem AF, Pol A, Jetten MSM, Op Den Camp HJM (2010) Nitrogen fixation by the verrucomicrobial methanotroph "*Methylacidiphilum fumariolicum*" SolV. Microbiology 156:1052–1059

Khadem AF, Pol A, Wieczorek A, Mohammadi SS, Francoijs K-J, Stunnenberg HG, Jetten MSM, Op den Camp HJM (2011) Autotrophic methanotrophy in verrucomicrobia: Methylacidiphilum fumariolicum SolV uses the calvin-benson-bassham cycle for carbon dioxide fixation. J Bacteriol 193:4438–4446

Khadem AF, van Teeseling MCF, van Niftrik L, Jetten MSM, Op den Camp HJM, Pol A (2012a) Genomic and physiological analysis of carbon storage in the verrucomicrobial methanotroph Ca. *Methylacidiphilum fumariolicum* SolV. Front Microbiol 3:345

Khadem AF, Van Teeseling MCF, Van Niftrik L, Jetten MSM, Op Den Camp HJM, Pol A (2012b) Genomic and physiological analysis of carbon storage in the verrucomicrobial methanotroph "Ca. *Methylacidiphilum fumariolicum*" SolV. Front Microbiol 3:1–10

Klotz A, Georg J, Bučinská L, Watanabe S, Reimann V, Januszewski W, Sobotka R, Jendrossek D, Hess WR, Forchhammer K (2016) Awakening of a dormant cyanobacterium from nitrogen chlorosis reveals a genetically determined program. Curr Biol 26:2862–2872

Koliwer-Brandl H, Syson K, van de Weerd R, Chandra G, Appelmelk B, Alber M, Ioerger TR, Jacobs WR, Geurtsen J, Bornemann S et al (2016) Metabolic network for the biosynthesis of intra- and extracellular α-glucans required for virulence of *Mycobacterium tuberculosis*. PLoS Pathog 12:1–26

Kopp RE, Kirschvink JL, Hilburn IA, Nash CZ (2005) The Paleoproterozoic snowball Earth: a climate disaster triggered by the evolution of oxygenic photosynthesis. Proc Natl Acad Sci USA 102:11131–11136

Laderman KA, Asada K, Uemori T, Mukai H, Taguchi Y, Kato I, Anfinsen CB (1993) α-Amylase from the hyperthermophilic archaebacterium Pyrococcus furiosus. Cloning and sequencing of the gene and expression in *Escherichia coli*. J Biol Chem 268:24402–24407

Leiba J, Syson K, Baronian G, Zanella-Cléon I, Kalscheuer R, Kremer L, Bornemann S, Molle V (2013) *Mycobacterium tuberculosis* maltosyltransferase GlgE, a genetically validated antituberculosis target, is negatively regulated by Ser/Thr phosphorylation. J Biol Chem 288:16546–16556

Leloir LF, Cardini CE (1957) Biosynthesis of glycogen from uridine diphosphate glucose. J Am Chem Soc 79:6340–6341

Lemassu A, Daffé M (1994) Structural features of the exocellular polysaccharides of *Mycobacterium tuberculosis*. Biochem J 297:351–357

Liberton M, Bandyopadhyay A, Pakrasi HB (2019) Enhanced nitrogen fixation in a glgX-deficient strain of Cyanothece sp. strain ATCC 51142, a unicellular nitrogen-fixing cyanobacterium. Appl Environ Microbiol. https://doi.org/10.1128/AEM.02887-18

Lindner AB, Madden R, Demarez A, Stewart EJ, Taddei F (2008) Asymmetric segregation of protein aggregates is associated with cellular aging and rejuvenation. Proc Natl Acad Sci USA 105:3076–3081

Lou J, Dawson KA, Strobel HJ (1997) Glycogen biosynthesis via UDP-glucose in the ruminal bacterium *Prevotella bryantii* B14. Appl Environ Microbiol 63:4355–4359

Mattoo RL, Waygood EB (1983) Determination of the levels of HPr and enzyme I of the phosphoenolpyruvate-sugar phosphotransferase system in *Escherichia coli* and *Salmonella typhimurium*. Can J Biochem Cell Biol 61:29–37

Meléndez R, Meléndez-Hevia E, Cascante M (1997) How did glycogen structure evolve to satisfy the requirement for rapid mobilization of glucose? A problem of physical constraints in structure building. J Mol Evol 45:446–455

Meléndez R, Meléndez-Hevia E, Canela EI (1999) The fractal structure of glycogen: a clever solution to optimize cell metabolism. Biophys J 77:1327–1332

Melendez-hevia E, Waddellt TG, Sheltont ED (1993) Optimization of molecular design in the evolution of metabolism : the glycogen molecule. Biochem J 295:477–483

Miao X, Wu Q, Wu G, Zhao N (2003) Changes in photosynthesis and pigmentation in an agp deletion mutant of the cyanobacterium *Synechocystis* sp. Biotechnol Lett 25:391–396

Monod J, Torriani A-M (1950) De l'amylomaltase d'*Escherichia coli*. Ann Inst Pasteur 78:65–77

Mouille G, Maddelein ML, Libessart N, Talaga P, Decq A, Delrue B, Ball S (1996) Preamylopectin processing: A mandatory step for starch biosynthesis in plants. Plant Cell 8:1353–1366

Murakami T, Kanai T, Takata H, Kuriki T, Imanaka T (2006) A novel branching enzyme of the GH-57 family in the hyperthermophilic archaeon *Thermococcus kodakaraensis* KOD1. J Bacteriol 188:5915–5924

Nakamura Y, Takahashi JI, Sakurai A, Inaba Y, Suzuki E, Nihei S, Fujiwara S, Tsuzuki M, Miyashita H, Ikemoto H et al (2005) Some cyanobacteria synthesize semi-amylopectin type α-polyglucans instead of glycogen. Plant Cell Physiol 46:539–545

Newton ILG, Woyke T, Auchtung TA, Dilly GF, Dutton RJ, Fisher MC, Fontanez KM, Lau E, Stewart FJ, Richardson PM et al (2007) The Calyptogena magnifica chemoautotrophic symbiont genome. Science 315:998–1001

Nikoh N, Hosokawa T, Oshima K, Hattori M, Fukatsu T (2011) Reductive evolution of bacterial genome in insect gut environment. Genome Biol Evol 3:702–714

Palomo M, Kralj S, Van Der Maarel MJEC, Dijkhuizen L (2009) The unique branching patterns of Deinococcus glycogen branching enzymes are determined by their N-terminal domains. Appl Environ Microbiol 75:1355–1362

Park JT, Shim JH, Tran PL, Hong IH, Yong HU, Oktavina EF, Nguyen HD, Kim JW, Lee TS, Park SH et al (2011) Role of maltose enzymes in glycogen synthesis by *Escherichia coli*. J Bacteriol 193:2517–2526

Piast RW, Wieczorek RM (2017) Origin of life and the phosphate transfer catalyst. Astrobiology 17:277–285

Pol A, Heijmans K, Harhangi HR, Tedesco D, Jetten MSM, Op Den Camp HJM (2007) Methanotrophy below pH 1 by a new Verrucomicrobia species. Nature 450:874–878

Porchia AC, Curatti L, Salerno GL (1999) Sucrose metabolism in cyanobacteria: sucrose synthase from *Anabaena* sp. strain PCC 7119 is remarkably different from the plant enzymes with respect to substrate affinity and amino-terminal sequence. Planta 210:34–40

Preiss J (2014) Glycogen: biosynthesis and regulation. EcoSal Plus. 6:1–28

Puszynska AM, O'Shea EK (2017) Switching of metabolic programs in response to light availability is an essential function of the cyanobacterial circadian output pathway. elife. https://doi.org/10.7554/eLife.23210

Rahimpour M, Montero M, Almagro G, Viale AM, Sevilla Á, Cánovas M, Muñoz FJ, Baroja-Fernández E, Bahaji A, Eydallin G et al (2013) GlgS, described previously as a glycogen synthesis control protein, negatively regulates motility and biofilm formation in *Escherichia coli*. Biochem J 452:559–573

Rao NN, Gómez-García MR, Kornberg A (2009) Inorganic polyphosphate: essential for growth and survival. Annu Rev Biochem 78:605–647

Reysenbach AL, Ravenschroft N, Long S (1986) Characterization, biosynthesis, and regulation of granulose in *Clostridium acetobutylicum*. Appl Environ Microbiol 52:185–190

Robson RL, Robson RM, Morris JG (1974) The biosynthesis of granulose by *Clostridium pasteurianum*. Biochem J 144:503–511

Roy R, Usha V, Kermani A, Scott DJ, Hyde EI, Besra GS, Alderwick LJ, Fütterer K (2013) Synthesis of α-glucan in mycobacteria involves a hetero-octameric complex of trehalose synthase tres and maltokinase Pep2. ACS Chem Biol 8:2245–2255

Sabnis NA, Yang H, Romeo T (1995) Pleiotropic regulation of central carbohydrate metabolism in *Escherichia coli* via the gene csrA. J Biol Chem 270:29096–29104

Sambou T, Dinadayala P, Stadthagen G, Barilone N, Bordat Y, Constant P, Levillain F, Neyrolles O, Gicquel B, Lemassu A et al (2008) Capsular glucan and intracellular glycogen of *Mycobacterium tuberculosis* : biosynthesis and impact on the persistence in mice. Mol Microbiol 70:762–774

Sánchez-Baracaldo P, Ridgwell A, Raven JA (2014) A neoproterozoic transition in the marine nitrogen cycle. Curr Biol 24:652–657

Santos CR, Tonoli CCC, Trindade DM, Betzel C, Takata H, Kuriki T, Kanai T, Imanaka T, Arni RK, Murakami MT (2011) Structural basis for branching-enzyme activity of glycoside hydrolase family 57: structure and stability studies of a novel branching enzyme from the hyperthermophilic archaeon *Thermococcus Kodakaraensis* KOD1. Proteins 79:547–557

Sawada T, Nakamura Y, Ohdan T, Saitoh A, Francisco PB, Suzuki E, Fujita N, Shimonaga T, Fujiwara S, Tsuzuki M et al (2014) Diversity of reaction characteristics of glucan branching enzymes and the fine structure of α-glucan from various sources. Arch Biochem Biophys 562:9–21

Schneegurt MA, Sherman DM, Nayar S, Sherman LA (1994) Oscillating behavior of carbohydrate granule formation and dinitrogen fixation in the cyanobacterium *Cyanothece* sp. strain ATCC51142. J Bacteriol 176:1586–1597

Schneegurt MA, Sherman DM, Sherman LA (1997) Composition of the carbohydrate granules of the cyanobacterium, *Cyanothece* sp. strain ATCC 51142. Arch Microbiol 167:89–98

Schwartz M (1967) Expression phenotypique et localisation génétique de mutations affectant le métabolism du maltose chez *Escherichia coli* K12. Ann Inst Pasteur (Paris) 112:673–700

Seibold GM, Wurst M, Eikmanns BJ (2009) Roles of maltodextrin and glycogen phosphorylases in maltose utilization and glycogen metabolism in *Corynebacterium glutamicum*. Microbiology 155:347–358

Seok YJ, Sondej M, Badawi P, Lewis MS, Briggs MC, Jaffe H, Peterkofsky A (1997) High affinity binding and allosteric regulation of *Escherichia coli* glycogen phosphorylase by the histidine phosphocarrier protein, HPr. J Biol Chem 272:26511–26521

Seok YJ, Koo BM, Sondej M, Peterkofsky A (2001) Regulation of *E. coli* glycogen phosphorylase activity by HPr. J Mol Microbiol Biotechnol 3:385–393

Singh PK (1973) Nitrogen fixation by the unicellular green-alga *Aphanothece*. Arch Mikrobiol 92:59–62

Stam MR, Danchin EGJ, Rancurel C, Coutinho PM, Henrissat B (2006) Dividing the large glycoside hydrolase family 13 into subfamilies: towards improved functional annotations of α-amylase-related proteins. Protein Eng Des Sel 19:555–562

Stewart EJ, Madden R, Paul G, Taddei F (2005) Aging and death in an organism that reproduces by morphologically symmetric division. PLoS Biol 3:0295–0300

Streb S, Delatte T, Umhang M, Eicke S, Schorderet M, Reinhardt D, Zeeman SC (2008) Starch granule biosynthesis in arabidopsis is abolished by removal of all debranching enzymes but restored by the subsequent removal of an endoamylase. Plant Cell 20:3448–3466

Summons RE, Jahnke LL, Hope JM, Logan GA (1999) 2-Methylhopanoids as biomarkers for cyanobacterial oxygenic photosynthesis. Nature 400:554–557

Suzuki E, Suzuki R (2016) Distribution of glucan-branching enzymes among prokaryotes. Cell Mol Life Sci 73:2643–2660

Suzuki E, Umeda K, Nihei S, Moriya K, Ohkawa H, Fujiwara S, Tsuzuki M, Nakamura Y (2007) Role of the GlgX protein in glycogen metabolism of the cyanobacterium, *Synechococcus elongatus* PCC 7942. Biochim Biophys Acta Gen Subj 1770:763–773

Suzuki E, Ohkawa H, Moriya K, Matsubara T, Nagaike Y, Iwasaki I, Fujiwara S, Tsuzuki M, Nakamura Y (2010) Carbohydrate Metabolism in Mutants of the Cyanobacterium

Synechococcus elongatus PCC 7942 Defective in Glycogen Synthesis. Appl Environ Microbiol 76:3153–3159

Suzuki E, Onoda M, Colleoni C, Ball S, Fujita N, Nakamura Y (2013) Physicochemical variation of cyanobacterial starch, the insoluble α-glucans in cyanobacteria. Plant Cell Physiol 54:465–473

Suzuki R, Koide K, Hayashi M, Suzuki T, Sawada T, Ohdan T, Takahashi H, Nakamura Y, Fujita N, Suzuki E (2015) Functional characterization of three (GH13) branching enzymes involved in cyanobacterial starch biosynthesis from Cyanobacterium sp. NBRC 102756. Biochim Biophys Acta Proteins Proteomics 1854:476–484

Syson K, Stevenson CEM, Rashid AM, Saalbach G, Tang M, Tuukkanen A, Svergun DI, Withers SG, Lawson DM, Bornemann S (2014) Structural insight into how *Streptomyces coelicolor* maltosyl transferase GlgE binds α-maltose 1-phosphate and forms a maltosyl-enzyme intermediate. Biochemistry 53:2494–2504

Takata H, Takaha T, Okada S, Takagi M, Imanaka T (1998) Purification and characterization of α-glucan phosphorylase from bacillus stearothermophilus. J Ferment Bioeng 85:156–161

Testoni G, Duran J, García-Rocha M, Vilaplana F, Serrano AL, Sebastián D, López-Soldado I, Sullivan MA, Slebe F, Vilaseca M et al (2017) Lack of glycogenin causes glycogen accumulation and muscle function impairment. Cell Metab 26:256–266

Torija MJ, Novo M, Lemassu A, Wilson W, Roach PJ, François J, Parrou JL (2005) Glycogen synthesis in the absence of glycogenin in the yeast Saccharomyces cerevisiae. FEBS Lett 579:3999–4004

Tracy BP, Gaida SM, Papoutsakis ET (2008) Development and application of flow-cytometric techniques for analyzing and sorting endospore-forming clostridia. Appl Environ Microbiol 74:7497–7506

Ugalde JE, Parodi AJ, Ugalde RA (2003) *De novo* synthesis of bacterial glycogen: *Agrobacterium tumefaciens* glycogen synthase is involved in glucan initiation and elongation. Proc Natl Acad Sci U S A 100:10659–10663

Vamadevan V, Bertoft E (2014) Structure-function relationships of starch components. Starch/Stärke 66:1–14

Wang L, Regina A, Butardo VM, Kosar-Hashemi B, Larroque O, Kahler CM, Wise MJ (2015) Influence of *in situ* progressive N-terminal is still controversial truncation of glycogen branching enzyme in Escherichia coli DH5α on glycogen structure, accumulation, and bacterial viability. BMC Microbiol. 15:96

Wilson WA, Roach PJ, Montero M, Baroja-Fernández E, Muñoz FJ, Eydallin G, Viale AM, Pozueta-Romero J (2010) Regulation of glycogen metabolism in yeast and bacteria. FEMS Microbiol Rev 34:952–985

Wyatt JT, Silvey JKG (1969) Nitrogen fixation by Gloeocapsa. Science. 165:908–909

Yoo S-H, Lee B-H, Moon Y, Spalding MH, Jane J (2014) Glycogen synthase isoforms in Synechocystis sp. PCC6803: identification of different roles to produce glycogen by targeted mutagenesis. PLoS One 9:e91524

Zona R, Chang-Pi-Hin F, O'Donohue MJ, Janeček Š (2004) Bioinformatics of the glycoside hydrolase family 57 and identification of catalytic residues in amylopullulanase from Thermococcus hydrothermalis. Eur J Biochem 271:2863–2872

Wax Ester and Triacylglycerol Inclusions

Alexander Steinbüchel and Marc Wältermann

Contents

1 Introduction 212
2 Occurrence of Storage Lipids in Prokaryotes and Their Chemical Structure 213
2.1 TAGs in Bacteria 214
2.2 WEs in Bacteria 215
3 Metabolism of Storage Lipids in Bacteria 216
3.1 TAG Metabolism 216
3.2 WE Metabolism 218
4 Structure and Morphology of Prokaryotic Lipid Inclusions 219
4.1 TAG Inclusions 219
4.2 WE Inclusions 222
5 Formation of TAG and WE Inclusions and PHA Granules in Prokaryotes 223
5.1 TAG Inclusions 223
5.2 WE Inclusions 225
5.3 PHA Granules 227
6 Structure, Formation, and Protein Equipment of Eukaryotic Lipid Inclusions 229
6.1 Plants 230
6.2 Mammals 231
6.3 Eukaryotic Microorganisms 233
7 Microbial Lipids as Sources for Biofuels 234
8 Concluding Remarks and Future Perspectives 234
References 235

Abstract Neutral lipids comprising wax esters (WEs) and triacylglycerols (TAGs) occur frequently as energy and carbon stores in several groups of bacteria.

A. Steinbüchel (✉)
Institut für Molekulare Mikrobiologie und Biotechnologie, Westfälische Wilhelms-Universität, Münster, Germany

Environmental Science Department, King Abdulaziz University, Jeddah, Saudi Arabia
e-mail: steinbu@uni-muenster.de

M. Wältermann
Institut für Molekulare Mikrobiologie und Biotechnologie, Westfälische Wilhelms-Universität, Münster, Germany

D. Jendrossek (ed.), *Bacterial Organelles and Organelle-like Inclusions*, Microbiology Monographs 34, https://doi.org/10.1007/978-3-030-60173-7_9

Biosynthesis of these lipids are promoted in cells grown during unbalanced growth, if an essential nutrient is limiting growth and if a surplus of a carbon source is available. They are mobilized under conditions of carbon and energy deficiency. In general, neutral lipids are stored as insoluble inclusions with different shapes and sizes inside the cytoplasm, depending on the lipid, strain, and culture conditions. The structure, morphology, and biogenesis of these inclusions are the main topics of this chapter. A short overview of the metabolic pathways leading to the biosynthesis of TAGs and WEs and the enzymes involved therein is given. Parallels to and differences from the structure and formation of polyhydroxyalkanoate granules in bacteria and neutral lipid bodies in multicellular organisms such as plants and animals and unicellular eukaryotes are described. The biotechnological potential of microbial lipids to satisfy the need for fuels is also shortly reviewed.

1 Introduction

Accumulation of storage lipids is a widespread property among prokaryotes. Three microorganisms have been studied in much detail regarding various aspects of wax esters (*Acinetobacter baylyi*), triacylglycerols (*Rhodococcus opacus*), and poly (3-hydroxybutyrate) (*Ralstonia eutropha*). All accumulated lipids consist of esterified fatty or hydroxy fatty acids. These lipids serve as a depot for energy and carbon needed for maintenance of metabolic activity and synthesis of cellular metabolites during starvation. By far most prokaryotic species are capable of synthesizing and accumulating hydrophobic polymeric substances like poly(3-hydroxybutyric acid) (PHB) (Fig. 1) or other polyhydroxyalkanoates (PHAs) such as poly (3-hydroxyoctanoate) (Fig. 1) (Steinbüchel 2001; Steinbüchel and Valentin 1995). A variety of species, mainly belonging to the actinomycetes group and some Gram-negative species, are capable of synthesizing and accumulating neutral lipids like triacylglycerols (TAGs) (Fig. 1) and wax esters (WEs) (Fig. 1) as intracellular storage compounds (Alvarez and Steinbüchel 2002; Bryn et al. 1977; Russel and Volkman 1980; Bredemeier et al. 2003). In contrast to their restricted occurrence in prokaryotes, lipids are more frequently accumulated in eukaryotes. Whereas TAGs occur often as the primary energy store in eukaryotes (Zweytick et al. 2000; Murphy 2001), intracellular accumulation of WEs occurs only in the seeds of the jojoba (*Simmondsia chinensis*) plant (Yermanos 1975). These lipids represent ideal storage compounds, since they exhibit a low biological toxicity compared with free fatty or hydroxy fatty acids and are water-insoluble and osmotically inert; therefore, they do not affect the water balance of the cells. Furthermore, accumulation of lipids is much more advantageous than that of proteins or carbohydrates, owing to their much higher caloric value and their relative compactness (Murphy 1993). Storage lipid biosynthesis in bacteria and unicellular eukaryotes is induced in response to stress imposed on the cells, for example, when the cells find themselves in a nutrient-limited environment and when a carbon source is available that can be used for

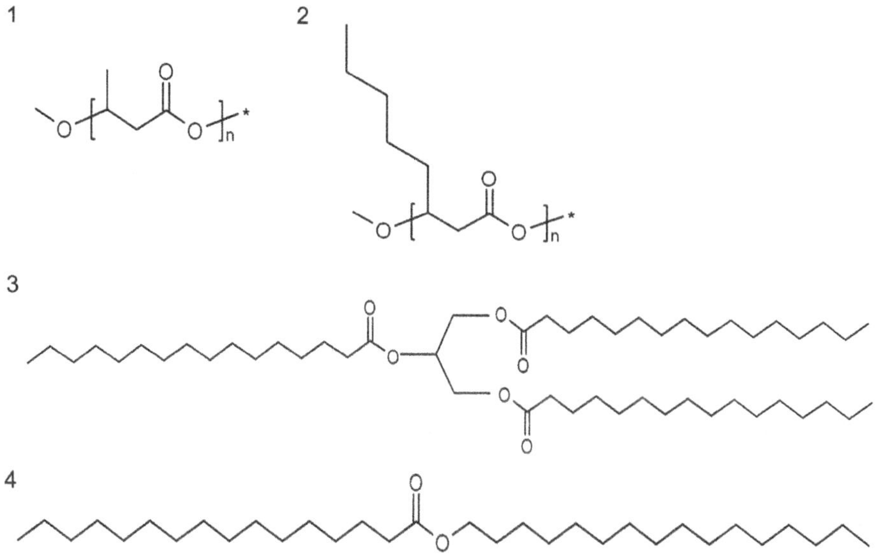

Fig. 1 Chemical structures of lipid storage compounds in bacteria. (**1**) Poly(3-hydroxybutyric acid) (*PHB*) as an example of a short-chain polyhydroxyalkanoate (*PHA*); (**2**) poly(3-hydroxyoctanoate) as an example of a medium-chain PHA; (**3**) tripalmitoylglycerol as an example of triacylglycerols (*TAGs*); (**4**) cetyl palmitate as an example of bacterial wax esters (*WEs*)

storage lipid biosynthesis. Owing to their hydrophobicity and water insolubility, storage lipids are always deposited as intracellular inclusions.

In general, these inclusions are composed of a core of the respective storage lipid (TAGs, WEs, sterol esters, or PHAs) which is surrounded by a layer of phospholipids and/or special classes of proteins (Wältermann and Steinbüchel 2005; Zweytick et al. 2000; Murphy 2001). However, although prokaryotic and eukaryotic lipid inclusions seem structurally related and relatively simple, they exhibit striking differences. The occurrence, composition, morphology, and formation of prokaryotic TAG and WE inclusions and their relationships to PHA granules in bacteria will be an important topic of this treatise. In addition, a basic overview of the different situation of neutral lipid biosynthesis and lipid body structure and formation in plants, mammals, and eukaryotic microorganisms will be given.

2 Occurrence of Storage Lipids in Prokaryotes and Their Chemical Structure

Storage lipids occur in many prokaryotes. The role and functions of these intracellular lipid droplets in prokaryotes were critically summarized and reviewed by several authors (e.g., Murphy 2012). Only a few prokaryotes lack the ability to accumulate lipids. The species belonging to the latter group exist often in nutrient-rich habitats, in which biosynthesis of lipophilic storage compounds seems meaningless (e.g., lactobacilli, *Enterobacteriaceae*, and methanogens). Therefore, it is very likely that accumulation of lipophilic storage compounds is advantageous for survival in natural habitats usually providing little carbon and that it has evolved during evolution.

2.1 TAGs in Bacteria

TAGs (trioxoesters of glycerol and long-chain fatty acids) are abundant and important storage lipids in bacteria and have mainly been described in *Actinomycetales* like *Actinomyces, Arthrobacter, Mycobacterium* sp. (including *Mycobacterium tuberculosis*), *Nocardia* ssp., *Rhodococcus* sp., *Gordonia* sp., and *Dietzia* sp. (Koval'schuk et al. 1973; Wayman et al. 1984; Akao and Kusaka 1976; Alvarez et al. 1997a, b; Alvarez and Steinbüchel 2002; Barksdale and Kim 1977) and in streptomycetes such as *Streptomyces coelicolor*, *Streptomyces lividans*, and *Micromonospora echinospora* (Olukoshi and Packter 1994; Hoskisson et al. 2001). The amounts of TAGs in these species strongly vary from merely 10% up to an enormous 80% of the cellular dry matter, for example, in *Rhodococcus opacus* PD630, which is one of the best studied model organisms with respect to its neutral lipid metabolism (Alvarez et al. 1996). In general, species of the genus *Rhodococcus* seem to be all oleaginous and accumulate interesting lipids (Alvarez et al. 2019). It should be mentioned that a few actinomycetes are capable of accumulating significant amounts of PHAs simultaneously with TAGs. For example, *R. ruber* is capable of accumulating similar amounts of TAGs and the copolyester poly (3-hydroxybutyrate-*co*-3-hydroxyvalerate) from glucose (Alvarez et al. 1997b; Kalscheuer et al. 2001). TAGs are also frequently accumulated in the Gram-negative genera *Acinetobacter* and *Alcanivorax* (Makula et al. 1975; Scott and Finnerty 1976; Singer et al. 1985; Bredemeier et al. 2003). These species accumulate minor amounts of TAGs besides major amounts of WEs, which serve as the predominant storage lipid. Members of the genus *Alcanivorax* and others belonging to the hydrocarbonoclastic marine bacteria, like *Marinobacter hydrocarbonoclasticus*, form lipids from hydrocarbons (Manilla-Perez et al. 2010). Recently, lipids were detected in many bacteria belonging to the genera *Rhodococcus* and *Streptomyces* that were isolated from desert soils (Röttig et al. 2016; Hauschild et al. 2017).

There was also a report of TAG biosynthesis in *Pseudomonas aeruginosa* strain 44T1 (De Andrès et al. 1991). However, the occurrence of TAGs in this strain could not be confirmed in our laboratory (Wältermann and Steinbüchel, unpublished results). In addition, TAGs have so far never been reported to occur in Archaea.

Compositions and amounts of bacterial TAGs strongly depend on the cultivation conditions. In general, cells do not contain significant amounts of TAGs in their exponential growth phases or when cultivated in nutrient-rich complex media. The amount of TAGs in the cells increases when the cells enter the stationary growth phase, in particular when they are cultivated in mineral salt medium with a high carbon-to-nitrogen ratio with unbalanced composition and under unbalanced conditions (Packter and Olukoshi 1995; Wältermann et al. 2005). The fatty acid composition of the accumulated TAGs depends strongly on the carbon source used for growth and fatty acid biosynthesis as well as on its concentration. In general, bacterial TAGs are composed mainly of long-chain, even-numbered saturated, and monoenic fatty acids such as palmitic and oleic acids when grown on structurally unrelated carbon sources like fructose, glucose, or gluconate; however, significant variations also occur (Alvarez et al. 1997a, b). For example, TAGs from gluconate-grown cells of *R. opacus* PD630 contain considerable amounts of odd-numbered fatty acids such as margaric acid (11.4%) and heptadecenoic acid (10.6%) besides palmitic acid (36.4% of total fatty acid content) and oleic acid (19.4%). The fatty acid composition of TAGs in *R. opacus* varies strongly when the cells are cultivated on propionate or valerate. Under these conditions, the fraction of odd-numbered fatty acids strongly increases, whereas the fraction of even-numbered fatty acids decreases (Alvarez et al. 1997b). Biosynthesis of TAGs with polyunsaturated fatty acid residues from unrelated carbon sources has never been observed in bacteria. In addition, formation of TAGs does not only depend on fatty acid de novo synthesis, as was revealed by growth experiments using different hydrocarbons as carbon sources, which were directly converted into their respective acyl coenzyme A (acyl-CoA) and were incorporated into TAGs (Alvarez et al. 2001, 2002).

Similar to TAGs from plant or animal origin, the fatty acid distribution of the acyl residues at the glycerol backbone is not random as was revealed from stereospecific chemical analyses. For example, in *Mycobacterium smegmatis* and *Mycobacterium bovis*, fatty acids with acyl chains of more than 20 carbon atoms were predominantly esterified to the *sn*-3 position of the glycerol, whereas C16 fatty acids occupied the *sn*-2 position and either stearyl, oleyl, or tuberculostearyl residues were located at position *sn*-1 (Walker et al. 1970). In TAGs of gluconate-grown *R. opacus* PD630, shorter and saturated fatty acid residues occur more often at position *sn*-2 in comparison to position *sn*-1 or *sn*-3. The unsaturated fatty acid residues are predominantly found at position *sn*-3, whereas unsaturated fatty acids and acyl chains with more than 17 carbon atoms were not found at position *sn*-2 (Wältermann et al. 2000).

2.2 WEs in Bacteria

For more than 40 years, biosynthesis and accumulation of WEs (oxoesters of primary long-chain fatty alcohols and long-chain fatty acids) have often been reported in species of the genus *Acinetobacter* (Fixter and Fewson 1974; Fixter and McCormack 1976; Gallagher 1971; Scott and Finnerty 1976; Kalscheuer et al. 1999) and less frequently also in species of *Moraxella*, *Micrococcus*, and *Alcanivorax* (Bryn et al. 1977; Russel and Volkman 1980; Bredemeier et al. 2003). Also in actinomycetes, WEs have been described, for example, in *Corynebacterium*, *Mycobacterium tuberculosis*, *Mycobacterium ratisbonense*, and *Nocardia* (Bacchin et al. 1974; Raymond and Davies 1960; Silva et al. 2007;

Wang et al. 1972). Similar to the accumulation of PHAs and TAGs, WE accumulation occurs during growth limitation owing to a lack of a suitable nitrogen source. Under this condition, cells of *Acinetobacter* sp. also synthesize WEs from carbon sources like acetate, sugars, and sugar acids; however, the total amount of WEs is low in contrast to that of cells grown on alkanes, in which the total WE content can reach about 25% of the cellular dry matter (Fixter et al. 1986). The chemical structures of WEs are similar to those produced by jojoba and the sperm whale (*Physeter macrocephalus*). They consist mainly of 32–36 carbon atoms, with saturated and monoenic C16 and C18 fatty acid and fatty alcohol residues, respectively (Ervin et al. 1984; Ishige et al. 2002). *Acinetobacter* sp. also synthesizes WEs from long-chain alkanes, with fatty acid and fatty alcohol residues according to the chain length of the assimilated alkanes. Furthermore, WEs with chain lengths of two and four carbon atoms less were also synthesized (Dewitt et al. 1982). The degree of saturation of the accumulated WEs depends on the culture temperature, with higher ratios of monoenic and dienic WEs occurring at lower temperatures (Ervin et al. 1984). Interestingly, *A. baylyi* strain ADP1 and *Alcanivorax jadensis* are also capable of synthesizing unusual wax diesters in the presence of diols or even from hexadecane (Kalscheuer et al. 2003; Reers, Wältermann, Luftmann and Steinbüchel, unpublished results).

WEs do not exclusively occur as intracellular storage compounds, since some strains of *Acinetobacter* sp. and *Alcanivorax jadensis* were reported to also synthesize extracellular WEs from hydrocarbons, but the origin and possible function of these extracellular WEs remain to be elucidated (Dewitt et al. 1982; Makula et al. 1975; Singer et al. 1985; Bredemeier et al. 2003).

3 Metabolism of Storage Lipids in Bacteria

A very recent review about the biochemistry of triacylglycerol and wax ester accumulation in lipid-accumulating bacteria was published by Alvarez (Alvarez 2016). This and in addition physiological aspects were also reviewed by Athenstaedt and Daum (2006). As the interest of academia and industry in bacterial storage lipids is continuously rising, much research has been focused on the biochemical and molecular basics of storage lipid biosynthesis during recent years.

3.1 TAG Metabolism

In general, TAG biosynthesis in bacteria is carried out via the Kennedy pathway, involving three sequential acylation reactions (Fig. 2) (Lehner and Kuksis 1996). This pathway combines the formation of TAGs and phospholipids, since biosynthesis of phosphatidate and diacylglycerol (DAG) from glycerol-3-phosphate shares the same enzymes. Biosynthesis of DAG starts in the formation of lysophosphatidate

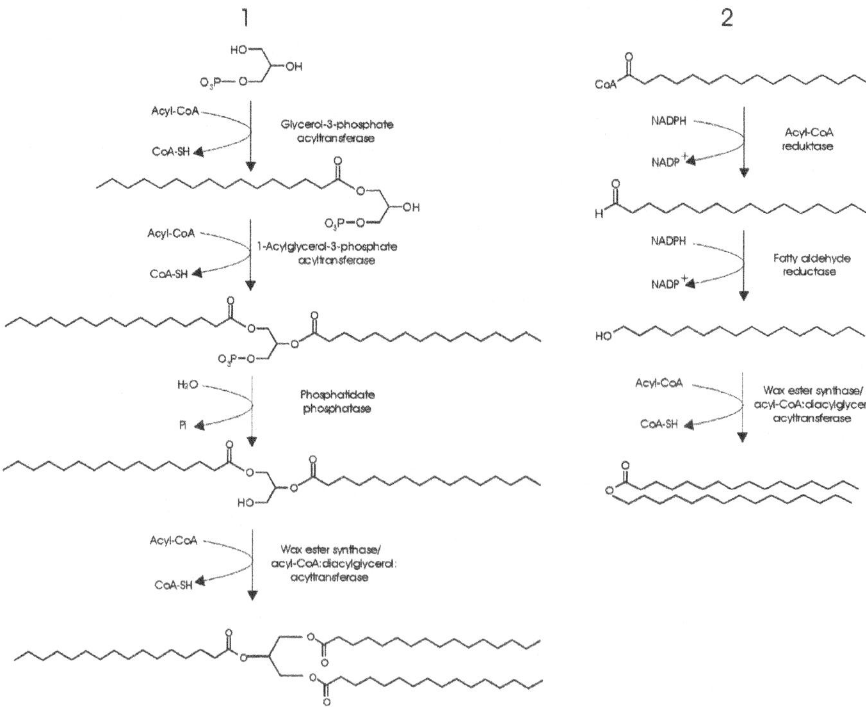

Fig. 2 Metabolic pathways for the biosynthesis of (**1**) TAGs and (**2**) WEs

(1-acylglycerol-3-phosphate) from glycerol-3-phosphate through catalysis of glycerol-3-phosphate acyltransferase. In *Escherichia coli,* this enzyme is able to use acyl-CoAs and also fatty acids bound to acyl carrier protein (ACP) as substrates, and it is localized at the cytoplasm membrane (Wilkinson and Bell 1997). Subsequently, lysophosphatidate is converted to phosphatidate by lysophosphatidate acyltransferase, which is also localized at the cytoplasm membrane (Coleman 1990). Finally, DAG is synthesized from phosphatidate through dephosphorylation catalyzed by phosphatidate phosphatase (Icho and Raetz 1983). In *A. baylyi* strain ADP1, the last step in biosynthesis of TAGs is mediated by a promiscuous WE synthase/DAG acyltransferase (WS/DGAT).

This WS/DGAT is responsible for biosynthesis of TAGs and WEs with acyl-CoA as an acyl donor and DAG or long-chain fatty alcohol as acyl acceptors, respectively (Kalscheuer and Steinbüchel 2003). These acyltransferases are the key enzymes for lipid biosynthesis in obviously all bacteria and have been studied in much detail (Röttig and Steinbüchel 2013). The activity of WS/DGAT using DAG as an acyl acceptor instead of fatty alcohols is one order of magnitude lower, with a high specificity using the *sn*-1 and *sn*-3 positions of glycerol as acyl acceptors, rather than the *sn*-2 position (Stöveken et al. 2005). The substrate specificities regarding fatty alcohols and DAG as acyl acceptors resemble the distribution of both storage lipids in *A. baylyi* strain ADP1 (Kalscheuer and Steinbüchel 2003) and in other bacteria. Furthermore, this enzyme exhibits also monoacylglycerol acyltransferase activity (Stöveken et al. 2005).

Although this WS/DGAT (also Atf1) is responsible for WE and TAG biosynthesis in *A. baylyi* strain ADP1, this type of long-chain acyl-CoA acyltransferase is not related to the known WS from jojoba, or the DAG:acyltransferase 1 (DGAT1) and DAG:acyltransferase 2 (DGAT2) families present in yeasts, plants, and animals, and the phospholipid:DGAT catalyzing TAG formation in yeasts and plants (Cases et al. 1998, 2001; Hobbs and Hills 1999; Routaboul et al. 1999; Zou et al. 1999; Bouvier-Navé et al. 2000; Sandager et al. 2002; Dahlquist et al. 2000). There is only one *atf1* gene present in *A. baylyi* strain ADP1, despite the fact that genes for Atf1 homologues seem to be widespread among the *Actinomycetales*, since an extensive group of related genes occurs in streptomycetes and mycobacteria, with, for example, 15 homologous genes in *Mycobacterium tuberculosis* H37Rv (Kalscheuer and Steinbüchel 2003; Daniel et al. 2004).

Similar to the degradation of WEs or PHAs, TAGs are degraded under conditions of carbon and energy deficiency. In this context, Alvarez et al. (2000) demonstrated mobilization of more than 90% of the accumulated TAGs in *R. opacus* and *R. ruber* within 120 h, when the cells were cultivated in the absence of a suitable carbon source and in the presence of ammonium. However, knowledge of the degradation of intracellular TAGs in other bacteria is very scarce, and the enzymes involved therein and their regulation are still unknown.

3.2 WE Metabolism

The metabolic pathway for biosynthesis of WEs in *Acinetobacter* sp. involves three enzymatic reactions (Fig. 2) (Ishige et al. 2003). First, long-chain acyl-CoAs, which are derived from fatty acid de novo synthesis or alkane oxidation, are reduced to the corresponding fatty aldehydes by an NADPH-dependent acyl-CoA reductase (Acr1). The corresponding gene (*acr1*) encodes a membrane-localized enzyme of 32.5 kDa and accepts acyl-CoAs with chain lengths from 14 to 22 carbon atoms (Reiser and Somerville 1997). The next reduction step is mediated by an NADPH-dependent fatty aldehyde reductase, yielding the corresponding fatty alcohols; however, the gene encoding this enzyme has not been identified yet. The crucial metabolic step, which condenses long-chain acyl-CoAs and long-chain fatty alcohols forming the respective WEs, is mediated by WS/DGAT encoded by *atfA* (Kalscheuer and Steinbüchel 2003). This 51.8-kDa enzyme comprises a putative membrane-spanning region and does not exhibit a very clear preference regarding its location. It was demonstrated to be mainly membrane-localized but seems also to be distributed throughout the cytoplasm and on the surface of intracellular lipid bodies (Wältermann et al. 2005; Stöveken et al. 2005; Kalscheuer et al. 2006a, b). WS/DGAT from *A. baylyi* strain ADP1 shows a broad ability to utilize fatty alcohols and also to some extent DAGs as acyl acceptors in vitro. Actually, this enzyme is also responsible for biosynthesis of the minor amounts of storage TAGs occurring in *A. baylyi* strain ADP1 (Kalscheuer et al. 2003; Stöveken et al. 2005).

Since WEs act as true storage lipids, degradation of the waxes has to occur in growth-accommodating environments and under carbon limitation. The initial step in the degradation of intracellular WEs is surely not the reverse reaction mediated by the WS/DGAT enzyme. Most probably, degradation of storage WEs is catalyzed by an intracellular esterase or lipase. Attempts to identify such an enzyme have not been successful so far.

4 Structure and Morphology of Prokaryotic Lipid Inclusions

All storage lipids in bacteria are stored in discrete, intracellular lipid inclusions. When observed by light microscopic techniques, these inclusions appear as intensively light-scattering particles inside the cytoplasm. In transmission electron microscopic studies, lipophilic inclusions appear as electron-transparent structures of different sizes and shapes and without a visible internal structure but with a single boundary layer. Although their structure seems relatively simple, interesting differences between the different types of inclusions occur.

4.1 TAG Inclusions

TAG inclusions in actinomycetes, whose formation has been completed, occur as spherical and electron-transparent storage droplets inside the cytoplasm, with diameters ranging from 50 to 400 nm (Fig. 3). The number and size of these inclusions vary depending on the respective strain, growth phase, and cultivation conditions. Similar to their plant, animal, and yeast analogues, TAG inclusions in bacteria were frequently reported to possess a thin boundary layer comprising half the width of the plasma membrane, which most probably shields the lipid core from the cytoplasm (Packter and Olukoshi 1995; Alvarez et al. 1996). However, in cells harboring large amounts of TAGs, inclusions show some indication of coalescence, most probably caused by tight direct interactions of the inclusions (Packter and Olukoshi 1995; Alvarez et al. 1996). Similar to lipid bodies in cultured mammalian cells, bacterial TAG inclusions split into numerous internal fracture planes when freeze-fractured, leading to a lamellar internal view of the hydrophobic core (Fig. 3) (Packter and Olukoshi 1995; Robenek et al. 2004; Wältermann et al. 2005). These fracture planes are very similar to that of membrane fracture faces. It is not known whether these fracture planes are caused by physical artifacts during the fracturing process or are a true series of tightly compressed lipid layers making up the TAG core (Wältermann et al. 2005). It should be mentioned that PHA inclusions inside the bacterial

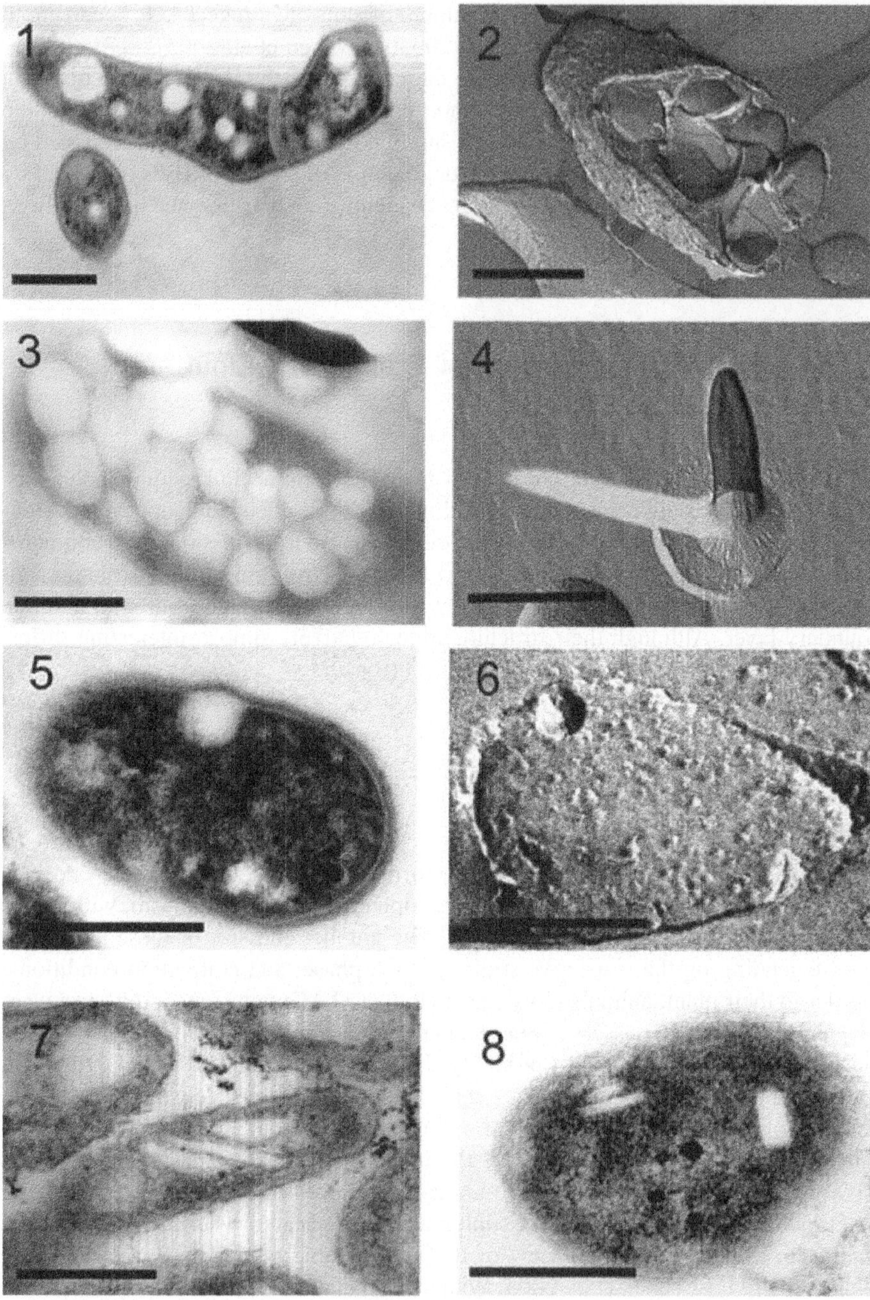

Fig. 3 Morphology of bacterial lipid inclusions. (**1**) Lipid inclusions and lipid prebodies in a cell of *Rhodococcus opacus* PD630; (**2**) freeze-fracture preparation of a cell of *R. opacus* PD630 exhibiting lamellar fracture planes on intracellular TAG inclusions; (**3**) cell of *R. eutropha* H16 with PHB granules; (**4**) needlelike deformation of a PHB granule in *R. eutropha* H16 in a freeze-fracture preparation; (**5**) cell of *Acinetobacter* sp. strain ADP1 with a nascent, spherical WE inclusion; (**6**) freeze-fractured cell of *Acinetobacter* sp. strain ADP1 accumulating WE inclusions; (**7**) cells of

cytoplasm can hardly be distinguished from WE and TAG inclusions by transmission electron microscopy (Fig. 3). However, they behave quite differently in freeze-fracture experiments. Depending on the chemical composition of PHA and the temperature at which the fracturing is performed, PHA granules exhibit needlelike or mushroomlike artifacts, which occur owing to a plastic deformation process. Fractured granules are often observed to be composed of a central, stretchable inner core and an outer core which withstands deformation while fracturing (Fig. 3) (Preusting et al. 1993).

A particularly interesting situation occurs in actinomycetes, which are capable of simultaneously accumulating TAGs and PHAs in similar amounts like in *R. ruber*. It remains to be elucidated whether TAGs and PHAs occur in discrete, separated inclusions, or if both types of storage lipids are deposited in one and the same type of inclusion. However, it could be demonstrated that the *R. ruber* phasin, a class of structural proteins thought to be specific for PHA granules, coated all intracellular inclusions, also the TAG inclusions (Pieper-Fürst et al. 1994). TAG inclusions from gluconate-grown cells of *R. opacus* PD630 were first isolated from cellular lysates by density-gradient ultracentrifugation, and the chemical composition was investigated by Alvarez et al. (1996). These inclusions were mainly composed of TAGs (87%), DAGs (approximately 5%), phospholipids (1.2%), and proteins (0.8%). TAG inclusions exhibited a very complex protein pattern strongly resembling that of total cell lysates when resolved by polyacrylamide gel electrophoresis. Our laboratory was unable to identify structural proteins specifically bound to the surface of these inclusions like proteins accomplishing similar functions as the oleosins on the surface of oil droplets in plants or phasins on the surface of PHA granules (Kalscheuer et al. 1999; Wältermann and Steinbüchel 2005). All proteins found to be strongly associated with the surface of isolated TAG inclusions could not be related to lipid metabolism or lipid body structure, indicating that these proteins were probably only bound unspecifically during cell disruption. Two proteins (ro02104 and PspA) exhibited the main proteins in the lipid droplets (Ding et al. 2012). Thus, the surface of bacterial TAG inclusions is most probably comparable to that of WE inclusions in *Acinetobacter* sp. and lipid bodies in yeasts or the mesocarb tissue from olive or avocado, which also lack specific structural proteins (Kalscheuer et al. 2006a, b; Murphy and Vance 1999; Murphy 2001). Recently, the proteomes of lipid droplets in microorganisms were analyzed (Yang et al. 2012).

Fig. 3 (continued) *Alcanivorax jadensis* exhibiting half-moon-like and disk-like WE inclusions; (**8**) rectangular WE inclusions in *Acinetobacter* sp. strain ADP1. *Bars* 1 μm

4.2 WE Inclusions

Since the amounts of WEs in the investigated bacteria are relatively low in comparison to the amounts of TAGs, only a limited number of WE inclusions occur in the cytoplasm (approximately three to five). WE inclusions in bacteria show a wide variety of different morphologies in electron microscopic images, ranging from (1) spherical and ellipsoid to (2) disk-like structures and to (3) even rectangular shapes:

1. Spherical inclusions in hexadecane-grown cells of *Acinetobacter* sp. strain HO1-N exhibited average diameters of 200 nm. These inclusions were first misinterpreted as depots of the unmodified carbon source. However, hexadecane was only a minor component, whereas WEs constituted the dominant part of these inclusions. On the basis of freeze-etching and transmission electron micrographs, the authors suggested that these inclusions were separated by a phospholipid monolayer membrane. Interestingly, also intracellular bilayer membranes were observed in these preparations (Scott and Finnerty 1976). Similarly, *A. baylyi* strain ADP1 accumulates spherical, electron-transparent WE inclusions when cultivated on carbohydrates or hexadecane (Fig. 3). These inclusions were also demonstrated to be surrounded by a phospholipid monolayer membrane (Wältermann et al. 2005). Freeze-fracturing of spherical WE inclusions leads to irregular internal fracture faces, with different fracture planes and sharply delineated edges, confirming a surrounding membrane (Fig. 3) (Wältermann et al. 2005; Scott and Finnerty 1976).
2. In *Acinetobacter* sp. strain M-1 cells, which were grown on hexadecane, WE inclusions occur as electron-transparent, smooth, and disk-like structures. On the basis of electron micrographs, these structures were reported to lack a surrounding membrane; however, the authors did not make further attempts to identify a possible membrane structure surrounding these inclusions. These WE inclusions grew to about the same diameter as the cell. The time course of their formation was also investigated. The authors described the sequential formation of WE disks, with one disk completion resulting in the formation of the following one (Ishige et al. 2002). However, the formation of disk-like WE inclusions in this strain seems not to be a curiosity or an artifact, since similar structures were also observed in *Alcanivorax* sp. cells accumulating WEs from hexadecane and to some extent also in *A. baylyi* strain ADP1 in parallel with spherical inclusions (Fig. 3) (Manilla, Wältermann and Steinbüchel, unpublished results).
3. Singer et al. (1985) reported on electron-transparent, rectangular inclusions with a typical bilayer structure at their surface in strain *Acinetobacter* sp. strain HO1-N, when cultivated on hexadecanol (Fig. 3). These rectangular structures were 100–200 nm in length, exhibited a width of approximately 30 nm, and consisted mainly of hexadecyl palmitate (85.6%) and minor amounts of hexadecanol (4.8%) and phospholipids (9.6%). The inclusions also occurred in parallel with cytoplasmic membrane structures. Whether the rectangular WE inclusions reported by Singer et al. (1985) and the disk-like WE inclusions reported by

Ishige et al. (2002) and our laboratory are just morphological variations of the more frequently observed spherical inclusions or whether their formation relies on mechanisms different from that described and discussed later remains to be elucidated.

Isolated WE inclusions from different *Acinetobacter* strains exhibited a complex pattern of copurified proteins, and no protein identified among them has been structurally related to these inclusions so far, reminding us of the situation of isolated TAG inclusions from *R. opacus* and *R. ruber* (Scott and Finnerty 1976; Alvarez et al. 1997a; Kalscheuer et al. 2001; Wältermann and Steinbüchel 2005). A very recent report on a recombinant *Escherichia coli*, which uses an artificial pathway for WE production and a bifunctional acyl-CoA reductase from jojoba and the *Acinetobacter* sp. strain ADP1 WS/DGAT, describes the formation of intracellular WE inclusions in this strain. The inclusions had the same size and spherical shape as those described for *A. baylyi* strain ADP1, which proves that structural proteins are not needed for the formation and structural integrity of WE inclusions (Kalscheuer et al. 2006a, b). Recently, it was demonstrated that WS/DGAT is partly localized on the surface of WE inclusions of *A. baylyi* strain ADP1 and also in recombinant *E. coli*; however, it remains questionable if the enzyme exhibits a significant biosynthesis activity at this location in vivo (Wältermann et al. 2005; Kalscheuer et al. 2006a, b).

5 Formation of TAG and WE Inclusions and PHA Granules in Prokaryotes

In eukaryotes, storage lipid synthesis and formation of lipid bodies are thought to arise most probably from the endoplasmic reticulum (ER) by a unique mechanism. Since prokaryotes do not possess an ER, the mechanism of the formation of neutral lipid inclusions must be quite different.

5.1 TAG Inclusions

The biogenesis of lipid droplets in cells was investigated with various methods and in many microorganisms (Wältermann et al. 2005; Kalantari et al. 2010). Interestingly the results of these studies supported each other and provided increasing insights into the onset of formation and ripening of lipid droplets in particular in bacteria.

Under appropriate experimental conditions (see before), strains like *R. opacus* PD630 or *Mycobacterium smegmatis* reach their maximum TAG content within a period of 72 h. During this time, nearly TAG and lipid inclusion free cells transform to cells in which the cytoplasm is substantially filled with lipid inclusions. Application of fluorescent dyes as lipid markers on TAG inclusion biogenesis indicated that

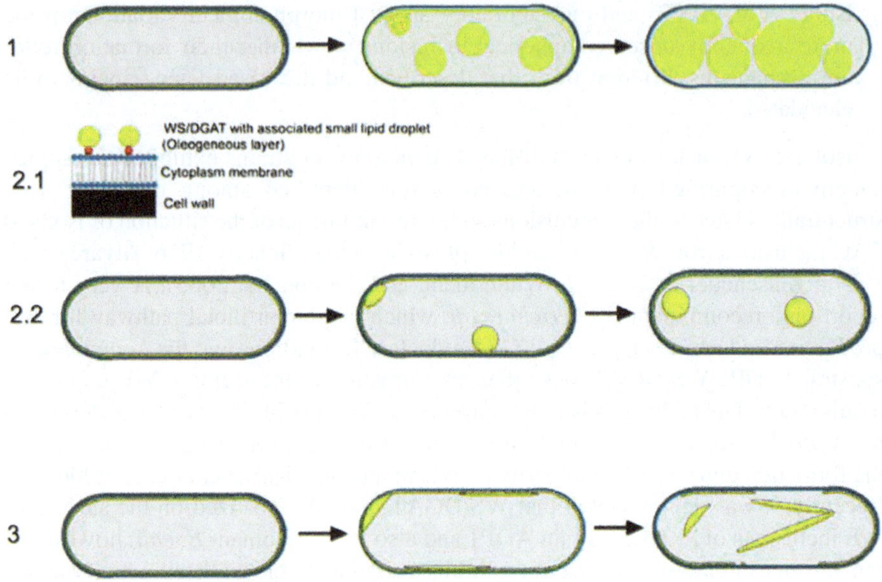

Fig. 4 Formation of intracellular neutral lipid inclusions in bacteria. (**1**) Formation of TAG inclusions; (**2.1**) formation of small lipid droplets at the cytoplasmic face of the cytoplasm membrane; (**2.2**) formation of spherical WE inclusions; (**3**) hypothetical model for the formation of nonspherical WE inclusions. *WS* WE synthase, *DGAT* diacylglycerol acyltransferase (updated from Wältermann and Steinbüchel 2005)

neutral lipid accumulation was initiated at peripheral lipid domains of the cells and that large lipid inclusions localized in the cytoplasm occur only at a later stage of lipid accumulation. When observed by phase-contrast microscopy, these early lipid domains appear as small, intensively light-scattering granules, whereas large lipid inclusions observed at later stages occur as clearly visible droplets filling almost the whole cytoplasm (Christensen et al. 1999; Wältermann et al. 2005). It was assumed that the formation of TAG inclusions in actinomycetes underlies the same mechanism as the formation of spherical WE inclusions in *A. baylyi* strain ADP1, which was investigated in more detail using purified WS/DGAT in in vitro experiments (Fig. 4; Sect. 5.2) (Wältermann et al. 2005). However, significant differences also occur.

Specimens of ultrathin sections of both *Mycobacterium smegmatis* and *R. opacus* exhibited a strong morphological diversification from low TAG-containing cells to cells harboring a maximum lipid content in the final stage. In early stages of lipid accumulation, flat-to-spherical structures of a higher electron density compared with matured TAG inclusions occurred close to the cytoplasm membrane (Fig. 3). These structures correspond to the peripheral lipid domains visible in fluorescence microscopy and were referred to as lipid prebodies. Lipid prebodies appear to be in conjunction with a distinct layer of varying thickness facing the cytoplasmic site of the cytoplasm membrane, which was referred to as an oleaginous layer. Lipid prebodies enlarged until they reached an average diameter of about 300 nm, before they left the oleaginous layer at the membrane and migrated into more central parts of the cytoplasm. Lipid prebodies bound to the cytoplasm membrane and cytoplasm-localized prebodies exhibit a relatively diffuse shape when compared with that of the electron-transparent lipid inclusions found in later stages of lipid accumulation and are not surrounded by a clearly visible outer layer in electron micrographs. It was assumed that lipid prebodies do not exhibit a homogeneous lipid core but are composed of an emulsion-like aggregation of very small lipid droplets (SLDs) in conjunction with phospholipids and proteins. How lipid prebodies convert to matured lipid inclusion is unknown. It was speculated that the emulsion-like aggregation of SLDs aggregate to structures with a homogeneous lipid core by proceeding, thermodynamically driven coalescence (Wältermann et al. 2005; Wältermann and Steinbüchel 2005).

5.2 WE Inclusions

Similar to the formation of TAG inclusions in the actinomycetes mentioned earlier, formation of spherical WE inclusions in *A. baylyi* strain ADP1 occurs at peripheral sites of the cells, as can be easily observed using lipophilic dyes and fluorescence microscopy techniques. According to this, nascent WE inclusions can be observed in proximity to the cytoplasm membrane in transmission electron microscopy preparations. They are often in conjunction with a thin layer facing the cytoplasmic site of the membrane (Fig. 3). The situation that WS/DGAT is mainly localized at the cytoplasm membrane indicates that this compartment is the natural site of WE biosynthesis and the origin of the formation of WE inclusions. As suggested by Wältermann et al. (2005), biosynthesis of WEs by membrane-bound WS/DGAT in *Acinetobacter* sp. strain ADP1 leads to the formation of very small WE droplets with diameters of only some nanometers at the cytoplasmic face of the cytoplasm membrane (Fig. 4). These SLDs remain associated with the enzyme and form a small, emulsion-like layer of WE biosynthesis. This layer is permeable for hydrophilic solutes and metabolites, thus enabling normal metabolic processes at the membrane. During continuing WE biosynthesis, the WS/DGAT:SLD complexes are released, and SLDs conglomerate or coalesce owing to hydrophobic interactions to larger structures at distinct sites of the membrane, which can be observed as nascent WE inclusions in transmission electron microscopy. This mechanism was confirmed by scanning force electron microscopy and quartz microbalance experiments on immobilized model phospholipid membranes using purified WS/DGAT for in vitro formation of WEs and TAGs. By these methods, the in vitro formation of small membrane-bound WE and TAG droplets and their conglomeration/coalescence to lipid prebodies were observed. Although not investigated in detail, it was assumed that phospholipids coat the nascent lipid inclusions owing to their proximity to the cytoplasmic membrane and phospholipid biosynthesis. However, in contrast to the situation of TAG inclusions in actinomycetes, nascent WE inclusions do not exhibit a more electron-dense inner structure in comparison to cytoplasmic WE inclusions. The cause of this phenomenon is unknown but might be due to an earlier coalescence of WEs inside the nascent inclusions in comparison to TAG prebodies (Fig. 4).

Although no detailed experiments on the formation of nonspherical WE inclusions, for example, in *Alcanivorax* sp. or *Acinetobacter* sp. strain M-1, have been performed so far, it may be speculated that their formation underlies principally the same mechanism. Electron micrographs of *Acinetobacter* sp. strain M-1, *Alcanivorax borkumensis*, and *Alcanivorax jadensis* revealed the occurrence of disk-, ellipsoid-, half-moon-, or rectangular-like WE inclusions of different sizes, of which a considerable amount was in direct conjunction with the cytoplasm membrane (Ishige et al. 2002; Reers, Wältermann and Steinbüchel, unpublished results). Since the first steps in the formation of WE inclusions in *A. baylyi* strain ADP1 are membrane-associated, and thus a nearly two-dimensional process, it is likely that nascent WE inclusions formed by this mechanism are not forced to become spherical structures (Wältermann et al. 2005). It might be speculated that WEs form inclusions, the shapes of which are molded from respective sites of the cytoplasm membrane. During continuing biosynthesis of small WE droplets at the membrane and without their conglomeration to spherical WE prebodies, a growing WE layer could be formed at the membrane; the WEs could occur as ellipsoid, disk-like, or otherwise irregular inclusions in transmission electron microscopy observations after they have left the membrane and migrated into the cytoplasm. However, although this assumption would not explain the occurrence of such large disk-like structures observed by Ishige et al. (2002), it would indicate that WEs in nonspherical inclusions occur in a nonliquid or crystalline state (Fig. 4, 3).

Although significant progresses in our knowledge of the formation of lipid inclusions and neutral lipid biosynthesis have been made in the last few years, the general mechanism of the formation of intracellular lipid inclusions is still a matter of debate. Although WS/DGAT is localized mainly at the cytoplasm membrane in *A. baylyi* strain ADP1, the enzyme exhibits also a distribution throughout the cytoplasm and at the surface of WE inclusions (Wältermann et al. 2005; Stöveken et al. 2005). This distribution was also reported to occur in bioengineered, WE-accumulating *E. coli* by Kalscheuer et al. (2006a, b). This strain is capable of synthesizing and accumulating small amounts of fatty acid butyl esters (FABE) and WE in media supplied with butanol. This strain was able to synthesize a small number of intracellular lipid inclusions of approximately 200 nm in diameter. Owing to its low lipid content, the authors suggested that this strain was virtually arrested in an early state of lipid accumulation, and that all lipids synthesized would be localized as a WS/DGAT:lipid complex at the cytoplasm membrane, and that no intracellular lipid inclusions could be formed according to the model suggested by Wältermann et al. (2005). This assumption is in contrast to the situation occurring in cells of *A. baylyi* strain ADP1 in early stages of WE accumulation, in which the WE content is also very low but in which WE inclusions can also be observed. On the basis of these findings, the authors suggested that the main location of FABE biosynthesis and also long-chain WE biosynthesis in this highly artificial system is associated with the surface of intracellular lipid inclusions, because both substrates for their formation are water-soluble. Although the possibility of substantial lipid biosynthesis at the surface of lipid inclusions cannot be excluded, experimental evidence for the latter mechanism is missing. In wild-type bacteria accumulating neutral lipids, TAGs and WEs are synthesized from acyl-CoAs and hydrophobic acyl acceptors, whose biosynthesis is associated with the cytoplasm membrane, and thus the putative situation of FABE synthesis from soluble butanol and acyl-CoA in the case of this recombinant *E. coli* strain does not reflect the natural situation in a lipid-accumulating strain and is not satisfactory to refute the current model for the formation of neutral lipid inclusions in bacteria at all. A significant biosynthesis activity at the surface of lipid inclusions would require that hydrophobic intermediates have to shuttle between the cytoplasm membrane and the surface of lipid inclusions. Furthermore, this would suggest that neutral lipid inclusions could arise de novo in the cytoplasm from soluble WS/DGAT starting from a small, hypothetical soluble enzyme:lipid complex continuously growing to a lipid inclusion. This would be similar to the formation of PHA inclusions from soluble substrates in PHA-accumulating bacteria (Jurasek and Marchessault 2004). However, such a situation would lead to many small lipid inclusions inside the cytoplasm, and larger inclusions would be formed with proceeding coalescence. However, such a situation has never been observed, not even in this recombinant strain. Furthermore, this model does not explain the incorporation of phospholipids into the surface of nascent lipid inclusions. Finally, although soluble WS/DGAT is capable of synthesizing lipids from soluble acyl-CoA and fatty alcohol emulsions, an in vitro formation of neutral lipid inclusions, similar to the in vitro formation of PHB granules reported several years ago (Jossek et al. 1998), has never been reported.

5.3 *PHA Granules*

PHA granules have been studied in much more detail than lipid inclusions in prokaryotes. These PHA granules are nowadays also referred to as carbonosomes (Jendrossek and Pfeiffer 2014; Jendrossek 2020). One chapter written by the editor of this volume (Dieter Jendrossek) provides a very concise review of our current knowledge about PHA granules (Jendrossek 2020).

Since PHAs represent a polymeric form of carbon and an energy store, the formation of PHA granules is quite different from the formation of prokaryotic neutral lipid inclusions. Biosynthesis of PHB is a template-independent homopolymerization process which forms an insoluble polymer from soluble substrates, mediated by the PHA synthase (PhaC) (Stubbe and Tian 2003). PhaC is a soluble enzyme and is found in the cytoplasm in cells harboring no PHB granules. In contrast, under conditions of PHA accumulation, PhaC binds to PhaM and becomes strictly bound to the surface of PHB granules. Not all aspects of the biogenesis of PHB granules are understood in detail. According to the most accepted model, cytoplasm-localized PHB-linked PhaC proteins generate micelle-like structures,

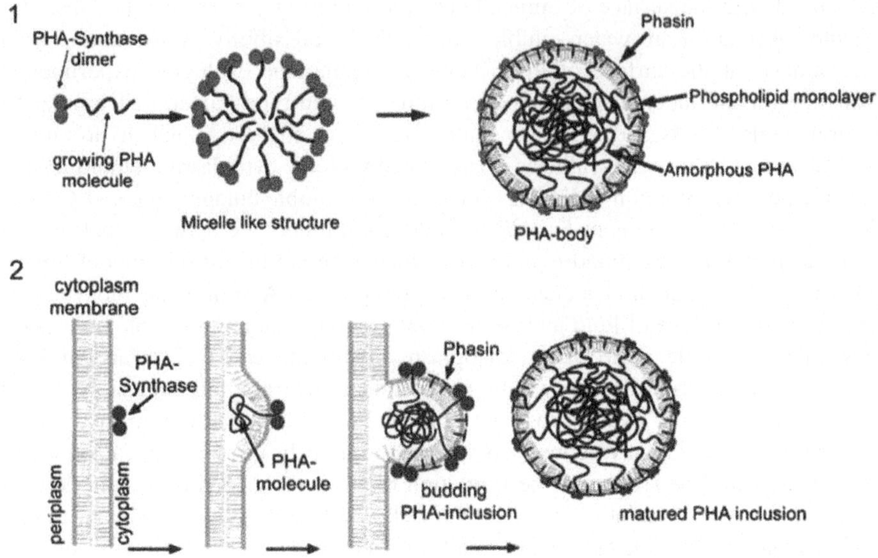

Fig. 5 Two models for the formation of PHA granules. (**1**) The micelle model; (**2**) the budding membrane model (Wältermann and Steinbüchel 2005). A third model, the scaffold model (Jendrossek and Pfeiffer 2014), is not shown

with the growing PHB chain representing the hydrophobic core and PhaC itself representing the hydrophilic head group. During the proceeding of PHB synthesis, PhaC remains covalently bound to the elongated PHB chain, while the resulting surface area of the nascent granule increases, thus dispersing PhaC on the surface of the granule. Probably at the very beginning of granule formation, the formation of phasin protein is induced, and the amphiphilic proteins invade into the empty space on the hydrophobic granule surface, thus preventing the granules from coalescing. The PHA granules do obviously not possess phospholipids at their surface (Bresan et al. 2016). This explains the existence and the need for phasins. In addition, formation of a large hydrophobic surface in the cells is prevented, which could lead to unspecific binding of other proteins. This function is similar to the situation of oleosins in plant seeds. There is also no plausible explanation how phospholipids could reach the surface of growing PHB inclusions from their site of synthesis if they are separated from the cytoplasm membrane. It may also be speculated whether they are required at all for maintenance of the structural integrity of the granules (Fig. 5) (Stubbe and Tian 2003; Jurasek and Marchessault 2004). However, the situation regarding the structure and composition of PHA granules is also discussed with some controversy. It may be possible that some other variants have evolved and may exist in some microorganisms.

One interesting finding concerns the binding of the PHB granules to the nucleoid region in the cells. There is conclusive evidence that this binding is mediated by the protein PhaM (Wahl et al. 2012). This could explain why the PHA granules are more or less evenly distributed to the daughter cells during cell division. Whether this is the case in all PHB-accumulating bacteria or whether it is a peculiarity of *Ralstonia eutropha* must be investigated in the future. However, there are probably also other effects or interactions of PhaM with other components of the PHB granules because the titer of PhaM has an impact on the size of the granules.

An alternative model for the formation of PHB granules has been proposed and shares some similarities with the formation of eukaryotic neutral lipid bodies. In this model, PhaC becomes membrane-bound and synthesizes a growing PHB molecule in the space between both leaflets of the plasma membrane, thus leading to a swelling of the membrane and a subsequent budding of a PHB granule which is coated by a monolayer membrane derived from the cytoplasmic face of the plasma membrane. Similar to the incorporation of oleosins into budding plant seed oil bodies along the cytoplasmic site of the ER membrane, incorporation of phasins could be facilitated by insertion of phasins into the plasma membrane (Jendrossek 2005; Stubbe and Tian 2003). Although this model provides a comprehensible explanation for the incorporation of phospholipids, it is contrary to the strict cytoplasmic localization of PhaC. Furthermore, phasins have never been reported to be membrane-associated in *R. eutropha* (Fig. 5).

6 Structure, Formation, and Protein Equipment of Eukaryotic Lipid Inclusions

Prokaryotic lipid inclusions and PHA granules are considered to be structurally related to eukaryotic lipid bodies, which are referred to as lipid particles, lipid globules, lipid bodies, or especially in plants oil bodies, spherosomes and oleosomes. Although the basic structure of eukaryotic lipid bodies is very similar to those of prokaryotes, these compartments are discussed whether they constitute a true organelle or not. The structure of eukaryotic lipid droplets and oil bodies is generally rather simple; a hydrophobic core consisting of the respective storage lipid is surrounded by a phospholipid monolayer with special proteins embedded therein. The main function of these inclusions is storage of neutral lipids as an energy source and as a source of components needed for membrane biogenesis or formation of other lipophilic components, like steroids. In higher eukaryotes, lipid storage occurs mainly in specialized tissues, for example, white and brown adipose tissue, steroidogenic cells, or the seeds of oleogenic plants (Zweytick et al. 2000). However, the formation and protein equipment of lipid bodies in eukaryotic cells are quite different from those in prokaryotes.

6.1 Plants

In some plant seeds, for example, rapeseed or sunflower, storage lipids account for more than half of the total weight. TAGs represent the dominant storage lipid of plants, whereas steryl esters generally constitute only a minor component. Related structures are the rubber particles of *Hevea brasiliensis* and other polyisoprene-producing plants in which the hydrophobic core is also surrounded by a phospholipid monolayer (Wood and Cornish 2000; Cherian et al. 2019). Although WEs are abundant in plants as a component of hydrophobic barriers, pathogen protection, and for light reflection, jojoba, a desert shrub native to southwest USA and Mexico, is the only exception among angiosperms known to accumulate WEs as energy stores (Yermanos 1975; Post-Beitenmiller 1996). In plants, TAGs are formed via the Kennedy pathway from glycerol-3-phosphate and acyl-ACP or acyl-CoA in the ER, plastids, and mitochondria. First, phosphatidic acid is formed from glycerol-3-phosphate through acylation at the *sn*-1 and *sn*-2 positions by glycerol-3-phosphate acyltransferase and 1-glycerol-3-phosphate acyltransferase, respectively. Subsequent dephosphorylation of phosphatidic acid is mediated by phosphatidate phosphatase (Frentzen 1996). Further acylation of the resulting DAG is catalyzed by either an enzyme of the DGAT1 or DGAT2 families and yields TAG. DGAT1 and DGAT2 are encoded by two unrelated genes in plants, from which one is closely related to acyl-CoA:cholesterol acyltransferase, whereas the other form does not resemble any other known genes (Zou et al. 1999; Lardizabal et al. 2000). As an acyl-CoA independent pathway, Dahlquist et al. (2000) described the formation of TAGs from DAG and phospholipid as an acyl donor in different plants and yeasts.

Plant oil bodies appear as electron-transparent structures in electron microscopy with an electron-dense phospholipid/protein layer. The average diameter of oil bodies in dessication-tolerant seeds varies from 0.6 to 2.5 μm, depending on the species (Tzen et al. 1993). This size is assumed to be optimal for TAG mobilization during germination of seeds, since it provides an ideal volume-to-surface ratio (Murphy 1993; Zweytick et al. 2000). In most plants, the main class of proteins embedded in the surrounding monolayer of seed oil bodies is oleosins; but they were also found in flowers, pollen, and tapetum. Oleosins are structural proteins of low molecular masses (15–26 kDa) (Murphy 1993; Roberts et al. 1991; Robert et al. 1994). It is assumed that these proteins form a meshwork-like organization on the surface of lipid bodies which protects oil bodies from unspecific attack of lipases and coalescence. This is particularly important in plant seeds, in which an optimal surface-to-volume ratio is essential for rapid mobilization of TAGs. Oleosins are composed of three structural domains, an expanded conserved α-helical region (approximately 70 hydrophobic amino acids) flanked by two amphiphilic regions of varying sizes. The central hydrophobic domain directly interacts with the lipid core of the oil body, and a proline knot motif in the central region of this domain was identified as a putative targeting signal to the lipid core (Chen et al. 1997; Lacey et al. 1998; Abell et al. 1997). However, although oleosins represent the majority of proteins on the surface of lipid bodies, it is unlikely that they play an essential role in their formation. This is confirmed by the situation of oil bodies in the seeds of olive or avocado, which do not possess oleosins (Murphy and Vance 1999; Ross et al. 1993). However, the lipid bodies in these species exhibit diameters of about 20 μm and are seemingly not degraded at all.

The budding model is the most accepted model for the assembling of plant oil bodies. In this model, TAG formation occurs between both phospholipid leaflets at special subdomains of the ER. This causes a swelling of the bilayer, and TAG droplets with a surface derived from the outer ER layer bud-off. This model is supported by ultrastructural observations and the fact that the plant ER is equipped with enzymes and structural proteins involved in biosynthesis of TAGs and oil bodies (Frey-Wyssling et al. 1963; Wanner et al. 1981; Wanner and Theimer 1978; Tzen et al. 1993; Galili et al. 1998). The coating of the nascent oil bodies with oleosins is mediated by a co-translational insertion of the oleosins into the ER membrane. Owing to their hydrophobic character mediated by their central hydrophobic domain, oleosins invade regions of TAG accumulation between the phospholipid leaflets and accumulate at the surface of nascent oil bodies (Hills et al. 1993; Loer and Herman 1993). The other hypothesis for plant oil body formation is the postencasement model. In this model, storage TAGs arise naked from the ER and form droplets inside the cytoplasm and associate subsequently with oleosins and phospholipids. However, although this model is supported by the observation that in some species oleosin formation lags temporarily TAG accumulation, it does not explain the formation of a phospholipid surface layer and insertion of oleosins (Murphy 1993; Ichihara 1982).

6.2 *Mammals*

Rather than being mere storage droplets for TAGs and sterol esters, which provide cells with fatty acids as an energy source, for membrane remodeling or with precursors for hormone synthesis, lipid bodies in mammals are recognized as indispensable and metabolically active organelles that participate in cell signaling, vesicle trafficking, and several important diseases, like obesity and arteriosclerosis (Umlauf et al. 2004; Liu et al. 2004; Wang et al. 1999). In mammals, TAGs, the primary bodily energy stores, are mainly synthesized by action of DGAT1 and DGAT2, which are associated with the microsomal fraction (Polokoff and Bell 1980). However, DGAT-deficient mice were reported to synthesize TAGs, indicating that an additional pathway for TAG biosynthesis must exist (Smith et al. 2000). Steryl esters, which constitute the second major lipid in mammalian lipid bodies, are synthesized by esterification of cholesterol with long-chain acyl-CoAs mediated by acyl-CoA:cholesterol acyltransferase. This enzyme is also located at the ER (Chang et al. 1997; Suckling and Stange 1985).

Lipid bodies occur in almost all mammalian cell types. However, the main location for TAGs is white adipose tissue. White adipose cells contain one or a few enormously large lipid bodies (10–100 μm in diameter), whereas steroidogenic cells and brown adipose cells, which occur in some hibernating animals and fetuses/neonates of many mammals, contain higher amounts of lipid droplets of much smaller size (2–10 μm) (Murphy and Vance 1999; Hammerson 1985; Weiss 1983; Weather et al. 1987). When mammalian lipid bodies are inspected by electron microscopic methods, they appear as distinct, electron-transparent structures surrounded by an electron-dense layer, often observed in a close association to the ER or mitochondria (Schlunk and Lombardi 1967). Blanchette-Mackie (1995) described a junction between the ER membrane and the lipid body surface on the basis of freeze-fracture observation, indicating that the formation of mammalian lipid bodies probably underlies a similar mechanism to that of oil bodies in plants, in which the lipids accumulate between the leaflets of the ER membrane. Distension of the membrane and budding of the membrane leaflet facing the cytoplasm are thought to leave a droplet surrounded by a monolayer consisting of the former cytoplasmic leaflet of the ER membrane.

The main proteins associated with mammalian lipid bodies are proteins of the PAT family, namely, perilipins, adipose differentiation-related protein (ADRP; orthologously termed adipophilin in humans) and tail-interacting protein of 47 kDa (TIP47) (Londos et al. 1999; Wolins et al. 2003; Wang et al. 2003). PAT family proteins are similar in their entire sequences, with the exception of an N-terminal extension of approximately 20 amino acids in TIP47 and a C-terminal extension of varying length in the different perilipins. Recently, caveolin, lipotransin, a 200-kDa capsular protein, and the hepatitis C virus core protein were reported as lipid body-associated proteins (Robenek et al. 2004; Syu and Saltiel 1999; Wang et al. 1997; Hope et al. 1998). Perilipins are localized at the surface of lipid bodies in adipocytes and steroidogenic cells but are virtually absent in all other types of cells. Perilipins are encoded by a single-copy gene yielding three isoforms, A, B, and C, by alternative splicing (Lu et al. 2001). Perilipin A is predominantly found in differentiated adipocytes but can also be found in steroidogenic cells. Isoform B can be found in lower amounts in adipocytes and steroidogenic cells, whereas isoform C coats exclusively the cholesteryl ester droplets in steroidogenic cells. Perilipins can be highly phosphorylated and seem to be involved in TAG lipolysis and lipid body stabilization (Londos et al. 1995). Both adipocytes and steroidogenic cells use a cyclic AMP (cAMP) stimulated process to activate a hormone-sensitive lipase for lipolysis of TAGs and cholesterol, respectively. After phosphorylation by protein kinase A, which is stimulated by hormonally controlled cAMP, perilipins are released from the lipid body surface, thus enabling the lipase to bind to the lipid core and hydrolyze the embedded lipids, whereas nonphosphorylated perilipin suppresses lipolysis (Sztalryd et al. 2003). ADRP and TIP47 are ubiquitously expressed in all other cells and early-differentiated adipocytes. The functions of ADRP and TIP47 in lipid metabolism are not well understood. Since both proteins exhibit 50% sequence identity and owing to their similarities to perilipin, both proteins might therefore have similar structural or metabolic roles. Similar to perilipins, ADRP has been found only associated with lipid droplets and not in any other subcellular compartment, whereas TIP47 is also abundant in the cytoplasm (Barbero et al. 2001; Miura et al. 2002). In later stages of adipocyte differentiation, ADRP is replaced by perilipins. Therefore, these enzymes are assumed to play an essential role in lipid droplet formation and during conversion of adipocytes from preadipocytes. In contrast to the oleosins in plants, which are assumed to be transferred to the lipid body surface along with the cytoplasmic leaflet of the ER, PAT family proteins are thought to be directed from cytosolic sites of synthesis to the surface of nascent lipid droplets (Londos et al. 1999).

6.3 Eukaryotic Microorganisms

Oleaginous microorganisms are defined as organisms able to accumulate more than 20% lipids of their cellular dry weight (Ratledge 1989). A wide variety of unicellular eukaryotes are capable of accumulating TAGs (Christiansen 1978; Leman 1997; Stahmann et al. 1994; Juanssilfero et al. 2018). The biochemistry of TAG and steryl ester biosynthesis and lipid body formation has been mostly investigated in oleaginous yeasts and filamentous fungi, particularly for biotechnological production of single-cell oils which are rich in diverse polyunsaturated fatty acids (Clausen et al. 1974; Ratledge 2002). Similar to the situation in prokaryotes, lipid accumulation in an eukaryotic microorganism starts upon depletion of a nutrient from the medium, for example, nitrogen, but when an excess of carbon still remains. The carbon source continues to be assimilated and is used for lipid synthesis. In eukaryotic microorganisms, accumulation of lipids parallels the presence of ATP:citrate lyase, which is localized in the cytoplasm and converts citrate to oxalacetate under hydrolysis of ATP and transfers the acetyl residue to CoA (Boulton and Ratledge 1981; Ratledge 1989). Acetyl-CoA is formed by oxidative decarboxylation of α-keto acids and β-oxidation of fatty acids in the mitochondrion. Because the inner membrane of the mitochondrion is impermeable for acetyl-CoA, it is condensed with oxalacetate in the mitochondrial matrix and converted to citrate, which is then transported into the cytoplasm where it is finally released by ATP:citrate lyase. The cytoplasm-localized acetyl-CoA enters fatty acid de novo synthesis and is incorporated into lipids (Leman 1997; Harwood 1994; Ratledge 1989). Because prokaryotes are not compartmented, this "acetyl-CoA shuttle" can only be a criterion for oleogenity in eukaryotic microorganisms.

Similar to plants and animals, lipids are stored in lipid bodies in yeasts. These lipid bodies exhibit a spherical shape and diameters between 0.3 and 2 μm and have a hydrophobic core constituted of similar amounts of TAGs and steryl esters (Schaffner and Matile 1981), which is surrounded by a phospholipid monolayer and proteins (Leber et al. 1994). Biosynthesis of TAGs in *Saccharomyces cerevisiae* is mediated mainly by two proteins, an acyl-CoA:DGAT (Dga1p; belonging to the DGAT2 family) and a phospholipid:DGAT (Lro1p), which utilize acyl-CoA and phosphatidylcholine as acyl donors, respectively. Lro1p activity was reported to occur exclusively in the microsomal fraction, whereas Dga1p activity was mainly localized in the lipid body fraction and to some extent in the microsomal fraction (Sorger and Daum 2003). In yeasts, steryl esters are formed by two acyl-CoA:sterol ester acyltransferases (ARE1 and ARE2), which are localized at the ER (Yang et al. 1996; Yu et al. 1996). Unlike higher eukaryotes, structural proteins comprising functions similar to that of perilipins or oleosins could not be identified on yeast lipid bodies, but most proteins identified so far are involved in sterol, fatty acid, and TAG metabolism (Athenstaedt and Daum 1997, 2003; Leber et al. 1994, 1998; Milla et al. 2002; Smid et al. 1995; Tsitsigiannis et al. 2004). Yeast lipid particles are assumed to originate from the ER, as was similarly described for plant and mammalian lipid bodies, because the ER and lipid droplets share certain proteins and are frequently observed in proximity to each other (Zweytick et al. 2000).

7 Microbial Lipids as Sources for Biofuels

Lipids produced by oleaginous microorganisms and photosynthetic microalgae were at the beginning of this century, when the oil prices became extremely high, transiently considered as sources for fuels. These fuels could then be produced autotrophically from carbon dioxide or heterotrophically from renewable resources, respectively. These "biofuels" could have make the economy more independent from fossil resources (Meng et al. 2009; Qadeer et al. 2017)

There were several strategies followed up. The one was to use triacylglycerol produced, for example, from microalgae directly as fuel or to convert them chemically or biochemically into fatty acid alkyl esters as it is done, for example, from rapeseed oil to produce "biodiesel" (Scott et al. 2010; Lennen and Pfleger 2013; Radakovits et al. 2010). Various attempts were pursued with microalgae. A different strategy was the conversion of carbohydrates into fatty acid acyl esters (FAAE) using recombinant bacteria. One direction was the use of a recombinant strain of *Escherichia coli* which harbored the p(Microdiesel) plasmid (Kalscheuer et al. 2006a, b; Elbahloul and Steinbüchel 2010). This plasmid harbored the genes encoding the pyruvate decarboxylase and the alcohol dehydrogenase from *Zymomonas mobilis* and the unspecific acyltransferase from *Acinetobacter baylyi*. The three additional enzymes catalyzed the efficient conversion of pyruvic acid into FAAE. The product was referred to as "microdiesel" because it was obtained from microorganisms. "Microdiesel" contained ethanol instead of methanol as for "biodiesel." This process has the potential to be engineered to utilize a wide range of different bacteria and other carbon sources like cellulose (Kalscheuer et al. 2007; Hetzler and Steinbüchel 2013).

8 Concluding Remarks and Future Perspectives

Research in the last few years on biosynthesis of neutral lipids and on the structure and formation of their deposits has revealed a tremendous amount of knowledge. Much research has still to be done to address unresolved problems, for example, to identify enzymes involved in the intracellular degradation of accumulated lipids, to identify the mechanism involved in excretion of neutral lipids in *Alcanivorax* sp., and the overall regulation of neutral lipid metabolism in bacteria. Furthermore, a biotechnological production of lipids, for example, the production of WEs as a cheaper substitute for jojoba oil, might be achievable in the near future. Furthermore, WE and TAG inclusions as well as PHA granules could be modified and used as a basis for the production of biodegradable, self-assembling nanoparticles, which could be useful for a wide range of sophisticated technical applications and in medicine.

Acknowledgments The authors are indebted to all coworkers at the Institut für Molekulare Mikrobiologie und Biotechnologie at the Westfälische Wilhelms-Universität Münster who have in the past contributed to lipid research. This manuscript is an update of the manuscript previously published in the corresponding first volumes of the "Microbiology Monographs" book series.

References

Abell BM, Holbrook LA, Abenes M, Murphy DJ, Hills MJ, Moloney MM (1997) Role of the proline knot motif in oleosin endoplasmic reticulum topology and oil body targeting. Plant Cell 9:1481–1493

Akao T, Kusaka T (1976) Solubilization of diglyceride acyltransferase from membrane of *Mycobacterium smegmatis*. J Biochem (Tokyo) 80:723–728

Alvarez HM (2016) Triacylglycrol and wax ester-accumulating machinery in prokaryotes. Biochemie 120:28–39

Alvarez HM, Steinbüchel A (2002) Triacylglycerols in prokaryotic microorganisms. Appl Microbiol Biotechnol 60:367–376

Alvarez HM, Mayer F, Fabritius D, Steinbüchel A (1996) Formation of intracytoplasmic lipid inclusions by *Rhodococcus opacus* strain PD630. Arch Microbiol 165:377–386

Alvarez HM, Pucci OH, Steinbüchel A (1997a) Lipid storage compounds in marine bacteria. Appl Microbiol Biotechnol 47:132–139

Alvarez HM, Kalscheuer R, Steinbüchel A (1997b) Accumulation of storage lipids in species of *Rhodococcus* and *Nocardia* and effects of inhibitors and polyethylene glycol. Fett/Lipid 99:239–246

Alvarez HM, Kalscheuer R, Steinbüchel A (2000) Accumulation and mobilization of storage lipids by *Rhodococcus opacus* PD630 and *Rhodococcus ruber* NCIMB40126. Appl Microbiol Biotechnol 47:132–139

Alvarez HM, Souto MF, Viale A, Pucci OH (2001) Biosynthesis of fatty acids and triacyl-glycerols by 2,6,10,14-tetramethyl pentadecane-grown cells of *Nocardia globerula* 432. FEMS Microbiol Lett 200:195–200

Alvarez HM, Luftmann H, Silva A, Cesari AC, Viale A, Wältermann M, Steinbüchel A (2002) Identification of phenyldecanoic acid as a constituent of triacylglycerols and wax ester produced by *Rhodococcus opacus* PD630. Microbiology 148:1407–1412

Alvarez HM, Herrero OM, Silva RA, Hernandez MA, Lanfranconi MP, Villalba MS (2019) Insights into the metabolism of oleaginous *Rhodococcus* spp. Appl Environ Microbiol 85: e00498–e00419

Athenstaedt K, Daum G (1997) Biosynthesis of phosphatidic acid in lipid particles and endoplasmic reticulum of *Saccharomyces cerevisiae*. J Bacteriol 179:7611–7616

Athenstaedt K, Daum G (2003) YMR313c/TGL3 encodes a novel triacylglycerol lipase located in lipid particles of *Saccharomyces cerevisiae*. J Biol Chem 278:23317–23323
Athenstaedt K, Daum G (2006) The life cycle of neutral lipids: synthesis, storage and degradation. Cell Mol Sci 63:1355–1369
Bacchin P, Robertiello A, Viglia A (1974) Identification of *n*-decane oxidation products in *Corynebacterium* cultures by combined gas chromatography mass spectrometry. Appl Microbiol 28:737–741
Barbero P, Buell E, Zulley S, Pfeffer SR (2001) TIP47 is not a component of lipid droplets. J Biol Chem 276:25324–25335
Barksdale L, Kim KS (1977) Mycobacterium. Bacteriol Rev 41:217–372
Blanchette-Mackie EI, Dwyer NK, Barber T, Coxey RA, Takeda T, Rondinone CM, Theodorakis JL, Grennberg AS, Londos C (1995) Perilipin is located on the surface-layer of intracellular lipid droplets in adipocytes. J Lipid Res 36:1211–1226
Boulton CA, Ratledge C (1981) Correlation of lipid accumulation in yeasts with possession of ATP: citrate lyase. J Gen Microbiol 102:33–43
Bouvier-Navé P, Benveniste P, Oelker P, Sturley SL, Schaller H (2000) Expression in yeast and tobacco of plant cDNAs encoding acyl CoA:diacylglycerol acyltransferase. Eur J Biochem 267:85–96
Bredemeier R, Hulsch R, Metzger JO, Berthe-Corti L (2003) Submersed culture production of extracellular wax esters by the marine bacterium *Fundibacter jadensis*. Mar Biotechnol 5:579–583
Bresan S, Sznajder A, Hauf W, Forchhammer K, Pfeiffer D, Jendrossek D (2016) Polyhydroxyalkanoate (PHA) granules have no phospholipids. Scientific Reports 6:26612
Bryn K, Jantzen E, Bovre K (1977) Occurrence and patterns of waxes in *Neisseriaceae*. J Gen Microbiol 102:33–43
Cases S, Smith SJ, Zheng YW, Myers HM, Lear SR, Sande E, Novak S, Collins C, Welch CB, Lusis AJ, Erickson SK, Farese RV (1998) Identification of a gene encoding as acyl CoA: diacylglycerol acyltransferase, a key enzyme in triacylglycerol synthesis. Proc Natl Acad Sci USA 95:13018–13023
Cases S, Stone SJ, Zhou P, Yen E, Tow B, Lardizabal KD, Voelker T, Farese RV (2001) Cloning of DGAT2, a second mammalian diacylglycerol acyltransferase, and related family members. J Biol Chem 276:38870–38876
Chang TY, Chang CCY, Cheng D (1997) Acyl-coenzyme A: cholesterol acyltransferase. Annu Rev Biochem 66:613–638
Chen JCF, Lin RH, Huang AHC, Tzen JTC (1997) Cloning, expression and isoform classification of a minor oleosin in sesame oil bodies. J Biochem (Tokyo) 122:819–824
Cherian S, Ryu SB, Cornish K (2019) Natural rubber biosynthesis in plants, the rubber transferase complex, and metabolic engineering progress and prospects. Plant Biotechnol J 17:2041–2061
Christensen H, Garton NJ, Horobin RW, Minnikin DE, Barer MR (1999) Lipid domains of mycobacteria studied fluorescent molecular probes. Mol Microbiol 31:1561–1572
Christiansen K (1978) Triacylglycerol synthesis in lipid particles of baker's yeast (*Saccharomyces cerevisiae*). Biochem Biophys Acta 530:78–90
Clausen MK, Christia K, Jensen PK, Behke O (1974) Isolation of lipid particles from bakers-yeast. FEBS Lett 43:176–179
Coleman J (1990) Characterization of *Escherichia coli* cells deficient in 1-acyl-sn-glycerol-3-phosphate acyltransferase activity. J Biol Chem 265:17215–17221
Dahlquist A, Stahl U, Lenman M, Banas A, Lee M, Sandager L, Ronne H, Stymne S (2000) Phospholipid:diacylglycerol acyltransferase: an enzyme that catalyzes the acyl-CoA-independent formation of triacylglycerol in yeast and plants. Proc Natl Acad Sci USA 97:6487–6492
Daniel J, Deb C, Dubey VS, Sirakova TD, Abomoelak B, Morbidoni HR, Kolattukudy PE (2004) Induction of a novel class of diacylglycerol acyltransferases in *Mycobacterium tuberculosis* as it goes into a dormancy-like state of culture. J Bacteriol 186:5017–5030

De Andrès C, Espuny MJ, Robert M, Mercadè ME, Manresa A, Guinea J (1991) Cellular lipid accumulation by *Pseudomonas aeruginosa* 44T1. Appl Microbiol Biotechnol 35:813–816

Dewitt S, Ervin JL, Howesorchison D, Dalietos D, Neidleman SL, Geigert J (1982) Saturated and unsaturated wax esters produced by *Acinetobacter* sp. HO1-N grown on C16-C20 *n*-alkanes. J Am Oil Chem Soc 59:69–74

Ding Y, Yang L, Zhang SY, Wang YL, Du J, Pu G, Peng G, Chen Y, Zhang HN, Yu JH, Hang HY, Wu P, Yang FQ, Yang HY, Steinbüchel A, Liu PS (2012) Identification of the major functional proteins of prokaryotic lipid droplets. J Lip Res *53*:399–411

Elbahloul Y, Steinbüchel A (2010) Pilol-scale production of fatty acid ethyl esters by an engineered *Escherichia coli* strain harboring the p(Microdiesel) plasmid. Appl Environ Microbiol 76:4560–4565

Ervin JL, Geigert J, Neidleman SL, Wadsworth J (1984) Substrate-dependent and growth temperature dependent changes in the wax ester compositions produced by *Acinetobacter* sp. HO1-N. In: Ratledge C, Dawson P, Rattray L (eds) Biotechnology of the oil and fats industry. American Oil Chemists Society, Champaign, pp 217–222

Fixter LM, Fewson CA (1974) Accumulation of waxes by *Acinetobacter calcoaceticus* NCIB-8250. Biochem Soc Trans 2:944–945

Fixter LM, McCormack JG (1976) Effect of growth conditions on wax-content of various strains of *Acinetobacter*. Biochem Soc Trans 4:504–505

Fixter LM, Nagi M, McCormack JG, Fewson CA (1986) Structure, distribution and function of wax esters in *Acinetobacter calcoaceticus*. J Gen Microbiol 132:3147–3157

Frentzen M (1996) Acyltransferases from basic science to modified seed oils. Fett/Lipid 100:161–166

Frey-Wyssling A, Muhlethaler K, Grieshaber E (1963) Origin of spherosomes in plant cells. J Ultrastruct Res 8:506–516

Galili G, Sengupta-Gopalan C, Ceriotti A (1998) The endoplasmic reticulum of plant cells and its role in protein maturation and biogenesis of oil bodies. Plant Mol Biol 38:1–19

Gallagher IHC (1971) Occurrence of waxes in *Acinetobacter*. J Gen Microbiol 68:245–247

Hammerson F (1985) Histology: color atlas of microscopic anatomy, 2nd edn. Urban and Schwarzenberg, Baltimore

Harwood JL (1994) Lipid metabolism. In: Gunstone FD, Harwood JL, Padley FB (eds) The lipid handbook. Chapman and Hall, London, pp 605–664

Hauschild P, Röttig A, Madkour MH, Al Ansari AM, Almakisha NH, Steinbüchel A (2017) Lipid accumulation in prokaryotic microorganisms from arid habitats. Appl Microbiol Biotechnol 71:942–952

Hetzler S, Steinbüchel A (2013) Establishment of cellulose utilization for lipid production in *Rhodococcus opacus* PD630. Appl Environ Microbiol 79:3122–3125

Hills MJ, Watson MD, Murphy DJ (1993) Targeting of oleosins to the oil bodies of oilseed rape (*Brassica napus* L.). Planta 189:24–29

Hobbs PJ, Hills MJ (1999) Expression and characterization of diacylglycerol acyltransferase from *Arabidopsis thaliana* in insect cell cultures. FEBS Lett 452:145–149

Hope HW, Moll R, Schwetlick J, Rackwitz HR, Keenan T (1998) Adipophilin is a specific marker of lipid accumulation in diverse cell types and diseases. Cell Tissue Res 294:309–321

Hoskisson PA, Hobbs G, Sharples GP (2001) Antibiotic production, accumulation of intracellular carbon reserves, and sporulation in *Micromonospora echinospora* (ATCC 15837). Can J Microbiol 47:148–152

Ichihara K (1982) Lipid-metabolism in safflower/formation of oleosomes in maturing safflower seeds. Agric Biol Chem 46:1767–1773

Icho T, Raetz CR (1983) Multiple genes for membrane-bound phosphatases in *Escherichia coli* and their action on phospholipid precursors. J Bacteriol 153:722–730

Ishige T, Tani A, Takabe K, Kawasaki K, Sakai Y, Kato N (2002) Wax ester production from *n*-alkanes by *Acinetobacter* sp. strain M-1: ultrastructure of cellular inclusions and role of acyl coenzyme A reductase. Appl Environ Microb 68:1192 1195

Ishige T, Tani A, Sakai JR, Kato N (2003) Wax ester production by bacteria. Cur Opin Microbiol 6:244–250
Jendrossek D (2005) Fluorescence microscopical investigation of poly(3-hydroxybutyrate) granule formation in bacteria. Biomacromolecules 6:598-603.
Jendrossek D (2020) Carbonosomes. In Jendrossek D (ed) Bacterial organelles and organelle-like inclusions. Springer Heidelberg
Jendrossek D, Pfeiffer D (2014) New insights in the formation of polyhydroxyalkanoate granules (carbonosomes) and novel functions of poly(3-hydroxybutyrate). Environ Microbiol 16:2357–2373
Jossek R, Reichelt R, Steinbüchel A (1998) In vitro biosynthesis of poly(3-hydroxybutyric acid) by using purified poly(hydroxyalkanoic acid) synthase of *Chromatium vinosum*. Appl Microbiol Biotechnol 49:258–266
Juanssilfero AB, Kahar P, Amza RL, Miyamoto N, Otsuka H, Matsumoto H, Kihira C, Thontowi A, Yopi OC (2018) Selection of oleaginous yeasts capable of high lipid accumulation during challenges from inhibitory chemical compounds. Biochem Eng J 137:182–191
Jurasek L, Marchessault RH (2004) Polyhydroxyalkanoate (PHA) granule formation in *Ralstonia eutropha* cells: a computer simulation. Appl Microbiol Biotechnol 64:611–617
Kalantari F, Bergeron JJM, Nilsson T (2010) Biogenesis of lipid droplets – how cells get fatter. Mol Membr Biol 27:462–468
Kalscheuer R, Wältermann M, Steinbüchel A (1999) Biosynthese und Speicherung von Triglyceriden und Wachsen in Bakterien. In: Biokonversion nachwachsender Rohstoffe 15. Landwirtschaftsverlag Münster, Münster, pp 253–261
Kalscheuer R, Wältermann M, Alvarez HM, Steinbüchel A (2001) Preparative isolation of lipid inclusions from *Rhodococcus opacus* and *Rhodococcus ruber* and identification of granule-associated proteins. Arch Microbiol 177:20–28
Kalscheuer R, Uthoff S, Luftmann H, Steinbüchel A (2003) In vitro and in vivo biosynthesis of wax diesters by an unspecific bifunctional wax ester synthase/acyl-CoA:diacylglycerol acyltransferase from *Acinetobacter calcoaceticus* ADP1. Eur J Lipid Sci Technol 105:578–584
Kalscheuer R, Stöveken T, Luftmann H, Malkus U, Reichelt R, Steinbüchel A (2006a) Neutral lipid biosynthesis in engineered *Escherichia coli*: Jojoba oil-like wax esters and fatty acid butyl esters. Appl Environ Microbiol 72:1373–1379
Kalscheuer R, Steinbüchel A (2003) A novel bifunctional wax ester synthase/acyl-CoA: diacylglycerol acyltransferase mediates wax ester and triacylglycerol biosynthesis in Acinetobacter calcoaceticus (italize) ADP1. J Biol Chem 278:8075-8082.
Kalscheuer R, Stoelting T, Steinbüchel A (2006b) Microdiesel: *Escherichia coli* engineered for fuel production. Microbiology (SGM) 152:2529–2536
Kalscheuer R, Stöveken T, Steinbüchel A (2007) Engineered microorganisms for sustainable production of diesel fuel and other oleochemicals from renewable plant biomass. Int Sugar J 109:16–19
Koval'schuk LP, Donets AP, Razumovskii PN (1973) Lipid biosynthesis by actinomycetes cultivated on different media. Mikrobiologiya Int Ed 42:567–571
Lacey DJ, Wellner N, Beaudoin F, Napier NA, Shewry PR (1998) Secondary structure of oleosins in oil bodies isolated from seeds of safflower (*Carthamus tinctorius* L.) and sunflower (*Helianthus annuus* L.). Biochem J 334:469–477
Lardizabal KD, Metz JG, Sakamoto T, Hutton WC, Pollard MR, Lassner MW (2000) Purification of a jojoba embryo wax synthase, cloning of its cDNA, and production of high levels of wax in seeds of transgenic Arabidopsis. Plant Physiol 122:645–655
Leber R, Zinser E, Zellnig G, Paltauf F, Daum G (1994) Characterization of lipid particles of the yeast. Yeast 10:1421–1428
Leber R, Landl K, Zinser E, Ahorn A, Spok A, Kohlwein SD, Turnowsky F, Daum G (1998) Dual localization of squalene epoxidase, Erg1p, in yeast reflect a relationship between endoplasmic reticulum and lipid particles. Mol Biol Cell 9:375–386
Lehner R, Kuksis A (1996) Biosynthesis of triacylglycerols. Prog Lipid Res 35:169–201

Leman J (1997) Oleaginous microorganisms: an assessment of the potential. Adv Appl Microbiol 43:195–243

Lennen RM, Pfleger BF (2013) Microbial production of fatty acid-derived fuels and chemicals. Curr Opin Biotechnol 24:1044–1053

Liu P, Ying Y, Zhao Y, Mundy DJ, Zhu M, Anderson RG (2004) Chinese hamster ovary K2 cell lipid droplets appear to be metabolic organelles in membrane traffic. J Biol Chem 279:3787–3792

Loer DS, Herman EM (1993) Cotranslational integration of soybean (Glycine max) oil body membrane-protein oleosin into microsomal membranes. Plant Physiol 101:993–998

Londos C, Brasaemle DL, Gruia-Gray J, Servetnik DA, Schultz CJ, Levin DM, Kimmel AR (1995) Perilipin – unique proteins associated with intracellular neutral lipid droplets in adipocytes and steroidogenic cells. Biochem Soc Trans 23:611–615

Londos C, Brasaemle DL, Schultz CJ, Segrest JP, Kimmel AR (1999) Perilipins, ADRP, and other proteins that associate with intracellular neutral lipid droplets in animal cells. Semin Cell Dev Biol 10:51–58

Lu XY, Gruia-Gray J, Copeland NG, Gilbert DJ, Jenkins NA, Londos C, Kimmel AR (2001) The murine perilipin gene: the lipid droplet-associated perilipins derive from tissue-specific, mRNA splice variants and define a gene of ancient origin. Mamm Genome 12:741–749

Makula RA, Lockwood PJ, Finnerty WR (1975) Comparative analysis of lipids of *Acinetobacter* species grown on hexadecane. J Bacteriol 121:303–312

Manilla-Perez E, Lange AB, Hetzler S, Steinbüchel A (2010) Occurrence, production, and export of lipophilic compounds by hydrocarbonoclastic marine bacteria and their potential use to produce bulk chemicals from hydrocarbons. Appl Microbiol Biotechnol 86:1693–1706

Meng X, Yang JM, Xu X, Zhang L, Nie QJ, Xian M (2009) Biodiesel production from oleaginous microorganisms. Renew Energ 34:1–5

Milla P, Athenstaedt K, Viola F, Oliaro-Bosso S, Kohlwein SD, Daum G, Balliano G (2002) Yeast oxidosqualene cyclase (Erg7p) is a major component of lipid particles. J Biol Chem 277:2406–2412

Miura S, Gan JW, Brzostowski J, Parisi MJ, Schultz CJ, Londos C, Oliver B, Kimmel AR (2002) Functional conservation for lipid storage droplet association among perilipin, ADRP, and TIP47 (PAT)-related proteins in mammals, *Drosophila*, and *Dictyostelium*. J Biol Chem 277:32253–32257

Murphy DJ (1993) Structure, function and biogenesis of storage lipid bodies and oleosins in plants and other organisms. Prog Lipid Res 29:299–324

Murphy DJ (2001) The biogenesis and functions of lipid bodies in animals, plants and microorganism. Prog Lipid Res 40:325–438

Murphy DJ (2012) The dynamic roles of intracellular lipid droplets: from archaea to mammals. Protoplasma 249:541–585

Murphy DJ, Vance J (1999) Mechanism of lipid body formation. Trends Biochem Sci 24:109–115

Olukoshi ER, Packter NM (1994) Importance of stored triacylglycerols in *Streptomyces* — possible carbon source for antibiotics. Microbiology 140:931–943

Packter NM, Olukoshi ER (1995) Ultrastructural studies of neutral lipid localisation in *Streptomycetes*. Arch Microbiol 164:420–427

Pieper-Fürst U, Madkour MH, Mayer F, Steinbüchel A (1994) Purification and characterization of a 14-kilodalton protein that is bound to the surface of polyhydroxyalkanoic acid granules in *Rhodococcus ruber*. J Bacteriol 176:4328–4337

Polokoff MA, Bell RM (1980) Solubilization, partial-purification and characterization of rat-liver microsomal diacylglycerol acyltransferase. Biochim Biophys Acta 618:129–142

Post-Beitenmiller D (1996) Biochemistry and molecular biology of wax production in plants. Annu Rev Plant Physiol Mol Biol 47:405–430

Preusting H, Kingma J, Huisman G, Steinbüchel A, Witholt B (1993) Formation of polyester blends by a recombinant strain of *Pseudomonas oleovorans*: different poly(3-hydroxyalkanoates) are stored in separate granules. J Environ Polym Degrad 1:11–21

Qadeer S, Khalid A, Mahmood S, Anjum M, Ahmad Z (2017) Utilizing oleaginous bacteria and fungi for cleaner energy production. J Clean Prod 168:917–928
Radakovits R, Jinkerson RE, Darzins A, Posewitz MC (2010) Genetic engineering of algae for enhanced biofuel production. Eukary Cell 9:486–501
Ratledge C (1989) Biotechnology of oils and fats. In: Ratledge C, Wilkinson SG (eds) Microbial lipids. Academic, London, pp 567–650
Ratledge C (2002) Regulation of lipid accumulation in oleaginous microorganisms. Biochem Soc Trans 30:1047–1050
Raymond RL, Davies JB (1960) *n*-Alkane utilization and lipid formation by a *Nocardia*. Appl Microbiol 8:329–334
Reiser S, Somerville C (1997) Isolation of mutants of *Acinetobacter calcoaceticus* deficient in wax ester synthesis and complementation of one mutation with a gene encoding a fatty acyl coenzyme A reductase. J Bacteriol 179:2969–2975
Robenek MJ, Severs NJ, Schlattmann K, Plenz G, Zimmer KP, Troyer D, Robenek H (2004) Lipids partition caveolin-1 from ER membranes into lipid droplets: updating the model of lipid droplet biogenesis. FASEB J 18:866–868
Robert LS, Gerster J, Allard S, Cass L, Simmonds J (1994) Molecular characterization of 2 *Brassica napus* genes related to oleosins which are highly expressed in the tapetum. Plant J 6:927–933
Roberts MR, Robson F, Foster GD, Draper J, Scott RJ (1991) A *Brassica napus* messenger-RNA expressed specifically in developing microspores. Plant Mol Biol 17:295–299
Ross JHE, Sanchez J, Millan F, Murphy DJ (1993) Differential presence of oleosins in oleogenic seed and mesocarp tissues in olive (*Olea europaea*) and avocado (*Persea americana*). Plant Sci 93:203–210
Röttig A, Steinbüchel A (2013) Acyltransferases in bacteria. Microbiol Molec Biol Rev. 77:277–321
Röttig A, Hauschild P, Madkour MH, Al-Ansari AM, Almakishah NH, Steinbüchel A (2016) Analysis and optimization of triacylglycerol synthesis in novel oleaginous *Rhodococcus* and *Streptomyces* strains isolated from desert soil. J Biotechnol 225:48–56
Routaboul JM, Benninng C, Bechtold N, Caboche M, Lepiniec L (1999) The TAG1 locus of *Arabidopsis* encodes for a diacylglycerol acyltransferase. Plant Physiol Biochem 37:831–840
Russel NJ, Volkman JK (1980) The effect of growth temperature and wax ester composition in the psychrophilic bacterium *Micrococcus cryophilus* ATCC 15174. J Gen Microbiol 118:131–141
Sandager L, Gustavson MH, Stahl U, Dahlquist A, Wiberg E, Banas A, Lenman M, Ronne H, Stymne S (2002) Storage lipid biosynthesis is non-essential in yeast. J Biol Chem 277:6478–6482
Schaffner G, Matile P (1981) Structure and composition of baker's yeast lipid globules. Biochem Physiol Pflanzen 176:659–666
Schlunk FF, Lombardi B (1967) Liver liposomes/isolation and chemical characterization. Lab Invest 17:30–38
Scott CCL, Finnerty WR (1976) Characterization of intracytoplasmic hydrocarbon inclusions from hydrocarbon-oxidizing *Acinetobacter* species HO1-N. J Bacteriol 127:481–489
Scott SA, Davey MP, Dennis JS, Horst I, Howe CJ, Lea-Smith DJ, Smith AG (2010) Biodiesel from algae: challenges and prospects. Curr Opin Biotechnol 21:277–286
Silva RA, Grossi V, Alvarez M (2007) Biodegradation of phytane (2,6,10,14-tetramethylhexadecane) and accumulation of related isoprenoid wax esters by *Mycobacterium ratisbonense* strain SD4 under nitrogen-starved conditions. FEMS Microbiol Lett 272:220–228
Singer ME, Tyler SM, Finnerty WR (1985) Growth of *Acinetobacter* sp. strain HO1-N on *n*-hexadecanol: physiological and ultrastructural characteristics. J Bacteriol 162:162–169
Smid A, Riva M, Bouet F, Sentenac A, Carles C (1995) The association of 3 subunits with yeast RNA-polymerase is stabilized by A14. J Biol Chem 270:13534–13540
Smith SJ, Cases S, Jensen DR, Chen HC, Sande E, Tow B, Sana DA, Raber J, Eckel RH, Farese RV Jr (2000) Obesity resistance and multiple mechanisms of triglyceride synthesis in mice lacking DGAT. Nat Genet 25:87–90

Sorger D, Daum G (2003) Triacylglycerol biosynthesis in yeast. Appl Microbiol Biotechnol 61:289–299
Stahmann KP, Kupp C, Feldmann SD, Sahm H (1994) Formation and degradation of lipid bodies found in the riboflavin-producing fungus *Ashbya gossypii*. Appl Microbiol Biotechnol 42:121–127
Steinbüchel A (2001) Perspectives for biotechnological production and utilization of biopolymers: metabolic engineering of polyhydroxyalkanoate biosynthesis pathways as a successful example. Macromol Biosci 1:1–24
Steinbüchel A, Valentin HE (1995) Diversity of bacterial polyhydroxyalkanoic acids. FEMS Microbiol Lett 128:219–228
Stöveken T, Kalscheuer R, Malkus U, Reichelt R, Steinbüchel A (2005) The wax ester synthase/acyl coenzyme A: diacylglycerol acyltransferase from *Acinetobacter* sp. strain ADP1: characterization of a novel type of acyltransferase. J Bacteriol 187:1369–1376
Stubbe JA, Tian J (2003) Polyhydroxyalkanoate (PHA) homeostasis: the role of the PHA synthase. Nat Prod Rep 20:445–457
Suckling KE, Stange EF (1985) Role of acyl-CoA:cholesterol acyltransferase in cellular cholesterol-metabolism. J Lipid Res 26:647–670
Syu LJ, Saltiel AR (1999) Lipotransin: a novel docking protein for hormone-sensitive lipase. Mol Cell 4:109–115
Sztalryd C, Xu GH, Dorward H, Tansey JT, Contreras JA, Kimmel AR, Londos C (2003) Perilipin A is essential for the translocation of hormone-sensitive lipase during lipolytic activation. J Cell Biol 161:1093–1103
Tsitsigiannis DI, Zarnowski R, Keller NP (2004) The lipid body protein, PpoA, coordinates sexual and asexual sporulation in *Aspergillus nidulans*. J Biol Chem 279:11344–11353
Tzen JT, Cao YZ, Laurent P, Ratnayake C, Huang AHC (1993) Lipids, proteins, and structure of seed oil bodies from diverse species. Plant Physiol 101:267–276
Umlauf E, Csaszar E, Moertelmaier M, Schuetz G, Parton RG, Prohaska R (2004) Association of stomatin with lipid droplets. J Biol Chem 280:23699–23709
Wahl A, Schuth N, Pfeiffer D, Nussberger S, Jendrossek D (2012) PHB granules are attached to the nucleoid via PhaM in *Ralstonia eutropha*. BMC Microbiol 12:262
Walker RW, Barakat H, Hung JGC (1970) Positional distribution of fatty acids in phospholipids and triglycerides of *Mycobacterium smegmatis* and *M. bovis* BCG. Lipids 5:684–691
Wältermann M, Steinbüchel A (2005) Neutral lipid bodies in prokaryotes: recent insights into structure, formation, and relationship to eukaryotic lipid depots. J Bacteriol 187:3607–3619
Wältermann M, Luftmann H, Baumeister D, Kalscheuer D, Steinbüchel A (2000) *Rhodococcus opacus* strain PD630 as a new source of high-value single cell oil? Isolation and characterization of triacylglycerols and other storage lipids. Microbiology 146:1143–1149
Wältermann M, Hinz A, Robenek H, Troyer D, Reichelt R, Malkus U, Galla HJ, Kalscheuer R, Stöveken T, von Landenberg P, Steinbüchel A (2005) Mechanism of lipid body formation in prokaryotes: how bacteria fatten up. Mol Microbiol 55:750–763
Wang L, Schnoes HK, Takayama K, Goldman DS (1972) Synthesis of alcohol and wax ester by a cell-free system in *Mycobacterium tuberculosis*. Biochem Biophys Acta 260:41–48
Wang SM, Fong T, Hsu S, Chien S, Wu J (1997) Reorganization of a novel vimentin-associated protein in 3T3-L1 cells during adipocyte conversion. J Cell Biochem 67:84–91
Wang X, Reape TJ, Li X, Rayner K, Webb C, Burnand KG, Lysko PG (1999) Induced expression of adipophilin mRNA in human macrophages stimulated with oxidized low-density lipoprotein in artheriosklerotic lesions. FEBS Lett 462:445–450
Wang SM, Hwang RD, Greenberg AS, Yeo HL (2003) Temporal and spatial assembly of lipid-droplet associated proteins in 3T3-L1 preadipocytes. Histochem Cell Biol 120:285–292
Wanner G, Theimer RR (1978) Membranous appendices of spherosomes (oleosomes)—possible role in fat utilization in germinating oil seeds. Planta 140:163–169
Wanner G, Formanek H, Theimer RR (1981) The ontogeny of lipid bodies (spherosomes) in plant cells—ultrastructural evidence. Planta 151:109–123

Wayman M, Jenkins AD, Kormendy AG (1984) Ratledge C, Dawson P, Rattray J (eds) Biotechnology for the oils and fat industry. American Oil Chemists' Society, monograph no 11. American Oil Chemists' Society, Champaign, pp 129–143

Weather PR, Burkiott HG, Daniels VG (1987) Functional histology: a text and color atlas, 2nd edn. Churchill Livingstone, New York

Weiss L (1983) Cell and tissue biology, 5th edn. Elsevier, New York

Wilkinson WO, Bell RM (1997) *sn*-Glycerol-3-phosphate acyltransferase from *Escherichia coli*. Biochem Biophys Acta 1348:3–9

Wolins NE, Skinner JR, Schoenfish MJ, Tzekow A, Bensch KG, Bickel PE (2003) Adipocyte protein S3–12 coats nascent lipid droplets. J Biol Chem 276:37713–37721

Wood D, Cornish K (2000) Microstructure of purified rubber particles. Int J Plant Sci 161:435–445

Yang H, Bard M, Bruner DA, Gleeson A, Deckelbaum RJ, Aljinovic G, Pohl TM, Rothstein R, Sturley SL (1996) Sterol esterification in yeast: a two-gene process. Science 272:1353–1356

Yang L, Ding YF, Chen Y, Zhang SY, Huo CX, Wang Y, Yu JH, Zhang P, Na HM, Zhang HN (2012) The proteomics of lipid droplets: structure, dynamics, and functions of the organelle conserved from bacteria to humans. J Lip Res 53:1245–1253

Yermanos DM (1975) Composition of jojoba seed during development. J Am Oil Chem Soc 52:115–117

Yu C, Kennedy NJ, Chang CC, Rothblatt JA (1996) Molecular cloning and characterization of two isoforms of *Saccharomyces cerevisiae* acyl-CoA:sterol acyltransferase. J Biol Chem 272:3980–3985

Zou JT, Wei YD, Jako C, Kumar A, Selvaraj G, Taylor DC (1999) The *Arabidopsis thaliana* TAG1 mutant has a mutation in a diacylglycerol acyltransferase gene. Plant J 19:645–653

Zweytick D, Athenstedt K, Daum G (2000) Intracellular lipids particles of eukaryotic cells. BBA Rev Biomembr 1469:101–120

Carbonosomes

Dieter Jendrossek

Contents

1 Introduction ... 244
2 Composition of the Carbonosome Core ... 244
3 Protein Composition of the Carbonosome Surface Layer ... 246
4 The Carbonosome of *Ralstonia eutropha* ... 248
5 Proteins of the *R. eutropha* Carbonosome ... 248
  5.1 PHB Synthases (PhaCs) ... 251
  5.2 Phasins (PhaPs) ... 252
  5.3 PHB Depolymerases (PhaZs) ... 253
  5.4 Proteins Binding to PHB Granules and to DNA ... 257
  5.5 Other Proteins Bound to PHB Granules ... 258
6 The Carbonosome of *Pseudomonas putida* ... 259
  6.1 mcl-PHA Synthases (mcl-PhaCs) ... 259
  6.2 mcl-PHA Depolymerase (mcl-PhaZs) ... 260
  6.3 Phasins ... 261
  6.4 Proteins Binding to mcl-PHA Granules and to DNA ... 261
  6.5 Other mcl-PHA-Bound Proteins ... 261
7 Are Carbonosomes Covered by a Phospholipid Monolayer? ... 262
8 Biogenesis of Carbonosomes ... 263
9 Novel Function of Carbonosomes ... 265
References ... 266

Abstract The understanding of the physiology of poly(3-hydroxybutyric acid) (PHB) and related polyhydroxyalkanoic acids (PHAs) has made great progress since the identification and cloning of the first PHB synthase gene from *Ralstonia eutropha* H16 in three laboratories in the late 1980s. In particular, the identification of many proteins as part of the proteinaceous surface layer of PHB and PHA granules and their classification into PHA synthases, PHA depolymerases, phasins, PHA- and DNA-binding proteins and others has led to the conclusion that PHB and PHA granules are functional supramolecular complexes for with the designation as

D. Jendrossek (✉)
Institute of Microbiology, University of Stuttgart, Stuttgart, Baden-Württemberg, Germany
e-mail: dieter.jendrossek@imb.uni-stuttgart.de

D. Jendrossek (ed.), *Bacterial Organelles and Organelle-like Inclusions*,
Microbiology Monographs 34, https://doi.org/10.1007/978-3-030-60173-7_10

carbonosomes has been suggested. In this chapter, the current knowledge on the functions of PHA granule associated proteins (PGAPs) and the composition of carbonosomes will be summarized at the example of PHB granules from *R. eutropha* and of PHA granules from *Pseudomonas putida*.

1 Introduction

Many prokaryotic species are able to synthesise short-chain-length polyhydroxyalkanoates (scl-PHAs) such as poly(3-hydroxybutyrate), PHB, or related medium-chain-length polyhydroxyalkanoates (mcl-PHAs). The bacteria accumulate these storage compounds in the form of ≈200 to ≈600 nm insoluble, granule-like inclusions (PHA granules) when they are cultivated under unbalanced growth conditions, e.g. at high carbon to nitrogen source ratios. PHA granules consist of a polyester core and a surrounding surface layer of several (up to 16) proteins. These so-called PHA granule-associated proteins (PGAPs) have specific, in parts, unknown functions in the synthesis, maintenance and mobilisation (intracellular reutilisation by depolymerisation) of PHAs.

The complexity of PHA granules with many PGAPs at the PHA surface indicates that PHA granules have more functions than that of a carbon and energy reserve material only. This is in agreement with data from several laboratories suggesting that PHA granules are important for the adaptation of cells to oxidative and other stresses. To indicate that PHA granules are supramolecular complexes and have organelle-like properties, the designation "carbonosomes" had been proposed as generic term for PHA granules (Jendrossek 2009). This review will provide an overview on the composition of the carbonosome core and of the carbonosome surface structure using *Ralstonia eutropha* and *Pseudomonas putida* as a model species for scl-PHA- and mcl-PHA-accumulating species, respectively. The function(s) of different PGAPs in the physiology of the accumulating species will be described and discussed. The biochemical pathways leading to the precursors of PHA monomers in different species have been extensively explored and are summarised in several previous reviews (for examples, see Anderson and Dawes 1990; Steinbüchel et al. 1993b; Steinbüchel and Valentin 1995; Madison and Huisman 1999; Pötter and Steinbüchel 2006) and will be not part of this contribution.

2 Composition of the Carbonosome Core

The term "carbonosome" in this review is used for all kinds of PHA granules in the native (in vivo) form in prokaryotes. Carbonosomes are defined as supramolecular complexes that consist of a polymer core with a backbone of mainly carbon atoms

Fig. 1 General structure of PHAs. (**a**) Polyoxoester, (**b**) Polythioester

and are surrounded by a surface layer of at least three types of proteins: (1) a polymer synthase that catalyses the polymerisation of the polymer core from the respective monomers, (2) a polymer depolymerase that catalyses the depolymerisation of the polymer to monomers and or oligomers and (3) one or more other types of proteins that contribute to or even define the properties of the surface layer. In the case of PHB or PHA granules, the phasins (PhaPs) fulfil this function. Granulose and glycogen that are synthesised, e.g. by many cyanobacteria, have proteins of type (1) and (2), but, at present, no evidence is available whether phasin homolog-like proteins of type (3) are present in vivo in granulose/glycogen-forming species. If such compounds will be detected, granulose/glycogen would represent another subtype of carbonosomes (polysaccharide-type) in comparison to polyester type of carbonosomes. For similar reasons cyanophycin granules at present are not (yet) considered as carbonosomes.

The core of polyester carbonosomes consists of the PHA polyester with the general formula shown in Fig. 1. In addition to these polyoxoesters, polythioesters such as poly(3-mercaptopropionate) can be also synthesised by some species if the bacteria are cultivated in the presence of thiol-containing precursors (Lütke-Eversloh et al. 2001; Thakor et al. 2005). Remarkably, and in sharp contrast to all PHAs, such polythioesters seem to be not biodegradable (Elbanna et al. 2004).

The molecular masses of the formed polymers in most species are in the range of $\approx 10^5$ to $\approx 10^7$ Da (Suzuki et al. 1988; Kawaguchi and Doi 1992; Myshkina et al. 2008; Adaya et al. 2018). Depending on the genotypic and metabolic capabilities of the PHA-accumulating species, the type of its PHA synthase and the available growth substrate, the monomeric constituents of the formed PHAs can verify to a large extent. Nevertheless, the most prominent and most common PHA is the homopolyester PHB consisting of ester-linked 3-hydroxybutyrate units only. However, due to the low substrate specificity of most PHA synthases with respect to the number of carbon atoms of the fatty acid monomer moiety and the position of the hydroxy group, PHAs can consist of monomers with 3 to at least 14 carbon atoms. The side group R might contain halogen atoms, unsaturated carbon-carbon bonds or even contain aromatic groups (Abraham et al. 2001; Olivera et al. 2010). Furthermore, PHA synthases can polymerise various substrates into the same polymer molecules if the metabolism of the accumulating strain is able to synthesise different CoA-activated monomeric precursors simultaneously. As a consequence, the resulting polyesters consist of two or more different hydroxyalkanoic acids and are copolyesters. The (bio)physical properties of PHA copolymers largely depend on the monomeric composition. For overviews on the complexity of possible compositions of microbially produced PHAs, see previous reviews (Steinbüchel and Valentin 1995; Meng and Chen 2018). For some applications of PHAs, it might be desirable to have homopolyesters. However, with the exception of PHB and polyhydroxyvalerate (PHV) (Steinbüchel et al. 1993a), most naturally produced PHAs are copolyesters. Homopolyester of PHAs can be produced by genetically modified organisms (GMOs), in which the β-oxidation pathway of fatty acids is the only pathway leading to CoA-activated monomers and the shortening of the substrate molecules by acetyl-CoA units is blocked by mutation (Liu et al. 2011). These GMOs accumulate PHA homopolyesters with a number of carbon atoms in the monomer that corresponds to the number of carbon atoms of the substrate fatty acid. For overview, see Meng and Chen (2018).

3 Protein Composition of the Carbonosome Surface Layer

PHA granules have a higher density than most other cellular components and form a distinct band after density gradient centrifugation of crude extracts obtained from PHA-containing cells with glycerol or sucrose gradients. PHA granules can be further purified by repeated centrifugation and washing steps. For details, see the pioneer works of J.W. Merrick and B. Witholt (Merrick and Doudoroff 1964; Griebel et al. 1968; de Smet et al. 1983); a methodical overview is given in Jendrossek (2007). The proteins of the isolated PHA granule fraction can be then analysed by SDS-PAGE and identified by proteome analysis. The first identification of PGAPs from purified carbonosomes was published in 1992 using *Chromatium vinosum* as a PHA-accumulating species (Liebergesell et al. 1992). Since then, numerous reports were published describing the isolation of PHA granules and identification of PGAPs from several species, e.g. *Pseudomonas oleovorans* (Fuller et al. 1992), *R. eutropha* (Wieczorek et al. 1995; Jendrossek and Pfeiffer 2014), *Herbaspirillum seropedicae* (Tirapelle et al. 2013), *Azotobacter vinelandii* (Adaya et al. 2018; Moreno et al. 2019) and many other species. In these publications, analysis of purified carbonosome fractions revealed the presence of several to many protein bands by SDS-PAGE. The number and intensities of these bands largely differed for carbonosome fractions obtained from different species and were also dependent on the staining method (Coomassie versus silver staining). In all cases it is not possible to decide whether a particular protein band of a carbonosome fraction represents a true PGAP in vivo or has become artificially bound to PHA granules during cell disruption and the carbonosome isolation process. That some proteins can artificially bind to PHB granules was already shown in the very first description of purified PHB granules from *C. vinosum* (Liebergesell et al. 1992) in which lysozyme, which had been added to the cells during isolation procedure, was tightly bound to purified PHA granules. Additional evidence for the artificial binding of proteins to carbonosomes comes from experiments, in which PHB granules were analysed that had been isolated from recombinant *E. coli* expressing the PHB biosynthetic *phaCAB* operon of *R. eutropha.* SDS-PAGE of this fraction revealed a large number of proteins that could not be removed even by repeated washing steps of the isolated PHB granules using buffers with different ionic strengths (Handrick et al. 2004b). This points to a high affinity of some *E. coli* proteins to the PHB granule surface. The number of proteins, which were identified in isolated carbonosome preparations, dramatically increased up to several hundreds with the improvement of the sensitivity of proteome analysis (Tirapelle et al. 2013; Jendrossek and Pfeiffer 2014). In conclusion, the identification of a specific protein in a PHB granule preparation does not proof whether this protein is a true PGAP in the PHB-accumulating strain; it may represent a false-positive protein that has some affinity to the (hydrophobic) surface of PHA granules and binds to isolated PHA granules in vitro. Therefore, additional evidence is necessary for the identification of a true PGAP in PHA-accumulating species. One possibility is the identification of PGAPs in transmission electron micrographs of thin sections of PHA-accumulating cells by the immuno-gold technique with a PGAP-specific antibody and a gold-labelled secondary antibody. This technique has been used to identify the PHB synthase and/or the phasin proteins at the surface of PHA granules in several PHA-accumulating species (Gerngross et al. 1993; Liebergesell et al. 1994; Pieper-Furst et al. 1994; Wieczorek et al. 1995; Follner et al. 1997; Pötter et al. 2002). A second possibility is the construction of gene fusions coding for fusion proteins of a PGAP candidate fused to a green fluorescent protein variant followed by fluorescence microscopical localisation of the fusion protein in living cells. A colocalisation of the fusion protein with PHA granules that can be visible in light microscopy if the granules have a sufficient diameter or that have been stained with another fluorophore (e.g. Nile red) will then indicate whether or not the test protein is indeed a PGAP in vivo. This technique was first applied in 1999 for the localisation of a PHB synthase and a phasin protein of *Bacillus megaterium* by McCool and Cannon (1999) and later became the method of choice for the identification of a considerable number of PGAPs in many PHA-accumulating species such as *R. eutropha* (Handrick et al. 2004a; Uchino et al. 2007; Neumann et al. 2008), *Rhodospirillum rubrum* (Jendrossek 2005; Sznajder and Jendrossek 2011), *Magnetospirillum gryphiswaldense* (Schultheiss et al. 2005), *Pseudomonas putida* (Ruth et al. 2008) or *Synechocystis* sp. (Hauf et al. 2015).

However, both methods require either a time-consuming purification of the protein of interest and generation of specific antibodies or substantial molecular biological cloning work for each test protein. Such procedures might be practicable to perform for one or two key enzymes of PHA metabolism if other indirect evidence is already available that the protein most likely is a true PGAP. However, these approaches can't be easily performed for the high number of proteins that are nowadays identified by the sensitive proteome analysis methods. For example, the number of identified proteins in two independently isolated batches of carbonosome preparations from *R. eutropha* amounted 404 and 268 proteins (Jendrossek and Pfeiffer 2014; Sznajder et al. 2015). In this case, comparative proteome analysis can be applied to reduce the number of false-positive proteins. To this end, the proteomes of different cell fractions (e.g. soluble fraction, membrane fraction and PHB granule fraction) can be determined, and those proteins that are specifically present in the PHB granule fraction only are likely candidates for in vivo PGAPs. Using comparative proteome analysis with *R. eutropha*, 14 proteins were identified to be specifically present in the PHB granule fraction (Sznajder et al. 2015). However, the two most abundant proteins of PHA granules (phasin PhaP1 and PHB synthase PhaC1) were also present in traces in the soluble fraction. The in vivo colocalisation of these 14 proteins with PHA granules was verified by fusion analysis of GFP variants. This approach enabled the identification of several new PGAPs (Sznajder et al. 2015).

4 The Carbonosome of *Ralstonia eutropha*

R. eutropha H16 was isolated in Hans G. Schlegel's lab in 1962 from the Weende spring in Göttingen, Germany, as a chemolithoautotrophic bacterium (Wilde 1962). It is also able to grow heterotrophically on some sugars, a large variety of fatty acids and many other carbon sources. It was first considered as a suitable species to produce single-cell protein from carbon dioxide and hydrogen in the 70th of the last century. However, the high content of (ribo)nucleic acids in prokaryotes can cause gout in mammals, and this prevented the use of *R. eutropha* as alternative food source. In the 1980s, the ability of *R. eutropha* to accumulate large amounts of PHB up to approximately 90% of its cellular dry weight from renewable resources came into focus (Schubert et al. 1988; Peoples and Sinskey 1989). PHB was considered as a promising biodegradable alternative to chemosynthetically produced polymers, and this was the starting point of worldwide increasing research activities on the molecular biological background of PHB biosynthesis. Although a variety of Gram-negative, Gram-positive and even some *Archaea* species were chosen to investigate the biosynthesis, maintenance and mobilisation of PHA, it was *R. eutropha* that became the model organism of PHB research, and meanwhile the PHB granules of *R. eutropha* H16 are the best characterised carbonosomes of all studied species. In the next paragraph, I will first summarise the current knowledge about the composition and properties of the PHB granules in *R. eutropha* and then contrast them with the properties of carbonosomes from mcl-PHA-accumulating species using *P. putida* as an example.

5 Proteins of the *R. eutropha* Carbonosome

Sixteen proteins have been identified as PGAPs by comparative proteome analysis of isolated PHB granules (Table 1, Fig. 2) (Sznajder et al. 2015). These are the following: the PHB synthases PhaC1 and PhaC2, the PHB depolymerase PhaZa1 and three other putative hydrolases (a putative phospholipase [H16_A0225], two putative PHB depolymerases PhaZe1 [H16_A0671] and PhaZe2 [H16_B1632]), the phasins PhaP1 to PhaP8 and PhaR and PhaM. All 16 proteins were identified in 2 separate batches of PHB granules that had been isolated from NB- or MSM-gluconate-grown cultures of *R. eutropha,* respectively. Interestingly, the abundance

Table 1 PHA depolymerases and/or PHB oligomer hydrolases

Species	Protein/ (KEGG)	Locus tag	Length (aa)	Mw [kDa]	Active site lipase box	Localisation/remark	Reference
R. eutropha H16	PhaZa1 (PhaZ1)	H16_A1150	419	47.3	$SVC_{183}QP$	nPHB	Handrick et al. (2000); Saegusa et al. (2001); Uchino et al. (2007); Eggers and Steinbüchel (2013)
R eutropha	PhaZa2 (PhaZ2)	H16_A2862	404	44.8	$AIC_{173}QP$	Cytoplasm[b]	Leuprecht (2012)
R. eutropha	PhaZa3 (PhaZ5)	H16_B1014	407	45.2	$AVC_{175}QP$	nPHB[b]	Leuprecht (2012)
R. eutropha	PhaZa4 (PhaZ4)	H16_PHG178	245[a]	27.4[a]	$GLC_{?}EG$[a]	nPHB[b]	Leuprecht (2012)
R. eutropha	PhaZa5 (PhaZ3)	H16_B0339	412	45.2	$AIC_{182}QA$	nPHB[b]	Leuprecht (2012)
R. eutropha	PhaZb (PhaY1)	H16_A2251	718	74.3	$SVS_{320}NG$	Cytoplasm/nPHB	Kobayashi et al. (2003) Leuprecht (2012)
R. eutropha	PhaZc (PhaY2)	H16_A1335	293	31.6	$GTS_{108}MG$	Cytoplasm/nPHB	Kobayashi et al. (2005); Leuprecht (2012)
R. eutropha	PhaZd1 (PhaZ6)	H16_B2073	362	39.2	$GMS_{190}AG$	High activity, capacity to bind to nPHB[b]	Sznajder and Jendrossek (2014)
R. eutropha	PhaZd2 (PhaZ7)	H16_B2401	365	38.4	$GLS_{193}AG$	High activity, capacity to hydrolyse nPHB[b]	Abe et al. (2005); Sznajder and Jendrossek (2014)
R. eutropha	PhaZe1	H16_A0671	761	83.9	$GNC_{190}QA$	nPHB	Sznajder et al. (2015)
R. eutropha	PhaZe2	H16_B1632	773	86.4	$GNC_{212}QG$	nPHB	Sznajder et al. (2015)
A. vinelandii OP	PhaZ1	Avin_03910	914	101	$GNC_{215}QA$	nPHB	Adaya et al. (2018)

(continued)

Table 1 (continued)

Species	Protein/ (KEGG)	Locus tag	Length (aa)	Mw [kDa]	Active site lipase box	Localisation/remark	Reference
P. putida GP01	PhaZ		283	31.5	$GVS_{102}WG$	mcl-PHA	Huisman et al. (1991); de Eugenio et al. (2007)
P. putida KT2440	PhaZ	PP_5004	283	31.4	$GVS_{102}WG$	mcl-PHA	Nelson et al. (2002)

The designation of PHB depolymerase proteins in the literature and in KEGG data base and the respective locus tags and important features are given. Question marks indicate that the position of the amino acid residue is questionable due to uncertain gene start

[a]Correct gene start and gene length questionable

[b]Data obtained by artificial overexpression, protein not found in isolated nPHB granules of wild-type cells

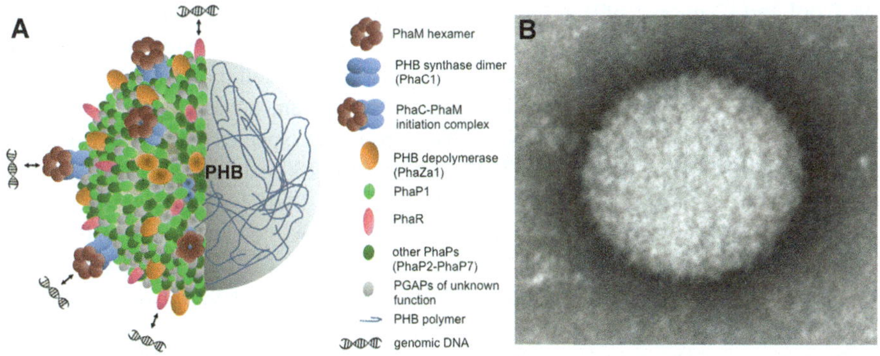

Fig. 2 PHB carbonosome. Schematic view of a PHB carbonosome in *R. eutropha* (**a**). PGAP proteins of the surface layer are symbolised in the left half of the PHB granule. Association of PhaM and of PhaR to regions on the genome is indicated by a symbol of a DNA helix. Image by F. Becker. The image in (**b**) shows a liberated PHB granule of *Caryophanon latum* after negative staining. Image modified from Jendrossek et al. (2007). The visible globular-like structures presumably represent PGAPs

of PGAPs in PHB granules of nutrient broth (NB)-grown cells was considerably higher than in PHB granules isolated from mineral salts medium (MSM)-grown cells. All identified proteins were associated with PHB granules in vivo as confirmed by the colocalisation of Nile red-stained PHB granules and fusions of these proteins with green fluorescent protein variants. The 16 identified PGAPs can be categorised into 4 subgroups: (1) PHB synthases (PhaCs), (2) phasins (PhaPs), (3) PHB depolymerases (PhaZs) and related hydrolases and (4) proteins with additional binding ability to DNA. Figure 2 shows a TEM image and a cartoon of a *R. eutropha* carbonosome.

5.1 PHB Synthases (PhaCs)

The PHA synthase is the key enzyme of PHA biosynthesis and catalyses the last step in PHA biosynthesis, i.e. the polymerisation of a hydroxyalkanoic acid from its CoA-activated monomer. In principle, the presence of a PHA synthase and a suited CoA-activated hydroxyalkanoic acid is sufficient to form a high molecular weight polymer (Gerngross and Martin 1995). In the case of *R. eutropha,* the PHB synthase is covalently linked to the growing polymer chain via the active site cysteine Cys319 (Gerngross et al. 1994; Wodzinska et al. 1996). The structure of the *R. eutropha* PHB synthase has been solved (Wittenborn et al. 2016; Kim et al. 2017a, b). The PHB-bound PHA synthase can be used as a marker protein for the identification and localisation of (native) PHA granules in vivo.

The first example of the use of fluorescent proteins for the *in vivo* co-localisation of PHA-associated proteins with PHB has been published in M. Cannon's lab

(McCool and Cannon 1999) at the example of the PHA synthase PhaC subunit and of the phasin PhaP in *Bacillus megaterium*. In this publication, fusions of PhaC and PhaP with the green fluorescent protein (GFP) were generated and formed fluorescent foci at the same position as PHB granules. Since then, several reports have been published in which PHA synthases in PHA-producing species were fused to fluorescent proteins (Jendrossek 2005; Peters and Rehm 2005; Jahns and Rehm 2009). An (almost forgotten) property of the PHB synthase PhaC1 of *R. eutropha* is the ability of the enzyme to catalyse the reverse reaction, i.e. the cleavage of PHB in the presence of free coenzyme A (CoA) to 3-hydroxybutyryl-CoA (3HB-CoA, thiolytic activity of PhaC1) (Uchino and Saito 2006). Thiolysis has been also detected for intracellular PHB depolymerase PhaZa1 (see below).

R. eutropha H16 has a second PHB synthase gene (*phaC2*). Transcription analysis and proteome determination of a PHB granule fraction and fusion analysis with eYFP confirmed that *phaC2* is expressed and that the gene product represents a true PGAP in *R. eutropha* although the degree of expression is very low (Peplinski et al. 2010; Brigham et al. 2012; Sznajder et al. 2015). PhaC2 is unable to synthesise PHB in a *ΔphaC1* background, but it can restore PHB accumulation in a *R. eutropha* background with a mutated, non-functional *phaC1* gene (strain PHB-04; for details, see Peplinski et al. (2010); Pfeiffer and Jendrossek (2012)). The function of PhaC2 in *R. eutropha*, if any, is obscure.

5.2 *Phasins (PhaPs)*

A common feature of carbonosomes in all investigated carbonosome-forming species is the presence of one or several so-called phasin proteins (PhaPs). Phasins are small amphiphilic proteins that cover a considerable part of the carbonosome surface and are thought to mediate between the hydrophobic polymer core and the hydrophilic cytoplasm (Steinbüchel et al. 1995). The first phasin (PhaP1) was discovered in A. Steinbüchel's laboratory by the characterisation of transposon-induced mutants that had a reduced but not completely abolished ability to accumulate PHB (so-called "PHB-leaky" phenotype) (Wieczorek et al. 1995). Phasins determine the surface-to-volume ratio of produced PHB granules: in the absence of PhaP1, *R. eutropha* synthesises less PHB compared to the wild type in the form of only one very big PHB granule, whereas an overproduction of PhaP1 leads to the formation of an increased number of unusually small PHB granules. The observation that phasins have a strong impact on the size of the produced PHB granules has been shown for many PHA-accumulating species (see Table 2 of Mezzina and Pettinari 2016) (Sun et al. 2019). Remarkably, eight phasin proteins (PhaP1–PhaP8) have been identified in PHB granules isolated from *R. eutropha* (Table 1) (Pötter et al. 2004; Kuchta et al. 2007; Pfeiffer and Jendrossek 2011; Pfeiffer and Jendrossek 2012; Sznajder et al. 2015). PhaP1 is by far the most abundant protein in PHB granule preparations, and deletion of the *phaP1* gene has a strong impact on PHB granule number and granule diameter. PhaP2, PhaP3 and PhaP4 are also present in substantial abundances in PHB granule preparations even though in much lower amounts than PhaP1 (Table 1); however, deletion of *phaP2, phaP3* and *phaP4* (single or in combination (Kuchta et al. 2007)) or of *phaP5, phaP6, phaP7* or *phaP8* (Pfeiffer and Jendrossek 2011; Pfeiffer and Jendrossek 2012; Schulze 2018) had no detectable phenotype in granule numbers or size. The overexpression of phasin genes (*phaP5, phaP6, phaP7* or *phaP8*) fused with *eyfp* led to a more polar localisation of clusters of PHB granules (Pfeiffer and Jendrossek 2011, 2012; Sznajder et al. 2015).

5.3 PHB Depolymerases (PhaZs)

Intracellular PHB depolymerases are the key enzymes for the reutilisation (mobilisation) of the storage polymer in the PHB-accumulating strain itself and catalyse the cleavage of PHAs to monomers and/or oligomers. Intercellular PHB depolymerases share only little homology to extracellular PHB depolymerases. The latter are secreted by PHA-degrading species that are able to utilise exogenous, denatured (partially crystalline) PHB as a carbon source but must not necessarily be able to accumulate PHB.

While no extracellular PHB depolymerase activity could be detected in culture fluids of *R. eutropha* (unpublished data), this species has a high number of intracellular enzymes with PHB depolymerase activity: the intracellular PHB depolymerase PhaZa1 is the main PHB depolymerase responsible for the mobilisation of PHB in carbon-starved cells (Handrick et al. 2000; Saegusa et al. 2001; Jüngert et al. 2017). It is bound to PHB granules in vivo as shown by fusion with GFP (Uchino et al. 2007; Sznajder and Jendrossek 2011; Sznajder et al. 2015). In contrast to most extracellular PHB depolymerases that hydrolyse extracellular (partially crystalline) PHB, i.e. PHB that has been released from lysed, dying bacteria, intracellular PHB depolymerases can cleave only the so-called native PHB granules (Merrick and Doudoroff 1964) that are in the amorphous state (for overviews on PHB degradation by extracellular PHA depolymerases, see Jendrossek (2001); Jendrossek and Handrick (2002)). Extracellular PHB depolymerases have a substrate-binding domain that enables them to bind to denatured (crystalline) PHB. The binding domain of intracellular PHB depolymerases has not been defined, yet. Native PHB granules harbouring PhaZa1 hydrolyse PHB to 3HB and oligomers of 3HB at a low rate in an aqueous in vitro system (Saegusa et al. 2001). However, in the presence of added CoA, the rate of hydrolysis decreased, and 3HB-CoA was formed (Uchino et al. 2007, 2008). This indicated that PhaZa1 in vivo most likely catalyses a thiolytic cleavage of PHB, thus saving the energy of the ester bonds in PHB in the thioester bond of 3HB-CoA. The thiolytic cleavage of PHB to crotonyl-CoA (in addition to (*R*)-3HB-CoA) was shown in a PHB synthesising recombinant *E. coli* strain if *phaZa1* was co-expressed with *phaCAB* and *phaP1* (Eggers and Steinbüchel 2013). Remarkably, when these experiments were performed with native PHB granules isolated from *R. eutropha*, the amount of formed crotonyl-CoA was reduced, and (*S*)-3HB-CoA was identified as main cleavage product. It was concluded that *R. eutropha* is able to distinguish between the biosynthesis of PHB via polymerisation of anabolically formed (*R*)-3HB-CoA to PHB and the (catabolic) mobilisation of PHB with PhaZa1 via (*S*)-3HB-CoA (Eggers and Steinbüchel 2013).

R. eutropha has four other putative PHB depolymerases (PhaZa2–PhaZa5) in addition to PhaZa1 (York et al. 2003; Pohlmann et al. 2006). All PhaZa isoenzymes share substantial amino acid similarity, and all have the same (unusual) cysteine-containing lipase box with the (putative) active site cysteine (Cys183 in PhaZa1) (Kobayashi and Saito 2003) (Table 2). The *phaZa2–phaZa5* genes are only poorly expressed under most tested culture conditions, and in agreement with transcript analysis, no evidence for the presence of detectable amounts of PhaZa2 to PhaZa5 was obtained by proteome analysis of NB and MSM cultures of *R. eutropha* (Lawrence et al. 2005; Brigham et al. 2012; Sznajder et al. 2015); the physiological functions of PhaZa2 to PhaZa5 remain unknown. Fusions of *phaZa2, phaZa3, phaZa4* and *phaZa5* with *eyfp* and overexpression in *R. eutropha* under PHB permissive conditions revealed that PhaZ3, PhaZ4 and PhaZ5 are true PGAPs in vivo (after artificial overexpression) (Leuprecht 2012). Surprisingly, PhaZa2-eYFP was detected only in the cytoplasm and did not colocalise with PHB granules.

Two 3-hydroxybutyrate oligomer hydrolases (PhaZb and PhaZc) (Kobayashi et al. 2003; Kobayashi et al. 2005) and two highly active intracellular PHB depolymerases (PhaZd1 and PhaZd2) have been additionally described for *R. eutropha* (Abe et al. 2005; Sznajder and Jendrossek 2014). However, PhaZb and PhaZc were located both in the cytosolic and PHB granule fraction in vitro (Kobayashi et al. 2003, 2005), and fusions of PhaZb or PhaZc with eYFP showed a soluble fluorescence and did not colocalise with accumulated PHB granules (Leuprecht 2012). Therefore, PhaZb and PhaZc can't be considered as PGAPs. Presumably, PhaZb and PhaZc are important for the hydrolysis of soluble 3-hydroxybutyrate dimers and/or other soluble ester compounds.

The PHB depolymerases PhaZd1 and PhaZd2—when expressed from a plasmid under control of the constitutive *phaC* promoter—hydrolysed PHB very efficiently (in vivo) so that free PHB granules were not detectable and a potential colocalisation of the PHB depolymerases with PHB granules could not been investigated. Accordingly, the purified PHB depolymerases PhaZd1 and PhaZd2 hydrolysed native PHB granules (in vitro) with extremely high specific activities (Sznajder and Jendrossek 2014). However, neither PhaZd1 nor PhaZd2 was detectably expressed in *R. eutropha* wild type, and the physiological function of these two PHB depolymerases therefore remains obscure. A colocalisation of a PhaZd1-eYFP fusion with PHB granules was observed if the catalytic site (S190) had been mutated to an alanine resulting in an inactive PhaZd1 protein. This suggested that PhaZd1 theoretically could be considered as a PGAP in *R. eutropha* if PhaZd1 would be expressed. Expression of a PhaZd2-eYFP fusion in *R. eutropha* showed that PhaZd2-eYFP did not colocalise with PHB granules and therefore a physiological function of PhaZd2 in PHB metabolism is unlikely.

Three other putative hydrolases are present in isolated PHB granules of *R. eutropha*. These are the H16_A0225, the H16_A0671 (PhaZe1) and the H16_B1632 (PhaZe2) gene products. All three proteins are putative hydrolases

Table 2 PHA granule-associated proteins (PGAPs)

Species	Protein/ (KEGG)	Locus tag	Length (aa)	Mw [kDa]	Localisation/ evidence	Reference
R. eutropha H16	PhaP1	H16_A1381	192	20	TEM-IG FM SDS-PAGE	Wieczorek et al. (1995)
R. eutropha	PhaP2	H16_PHG202	188	20	SDS-PAGE	Pötter et al. (2004)
R. eutropha	PhaP3	H16:2172	185	20	SDS-PAGE	Pötter et al. (2004)
R. eutropha	PhaP4	H16_B2021	189	20	SDS-PAGE	Pötter et al. (2004)
R. eutropha	PhaP5	H16_B1934	142	16	FM	Pfeiffer and Jendrossek (2011)
R. eutropha	PhaP6	H16_B1988	207	23	FM	Pfeiffer and Jendrossek (2012)
R. eutropha	PhaP7	H16_B2326	142	16	FM	Pfeiffer and Jendrossek (2012)
R. eutropha	PhaP8	H16_A2001	159	18	FM	Sznajder et al. (2015)
R. eutropha	PhaC1	H16_A1437	589	64	TEM-IG FM SDS-PAGE	Gerngross et al. (1993)
R. eutropha	PhaC2	H16_A2003	591	65	FM	Pfeiffer and Jendrossek (2012)
R. eutropha	PhaZa1	H16_A1150	419	47	FM	Uchino et al. (2007)
R. eutropha	PhaZe1	H16_B1632	773	86	FM	Sznajder et al. (2015)
R. eutropha	PhaZe2	H16_A0671	761	84	FM	Sznajder et al. (2015)
R. eutropha		H16_A0225	449	51	FM	Sznajder et al. (2015)
R. eutropha	PhaR	H16_A1440	183	21	SDS-PAGE[a]	Pötter et al. (2002)
R. eutropha	PhaM	H16_A0141	263	27	TH, FM	Pfeiffer et al. (2011)
A. vinelandii OP	PhaZ1	Avin_03910	914	101	SDS-PAGE	Adaya et al. (2018)
A. vinelandii UW136	OprI	Avin_23100		≈7	SDS-PAGE	Adaya et al. (2018); Moreno et al. (2019)
P. putida GP01	PhaC1		559	62	SDS-PAGE	Huisman et al. (1991); Ren et al. (2009)

(continued)

Table 2 (continued)

Species	Protein/ (KEGG)	Locus tag	Length (aa)	Mw [kDa]	Localisation/ evidence	Reference
P. putida GP01	PhaC2		560	63	SDS-PAGE	Huisman et al. (1991); Ren et al. (2009)
P. putida GPo1	PhaZ		283	31.5	SDS-PAGE	Huisman et al. (1991); Foster et al. (1996); Stuart et al. (1996); de Eugenio et al. (2007)
P. putida GPo1	PhaF		255	26	SDS-PAGE	Prieto et al. (1999)
P. putida GP01	PhaI		139	15	SDS-PAGE	Prieto et al. (1999)
P. putida GPo1	Acs1		565	62	FM	Ruth et al. (2008)
P. putida KT2440	PhaZ	PP_5004	283	31.4		Nelson et al. (2002)[b]

TH, two hybrid; FM, fluorescence microscopy; IG, immuno-gold
[a]Binding of PhaR to PHB shown in vitro using artificial PHB granules
[b]*P. putida* KT2440 and its derivative KT2442 strain have the same *pha* genes as *P. putida* GPo1

with H16_A0225 having a phospholipase signature. Fusions of each of the three genes with *eyfp* and expression in *R. eutropha* confirmed that all gene products colocalised with formed PHB granules (Sznajder et al. 2015). Purified H16_A0225 protein had no detectable PHB depolymerase activity with native PHB granules but could hydrolyse *p*-nitrophenyl-decanoic acid. The deletion of H16_A0225 had no detectable phenotype in PHB synthesis or in mobilisation of PHB (Schulze 2018). The physiological function of this PGAP remains therefore unknown. The deletion of each one or of both of the other two genes (*phaZe1* and *phaZe2*) did not substantially change the ability of the mutant to accumulate PHB. However, the two single and the double mutants without *phaZe1* and/or *phaZe2* had slightly higher PHB contents in the stationary growth phase (i.e. during PHB mobilisation), suggesting that the PhaZe1 and PhaZe2 could have PHB depolymerase activity although a biochemical proof with purified proteins is still lacking. Based on the presence of the putative cysteine containing lipase box and similarity to a thiolytic PHB depolymerase of *Azotobacter vinelandii* (Adaya et al. 2018) (Table 2), PhaZe1 and PhaZe2 are supposed to represent thiolytic PHB depolymerases (Sznajder et al. 2015).

Interestingly, H16_B1632 is the first gene of a putative four-genes operon coding for a putative phosphotransacetylase (H16_B1631, *pta1*), a putative acetate kinase (H16_B1630, *ackA*) and a putative enoyl-reductase (H16_B1629, *fabI2*). If indeed the H16_B1632 gene product encodes a thiolytic PHB depolymerase, the formed 3HB-CoA could be converted to 3-hydroxybutyryl-phosphate and subsequently to

3-hydroxybutyrate (3HB) via the action of (unspecific) Pta1 and AckA proteins. This would enable the cells to gain one ATP molecules and to save the energy of the cleaved ester bond. FabI2 potentially could interconvert 3HB-CoA and crotonyl-CoA; evidence that this reaction is catalysed by isolated native PHB granules from *R. eutropha* has been published, previously (Eggers and Steinbüchel 2013).

5.4 Proteins Binding to PHB Granules and to DNA

Two PGAP proteins have dual subcellular locations and can bind not only to PHB granules but also to genomic DNA in *R. eutropha*. These are the regulator (repressor) of phasin synthesis PhaR and the PhaM protein. PhaR binds to the operator regions of *phaP1*, *phaP3* and *phaR* and represses their transcription (York et al. 2002; Pötter et al. 2002, 2005). However, it also binds to the surface of PHB granules, and—if doing so—it cannot repress the transcription of *phasin* genes and its own transcription. As a consequence, the repression of transcription is released, and PhaP1, PhaP3 and PhaR can be formed. Since phasin proteins have a strong binding affinity to PHB, they displace PhaR molecules from the PHB granule surface, and this in turn enhances the probability that the liberated PhaR molecules bind to the operator regions of *phaP1* and *phaP3* (and *phaR*) leading to a stop of further transcription of *phaP1*, *phaP3* and *phaR*. Thus, the dual binding ability of PhaR to PHB granules and to DNA ensures that the production of PhaP1 and PhaP3 proteins (and of PhaR) is stopped only when the surface of PHB granules is covered with phasin proteins.

The multifunctional PhaM protein does not bind directly to PHB but efficiently interacts with the PHB synthase PhaC1 (Pfeiffer et al. 2011; Ushimaru and Tsuge 2016) and can be co-purified with PhaC (Cho et al. 2012; Schulze 2018). In vitro, purified PhaC1 and purified PhaM form a high molecular weight complex presumably consisting of a PhaC1 dimer and a PhaM hexamer (unpublished data). The specific activity of the PHB synthase in the presence of PhaM is approximately threefold higher, and the lag phase of the in vitro polymerisation reaction of PhaC1 (Gerngross and Martin 1995) is abolished by added PhaM (Pfeiffer and Jendrossek 2014). PHB granule-associated PhaM interacts with and binds to the nucleoid region in vivo. Electrophoretic gel mobility shift analysis (EMSA) confirmed that purified PhaM is able to bind to DNA in vitro, however, in a sequence-unspecific manner. The C-terminal part of PhaM is responsible for this DNA-binding ability because C-terminal truncations of PhaM have lost the ability to bind to DNA. The C-terminus of PhaM is highly basic and harbours histone-like PAKKA motifs. The lysine residues of these PAKKA motifs are essential for the interaction of PhaM with DNA in vivo and in vitro as revealed by site-directed mutagenesis of these residues (Bresan and Jendrossek 2017). As a consequence of the dual binding of PhaM to PhaC1 and to DNA, PHB granules are attached to the nucleoid region. Doubling of the genomic DNA and separation of the two daughter genomes during cell division distributes the genome-attached PHB granules to both daughter cells. Accordingly, mutants lacking *phaM* are unable to distribute formed PHB granules to both daughter cells during cell division (Pfeiffer et al. 2011; Wahl et al. 2012).

5.5 Other Proteins Bound to PHB Granules

In addition to the above-mentioned reports on true PGAPs of *R. eutropha*, several other proteins were described to be attached to PHB granules in various PHB-accumulating species (Tirapelle et al. 2013; Jendrossek and Pfeiffer 2014; Adaya et al. 2018; Moreno et al. 2019). Depending on the sensitivity of the staining procedure in SDS-PAGE of isolated PHB granule fractions (Coomassie brilliant blue or silver staining) or of the proteome analysis facility, several dozens to a few hundreds of proteins have been detected in purified PHB granule fractions. The annotated functions of these proteins (in *R. eutropha*) are diverse and include proteins involved in RNA and protein biosynthesis (RNA polymerase and ribosome components), energy metabolism (ATP synthase components), transport, C-metabolism (tri-carbonic acid cycle proteins) and others (Jendrossek and Pfeiffer 2014). It is unlikely that so many different proteins are truly bound to carbonosomes in vivo. More likely, the majority of these proteins represent false-positive PGAPs. However, it is difficult to exclude for sure that one or few of these in vitro carbonosome-attached proteins are true PGAPs in vivo. For example, in many reports on proteome analysis of isolated PHA granules, one or several proteins were identified that are annotated as outer membrane proteins. In case of *R. eutropha*, six Omp proteins were identified in the PHB granule fraction (Jendrossek and Pfeiffer 2014). Could it be that some of these Omp proteins are mis-annotated and in fact represent not yet identified PGAPs? A recent study, for example, showed that the outer membrane protein OprI of *Azotobacter vinelandii* is part of the PHB granule proteome and deletion of the o*prI* gene reduced the PHB content of cells grown on solid media (Adaya et al. 2018; Moreno et al. 2019). However, a direct in vivo proof of the presence of OprI in PHB granules, e.g. by fusion analysis with a fluorescent protein, is still missing. In conclusion, a function of Omp proteins in PHA metabolism is possible, but additional in vivo evidence is necessary to confirm this hypothesis.

In PHB-accumulating *Archaea*, two other proteins have been identified as PGAPs: a protein with a patatin-like phospholipase motif was identified in a PHB granule fraction of PHB mobilising *Haloferax mediterranei* cells (Liu et al. 2015). The purified enzyme was able to hydrolyse native PHB granules and thus represented a functional PHB depolymerase. Interestingly, the PHB-associated A0225 protein of *R. eutropha* (see above) also harbours a patatin-like phospholipase motif; however, a PHB depolymerase activity could not be shown for the purified protein (Schulze 2018). Another PHB-associated protein in *H. mediterranei* is the enoyl-CoA hydratase PhaJ1 (Liu et al. 2016).

6 The Carbonosome of *Pseudomonas putida*

The discovery that some bacteria can synthesise and accumulate polyesters consisting of mcl-hydroxyalkanoic acids (mcl-PHAs) goes back to 1983 when B. Witholt's lab published the formation of poly(3-hydroxyoctanoic acid) (PHO) by *P. putida* GPo1 (previously *Pseudomonas oleovorans* GPo1) (de Smet et al. 1983; Lageveen et al. 1988). Since then, numerous *Pseudomonas* species have been identified that are able to synthesise mcl-PHAs (for a selection of the first of those studies, see Huisman et al. (1989); Timm and Steinbüchel (1990); Huijberts et al. (1992)). Depending on the absence or presence of a 3-hydroxyacyl-CoA-acyl carrier protein transferase (*phaG* gene product that links β-oxidation with fatty acid biosynthesis (Rehm et al. 1998)), these species can synthesise mcl-PHAs only from related substrates, i.e. from medium- or long-chain fatty acids, or also from unrelated substrates such as glucose. Key enzyme for the synthesis of mcl-PHAs is the type II PHA synthase (PhaC) that differs from type I PHA synthases of PHB-accumulating bacteria (Rehm 2003; Rehm 2006; Mezzolla et al. 2018; Chek et al. 2019). A common feature of *P. putida* GP01 and other mcl-PHA-producing pseudomonads is the presence of two (related) *phaC* genes (*phaC1* and *phaC2*) on the chromosome that form a gene cluster together with a PHA depolymerase (*phaZ*) gene between *phaC1* and *phaC2* (Huisman et al. 1991; Kim et al. 2006). Meanwhile, two other related strains of *P. putida* (strain KT2440 and its derivative KT2442) have become the model species of mcl-PHA research, and most research activities on the identification of PGAPs in mcl-PHA-accumulating bacteria were performed with *P. putida* KT2440/KT2442. Below, a summary on the present knowledge on PGAPs of PHA granules in *Pseudomonas* species is provided (see also Fig. 3). It should be noted that the described results were obtained with either *P. putida* GPo1 or with *P. putida* KT2440/KT2442.

6.1 mcl-PHA Synthases (mcl-PhaCs)

All studied mcl-PHA-producing pseudomonads have two similar type II PHA synthase genes (*phaC1* and *phaC2*, separated by a PHA depolymerase *phaZ* gene) in the *pha* locus of their respective genomes. Both PHA synthases appear to be generally attached to mcl-PHA granules as revealed from proteome studies of isolated carbonosome preparations of several *Pseudomonas* species (Fuller et al. 1992; Stuart et al. 1996) although in vivo evidence by TEM-immuno-studies or by fusion with fluorescent proteins are rare (Peters and Rehm 2005). Purified PhaC1 and PhaC2 (either on isolated native PHA granules or isolated as hexa-histidine-tagged fusions) are able to catalyse the polymerisation of medium-chain-length hydroxyalkanoic acids (Qi et al. 2000); they differ in their K_m and V_{max} values (125 vs. 37 μM and 41 vs. 2.7 U/mg, respectively) with (*R*)-3-hydroxyoctanoyl-CoA as a substrate (Ren et al. 2009) as well as in their substrate specificities for mainly

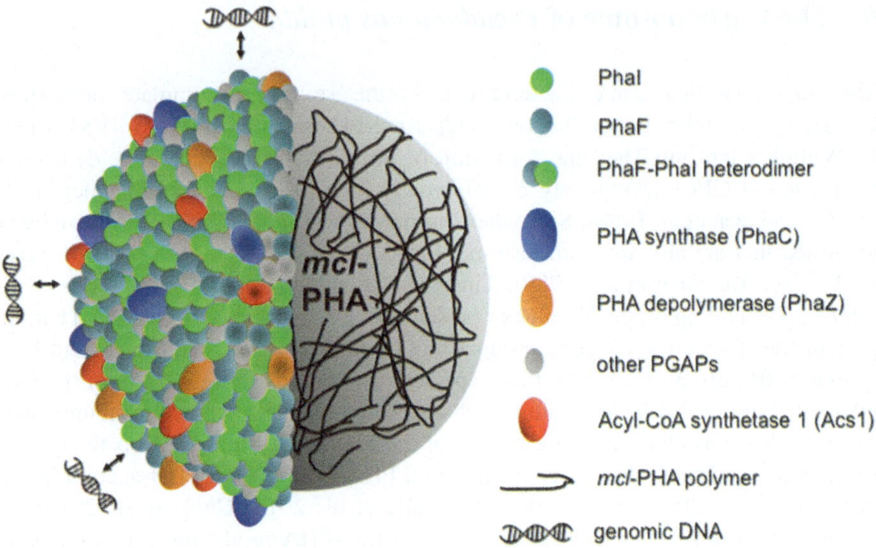

Fig. 3 mcl-PHA carbonosome. Schematic view of a PHA carbonosome in *P. putida*. PGAP proteins of the surface layer are symbolised in the left half of the PHA granule. Associations of PhaF and/or PhaF-PhaI hetero-dimers to the bacterial nucleoid are indicated. Image by F. Becker

mcl-3HA-monomers (PhaC1) or revealing a broad substrate specificity, respectively (short- to long-chain 3HA-monomers) (Chen et al. 2006).

6.2 *mcl-PHA Depolymerase (mcl-PhaZs)*

mcl-PHA-accumulating *Pseudomonas* species generally have a mcl-PHA depolymerase gene (mcl-*phaZ*) located between the *phaC1* and the *phaC1* genes of the respective *pha* locus. The conclusion that these mcl-PHA depolymerase proteins have indeed mcl-PHA depolymerase activity comes from *phaZ*-mutant analysis (Huisman et al. 1991; Solaiman et al. 2003; Cai et al. 2009), *phaZ* knock-out and *phaZ* overexpression analysis and by showing depolymerase activity of purified proteins with artificially prepared mcl-PHA granules (Foster et al. 1994, 1996; Jiang et al. 2004; de Eugenio et al. 2007, 2008; Isabel de Eugenio et al. 2010a; Arias et al. 2013). Proteome analysis of isolated mcl-PHA granule fractions from several *Pseudomonas* species had identified the mcl-PHA depolymerase protein, suggesting that mcl-PHA depolymerases are PGAPs in these species (Fuller et al. 1992; Stuart et al. 1996). However, in vivo studies that could unequivocally verify the subcellular location of the mcl-PHA depolymerase in vivo are missing. Intracellular mcl-PHA depolymerases of *Pseudomonas* species most likely are serine hydrolases and have a lipase box in the active centre (GxSxG in case of *P. putida* GPo1 (Huisman et al. 1991)) (Table 2). Therefore, mcl-PHA depolymerases are true hydrolases and are unlikely to have thiolytic activity as PhaZa1 and related isoenzymes of *R. eutropha* with a cysteine as the active site (see above).

6.3 Phasins

mcl-PHA-accumulating *Pseudomonas* species have two types of phasins, namely, PhaI and PhaF. While PhaI and the N-terminal part of PhaF share substantial amino acid similarity and have both coiled coil domains, neither PhaI nor PhaF have a pronounced amino acid similarity to phasins of PHB-accumulating species (Prieto et al. 1999; Maestro et al. 2013). PhaI and PhaF in vivo interact with each other and probably form hetero-oligomers (Tarazona et al. 2019). The N-terminal domain of PhaF is responsible for the attachment to mcl-PHA carbonosomes (Moldes et al. 2004). The function of the C-terminal domain of PhaF will be discussed below.

6.4 Proteins Binding to mcl-PHA Granules and to DNA

A homolog of the PhaR transcriptional repressor protein of *R. eutropha* that would be able to bind to mcl-PHA carbonosomes and to the operator region of phasin genes is not present in *P. putida* and has not been identified in any other mcl-PHA-accumulating species. Instead, a PhaD protein is present in *P. putida* that activates the transcription of PHA biosynthetic genes under appropriate conditions (Isabel de Eugenio et al. 2010b). A binding of PhaD to mcl-PHA carbonosomes has not been described for any mcl-PHA-accumulating species. However, the PhaF phasin protein has PAKKA-like motifs in the C-terminal domain of the protein (Prieto et al. 1999), and binding of the C-terminal domain to the nucleoid of *P. putida* has been demonstrated by fusion analysis with GFP (Galan et al. 2011). The properties of PhaF in *P. putida* and of PhaM in *R. eutropha* to bind to formed carbonosomes and to the nucleoid suggest that both proteins have similar functions and enable both species to distribute the storage granules more or less equally to both daughter cells during cell division (for more details, see below). Despite this similar function, PhaF and PhaM do not share significant amino acid homology to each other (except for the PAKKA motifs).

6.5 Other mcl-PHA-Bound Proteins

One of the two acyl-CoA synthetases (Acs1) of *P. putida* is in vivo attached to PHA granules as revealed by fusion analysis with GFP (Ruth et al. 2008) (Fig. 3). This finding suggests that monomeric hydroxyalkanoic acids can be converted to substrates of the PHA synthase in direct vicinity of PHA granule-bound PHA synthases supporting the view that PHA granules are functional units (carbonosomes).

7 Are Carbonosomes Covered by a Phospholipid Monolayer?

The presence of a phospholipid monolayer at the surface of carbonosomes has been assumed for more than 50 years: Merrick and co-workers published a work in 1967 in which they determined the constituents of a PHB granule fraction that had been purified from *Bacillus megaterium* (Griebel et al. 1968). PHB (97.7%) and protein (1.87%) were the two main components, but 0.46% lipids (including phosphatidic acid) were additionally present. Later, Horowitz and Sanders detected several phospholipids such as phosphatidyl-ethanolamine, phosphatidyl-glycerol, diphosphatidyl-glycerol and a fourth not identified compound (possibly phosphatidyl-serine) in PHB granules isolated from *R. eutropha* (Horowitz and Sanders 1994). These findings together with the determination of the thicknesses of the *R. eutropha* PHB granule boundary layer and of the *R. eutropha* cytoplasm membrane of 2.9–3.8 nm and 6.7–8.1 nm, respectively (Mayer and Hoppert 1997), were the background for the postulation of a phospholipid monolayer surrounding PHB granules. Based on these findings, most reviews on PHA metabolism and on PHA granule structure include the presence of a phospholipid monolayer in cartoons of PHB or PHA granules (Steinbüchel et al. 1995; Pötter and Steinbüchel 2006; Ruth et al. 2008; Grage et al. 2009; Rehm 2010; Dinjaski and Prieto 2013; Prieto et al. 2016). However, the assumed phospholipid layer of carbonosomes has never been shown in whole or living cells. Even in cryo-EM images of *R. eutropha* cells, no continuous surface layer could be detected at the surface of PHB granules (Beeby et al. 2012). Another remarkable point is that despite more than three decades of intensive research on PHB metabolism in *R. eutropha* and other PHB-accumulating species, no publication appeared in which a mutant in phospholipid metabolism was identified that resulted in a phenotype with altered PHB accumulation. The phasin protein PhaP1 in *R. eutropha* was discovered by the PHB-leaky phenotype in a screening approach in which strains with altered ability to accumulate PHB were screened in a transposon mutant library (Wieczorek et al. 1995). However, mutants with insertions in genes related to phospholipid biosynthesis were not detected in these studies. In 2008, the use of the phospholipid-binding C2 domain of bovine lactadherin fused to a fluorescent protein was described for its use to detect phosphatidyl-serine in eukaryotic cells (Yeung et al. 2008). This prompted Bresan et al. to use the same system in bacteria (Bresan et al. 2016). However, all attempts to find evidence for the presence of phospholipids at the surface of scl-PHA or mcl-PHA granules in different species failed, while the phospholipids of the cytoplasm membrane and the phospholipid bilayer membrane of magnetosomes in

Magnetospirillum gryphiswaldense were successfully detected by the reporter system (Dsred2EC-LactC2) (Bresan et al. 2016). Furthermore, the authors of that study could find evidence for the (artificial) in vitro binding ability of phospholipids to PHB granules in a crude cell extract of PHB-accumulating *R. eutropha* cells providing an explaining for the in vitro detection of phospholipids in earlier studies. The above-mentioned findings do not exclude—although not being likely—that single phospholipid molecules or phospholipid-related molecules differing chemically from conventional phospholipids and thus cannot be detected by phospholipid-binding proteins such as C2 domain or segment A of RNAse E (Schulze 2018) could be present in carbonosomes in vivo. In principle, phasin proteins can stably interact with phospholipid layers (Mato et al. 2019). Taken together, there is no in vivo evidence for the presence of a phospholipid (mono)-layer around PHB or PHA granules. Carbonosomes are covered by a surface layer consisting of only proteins, and the presence of a continuous monolayer of conventional phospholipids around carbonosomes is unlikely.

8 Biogenesis of Carbonosomes

The in vivo biogenesis of carbonosomes has been studied in *R. eutropha*. This species rapidly mobilises previously accumulated PHB in the stationary growth phase of cultures with nutrient broth (NB) as a carbon source. When such PHB-free cells are transferred to fresh medium and grown under PHB-permissive conditions (e.g. on NB medium with 0.2% gluconate to increase the C-to-N ratio), the cells start to synthesise one or two new PHB granules in the first 20 min after transfer. Remarkably, one of the newly formed PHB granules is always located close to one of the cell poles, while the second PHB granule originates near the other cell pole. A formation of two PHB granules in the same half of the rod-shaped cells has not been detected (as long as not more than two PHB granules have been formed). This indicates that the spatial formation of PHB granules is controlled in growing *R. eutropha* cells. The PhaM protein determines the subcellular localisation of newly formed PHB granules. PhaM forms oligomers (presumably hexamers) and is bound to the nucleoid of the cells in the absence of PHB (Pfeiffer et al. 2011). The lysine residues of the PAKKA motifs in the C-terminus of PhaM are responsible for the binding of PhaM to DNA in vitro and in vivo (Bresan and Jendrossek 2017). It is not known whether PhaM binds to a specific position on the nucleoid. DNA chip experiments revealed no binding preference of PhaM (unpublished data). Gel mobility shift experiments with purified PhaM showed that PhaM has no DNA sequence specificity in vitro. The N-terminus of PhaM interacts with the PHB synthase PhaC (Pfeiffer et al. 2011). These findings suggest that under PHB-permissive conditions (e.g. in NB-gluconate medium) nucleoid-bound PhaM oligomers recruit PHB synthase dimers and form a PHB granule initiation complex. This initiation complex is formed in each cell half near the cell pole. At the same time, the nucleoid has started to replicate and to separate. If one assumes that the genome has only one or few (not yet identified) binding sites for the attachment of the PhaM-PhaC initiation complex, one might speculate that genome replication and segregation of the daughter chromosomes during genome replication are the explanation why always one PHB granule in each cell half is formed at the beginning of PHB granule biosynthesis. As soon as the initiation complex is formed, PhaM-attached PHB synthase dimers start to synthesise PHB (Pfeiffer and Jendrossek 2014). The hydrophobic PHB chain aggregates and grows to a microscopically detectable granule. During this process, phasins and other PGAPs bind to the growing surface of the arising PHB granule. The PHB granules can be liberated from the covalently bound PhaM-PhaC complex at later stages of PHB accumulation. Evidence for this "aging phenomenon of carbonosomes" is the identification of non-fluorescent PHB granules at later stages of PHB accumulation in strains expressing *eyfp-phaC1* or *eyfp-phaM* (Bresan and Jendrossek 2017). If the cells with liberated (PhaM-PhaC1-free) PHB granules are further kept under PHB-permissive conditions, then free PhaC1-PhaM complexes can initiate the formation of additional, new PHB granules. The release of PHB granules from the PhaC1-PhaM complex ensures that PHB granules do not grow ad infinitum if the cells are permanently kept under PHB-permissive conditions.

The PhaM protein is not essential for PHB biosynthesis. Mutants with a deletion of the *phaM* gene accumulate slightly reduced amounts of PHB compared to the wild type, and the number of granules is only 0, 1 or 2 in most cells. However the diameters of the carbonosomes are substantially enlarged similar as in *ΔphaP1* mutants (Wieczorek et al. 1995). The distribution of the PHB granules during cell division in *ΔphaM* cells is disordered. Directly after cell division, one daughter cell harbours all (one or two) PHB granules, while the other daughter cell has no PHB granule. Only under ongoing PHB-permissive conditions, the PHB-free daughter cell can synthesise an own (new) PHB granule at later stages of growth. These data suggest that a major physiological function of PhaM is to ensure an equal distribution of the accumulated carbonosomes during cell division to both daughter cells. In this respect the function of PhaM seems to be similar to PhaF in *P. putida* (Galan et al. 2009, 2011; Prieto et al. 2016). PhaF has also C-terminal located PAKKA motifs as PhaM and enables PhaF to bind to DNA. However, the N-terminal domain of PhaF does not interact with the PHA synthase but has a phasin motif and interacts with phasin PhaI (Prieto et al. 1999). Therefore, PhaF is unlikely to be part of a PHA granule initiation complex in mcl-PHA-accumulating species. Nevertheless, *P. putida* strains, in which the *phaF* gene has been deleted, have a disordered localisation of accumulated PHA granules, and equal distribution of these granules to the daughter cells during cell division is blocked (Galan et al. 2011).

In summary, the biogenesis of carbonosomes in two model organisms, in scl-PHA-accumulating *R. eutropha* and in mcl-PHA-accumulating *P. putida*, follows via a scaffold model in which the scaffold is the bacterial nucleoid. Evidence for a scaffold model in PHB granule formation in *R. eutropha* was already provided in studies of J. Stubbe's laboratory (Tian et al. 2005a, b). In these studies, the scaffold transiently was detected as "dark-stained mediation elements" in *R. eutropha*. The data of Tian et al. are in good agreement with the assumption that the mediation elements represent the bacterial nucleoid. Previously proposed and discussed models of carbonosome biosynthesis (i.e. the budding model and the micelle model (Ellar et al. 1968; Stubbe and Tian 2003; Stubbe et al. 2005; Tian et al. 2005b; Pötter and Steinbüchel 2006)) are not supported by in vivo evidence.

9 Novel Function of Carbonosomes

The most obvious function of PHB and related carbonosomes is that of an insoluble and therefore osmotic inert reservoir for carbon and energy and has been experimentally verified for several species (for a selection of publications, see Dawes and Senior 1973; Matin et al. 1979; James et al. 1999; Handrick et al. 2000; Ratcliff et al. 2008; Wang et al. 2009). In addition, PHAs are relatively reduced compounds and can be therefore considered as sinks for electrons (Senior and Dawes 1971). In agreement with this, oxygen limitation in many PHA-producing species increases the amount of accumulated PHAs. Degradation of PHA and oxidation of the hydrolysis products can be used to promote consumption of oxygen in nitrogen-fixing species to protect the nitrogenase from inactivation by oxic conditions (Stam et al. 1986; Mandon et al. 1998). Another function of accumulated PHAs is its protective role in coping with various stresses such as oxidative stress, high/low temperatures or osmotic stress. For examples, see Kadouri et al. (2003); Zhao et al. (2007); Ayub et al. (2009); Obruca et al. (2016); Koskimäki et al. (2016); Nowroth et al. (2016); Obruca et al. (2017); Batista et al. (2018); and Sedlacek et al. (2019). It is not clear in which cases the enhanced stress resistance of PHB-containing cells is a direct protective effect of the polymer or is only an indirect effect. The function of PHB in enhancing stress resistance is a relatively young research field, and the molecular mechanisms, by which PHB can effectuate a higher level of stress resistance, are only poorly understood. In this context, an astonishing result was the finding that PHB degradation products such as 3HB and methyl esters of 3HB have strong radical scavenging activities and help PHB-accumulating plant-symbiotic bacteria (*Methylobacterium extorquens*) to tolerate oxygen radicals produced by the plant cells during contact with incoming symbionts (Koskimäki et al. 2016). Recently, metabolism of PHB was investigated in *Synechocystis* sp. PCC 6803 (Koch et al. 2019). The authors provided evidence that PHB is involved in glycogen metabolism but presumably has no (prominent) function for survival under (nitrogen) starving conditions. In summary, it is evident that, despite several decades of intensive research on PHB metabolism, still more research is necessary to understand the molecular basis of the protective roles of PHB in prokaryotic species.

Acknowledgements I thank the Deutsche Forschungsgemeinschaft for the continuous support on my work on carbonosomes. I deeply appreciate the work of my current and previous co-workers and students in the field of PHB metabolism: Libertad Adaya, Meike Andermann, Felix Becker, Astrid Behrends, Bernd-Holger Briese, Martina Backhaus, Stephanie Bresan (Schulze), Bernd-Holger Briese, Chiara Brück, Nicole Fischer, Letizia Focarete, Anja Frank, Andrea Frisse, Birgit Gebauer,

Rene Handrick, Eva-Maria Hüper, Ingrid Knoke, Siska Hermawan, Pia Kaufmann, Philipp Kimmig, Britta Klingbeil, Lea Kneissle, Janina Jüngert (Berg), Daniela Leuprecht, Heike Lill, Claudia Matz, Beate Müller, Verena Nowroth, Cameron Patterson, Elena Pluta, Thomas Reichert, Dominik Reis, Watsana Penkrue, Daniel Pfeiffer, Ute Schöber, Nora Schuth, Simone Reinhardt, Fabian Romen, Thomas Sander, Andreas Schirmer, Eva Slaninova, Thomas Stanislawski, Anna Sznajder (Sommer), Naihmeh Taheri Talesh, Ulricke Technow, Monika Tischbirek, Kay Terpe, Christian Thiel, Gianpaolo Tomasi, Keiichi Uchino, Manuela Unsinn, Tanja Veselinovic, Andreas Wahl and Lei Wang.

References

Abe T, Kobayashi T, Saito T (2005) Properties of a novel intracellular poly(3-hydroxybutyrate) depolymerase with high specific activity (PhaZd) in *Wautersia eutropha* H16. J Bacteriol 187:6982–6990. https://doi.org/10.1128/JB.187.20.6982-6990.2005

Abraham GA, Gallardo A, San Roman J, Olivera ER, Jodra R, García B, Miñambres B, Garcia JL, Luengo JM (2001) Microbial synthesis of poly(beta-hydroxyalkanoates) bearing phenyl groups from *Pseudomonas putida*: chemical structure and characterization. Biomacromolecules 2:562–567. https://doi.org/10.1021/bm010018h

Adaya L, Millán M, Peña C, Jendrossek D, Espín G, Tinoco-Valencia R, Guzmán J, Pfeiffer D, Segura D (2018) Inactivation of an intracellular poly-3-hydroxybutyrate depolymerase of *Azotobacter vinelandii* allows to obtain a polymer of uniform high molecular mass. Appl Microbiol Biotechnol 187:6982–6915. https://doi.org/10.1007/s00253-018-8806-y

Anderson AJ, Dawes EA (1990) Occurrence, metabolism, metabolic role, and industrial uses of bacterial polyhydroxyalkanoates. Microbiol Rev 54:450–472

Arias S, Bassas-Galia M, Molinari G, Timmis KN (2013) Tight coupling of polymerization and depolymerization of polyhydroxyalkanoates ensures efficient management of carbon resources in *Pseudomonas putida*. Microb Biotechnol 6:551–563. https://doi.org/10.1111/1751-7915.12040

Ayub ND, Tribelli PM, López NI (2009) Polyhydroxyalkanoates are essential for maintenance of redox state in the Antarctic bacterium *Pseudomonas* sp. 14-3 during low temperature adaptation. Extremophiles 13:59–66. https://doi.org/10.1007/s00792-008-0197-z

Batista MB, Teixeira CS, Sfeir MZT, Alves LPS, Valdameri G, Pedrosa F de O, Sassaki GL, Steffens MBR, de Souza EM, Dixon R, Müller-Santos M (2018) PHB biosynthesis counteracts redox stress in *Herbaspirillum seropedicae*. Front Microbiol 9:472. https://doi.org/10.3389/fmicb.2018.00472

Beeby M, Cho M, Stubbe J, Jensen GJ (2012) Growth and localization of polyhydroxybutyrate granules in *Ralstonia eutropha*. J Bacteriol 194:1092–1099. https://doi.org/10.1128/JB.06125-11

Bresan S, Jendrossek D (2017) New insights in PhaM-PhaC-mediated localization of PHB granules in *Ralstonia eutropha* H16. Appl Environ Microbiol AEM:00505–00517. https://doi.org/10.1128/AEM.00505-17

Bresan S, Sznajder A, Hauf W, Forchhammer K, Pfeiffer D, Jendrossek D (2016) Polyhydroxyalkanoate (PHA) Granules Have no Phospholipids. Sci Rep 6:26612. https://doi.org/10.1038/srep26612

Brigham CJ, Speth DR, Rha C, Sinskey AJ (2012) Whole-genome microarray and gene deletion studies reveal regulation of the polyhydroxyalkanoate production cycle by the stringent response in *Ralstonia eutropha* H16. Appl Environ Microbiol 78:8033–8044. https://doi.org/10.1128/AEM.01693-12

Cai L, Yuan M-Q, Liu F, Jian J, Chen G-Q (2009) Enhanced production of medium-chain-length polyhydroxyalkanoates (PHA) by PHA depolymerase knockout mutant of *Pseudomonas putida* KT2442. Bioresour Technol 100:2265–2270. https://doi.org/10.1016/j.biortech.2008.11.020

Chek MF, Hiroe A, Hakoshima T, Sudesh K, Taguchi S (2019) PHA synthase (PhaC): interpreting the functions of bioplastic-producing enzyme from a structural perspective. Appl Microbiol Biotechnol 103:1131–1141. https://doi.org/10.1007/s00253-018-9538-8

Chen J-Y, Song G, Chen G-Q (2006) A lower specificity PhaC2 synthase from *Pseudomonas stutzeri* catalyses the production of copolyesters consisting of short-chain-length and medium-chain-length 3-hydroxyalkanoates. Antonie Van Leeuwenhoek 89:157–167. https://doi.org/10.1007/s10482-005-9019-9

Cho M, Brigham CJ, Sinskey AJ, Stubbe J (2012) Purification of polyhydroxybutyrate synthase from its native organism, *Ralstonia eutropha*: implications for the initiation and elongation of polymer formation in vivo. Biochemistry 51:2276–2288. https://doi.org/10.1021/bi2013596

Dawes EA, Senior PJ (1973) The role and regulation of energy reserve polymers in micro-organisms. Adv Microb Physiol 10:135–266. https://doi.org/10.1016/s0065-2911(08)60088-0

de Eugenio LI, Garcia P, Luengo JM, Sanz JM, San Roman J, Garcia JL, Prieto MA (2007) Biochemical evidence that phaZ gene encodes a specific intracellular medium chain length polyhydroxyalkanoate depolymerase in *Pseudomonas putida* KT2442—Characterization of a paradigmatic enzyme. J Biol Chem 282:4951–4962. https://doi.org/10.1074/jbc.M608119200

de Eugenio LI, Garcia JL, Garcia P, Prieto MA, Sanz JM (2008) Comparative analysis of the physiological and structural properties of a medium chain length polyhydroxyalkanoate depolymerase from *Pseudomonas putida* KT2442. Eng Life Sci 8:260–267. https://doi.org/10.1002/elsc.200700057

de Smet MJ, Eggink G, Witholt B, Kingma J, Wynberg H (1983) Characterization of intracellular inclusions formed by *Pseudomonas oleovorans* during growth on octane. J Bacteriol 154:870–878

Dinjaski N, Prieto MA (2013) Swapping of phasin modules to optimize the in vivo immobilization of proteins to medium-chain-length polyhydroxyalkanoate granules in *Pseudomonas putida*. Biomacromolecules 14:3285–3293. https://doi.org/10.1021/bm4008937

Eggers J, Steinbüchel A (2013) Poly(3-Hydroxybutyrate) Degradation in *Ralstonia eutropha* H16 Is Mediated Stereoselectively to (S)-3-Hydroxybutyryl Coenzyme A (CoA) via Crotonyl-CoA. J Bacteriol 195:3213–3223. https://doi.org/10.1128/JB.00358-13

Elbanna K, Lütke-Eversloh T, Jendrossek D, Luftmann H, Steinbüchel A (2004) Studies on the biodegradability of polythioester copolymers and homopolymers by polyhydroxyalkanoate (PHA)-degrading bacteria and PHA depolymerases. Arch Microbiol 182:212–225. https://doi.org/10.1007/s00203-004-0715-z

Ellar D, Lundgren DG, Okamura K, Marchessault RH (1968) Morphology of poly-beta-hydroxybutyrate granules. J Mol Biol 35:489–502. https://doi.org/10.1016/s0022-2836(68)80009-9

Follner CG, Madkour M, Mayer F, Babel W, Steinbüchel A (1997) Analysis of the PHA granule-associated proteins GA20 and GA11 in *Methylobacterium extorquens* and *Methylobacterium rhodesianum*. J Basic Microbiol 37:11–21

Foster LJ, Lenz RW, Fuller RC (1994) Quantitative determination of intracellular depolymerase activity in *Pseudomonas oleovorans* inclusions containing poly-3-hydroxyalkanoates with long alkyl substituents. FEMS Microbiol Lett 118:279–282. https://doi.org/10.1111/j.1574-6968.1994.tb06841.x

Foster LJ, Stuart ES, Tehrani A, Lenz RW, Fuller RC (1996) Intracellular depolymerase and polyhydroxyoctanoate granule integrity in *Pseudomonas oleovorans*. Int J Biol Macromol 19:177–183

Fuller RC, O Donnel JP, Saulnier J, Redlinger TE, Foster J, Lenz RW (1992) The supramolecular architecture of the polyhydroxyalkanoate inclusions in *Pseudomonas-oleovorans*. FEMS Microbiol Lett 103:279–288. https://doi.org/10.1016/0378-1097(92)90321-E

Galan B, Maestro B, Garcia JL, Sanz JM, Prieto MA (2009) Role of the bi-modular PhaF protein as DNA binding factor in *Pseudomonas putida* KT2442. New Biotechnol 25:S62. https://doi.org/10.1016/j.nbt.2009.06.292

Galan B, Dinjaski N, Maestro B, de Eugenio LI, Escapa IF, Sanz JM, Garcia JL, Prieto MA (2011) Nucleoid-associated PhaF phasin drives intracellular location and segregation of polyhydroxyalkanoate granules in *Pseudomonas putida* KT2442. Mol Microbiol 79:402–418. https://doi.org/10.1111/j.1365-2958.2010.07450.x

Gerngross TU, Martin DP (1995) Enzyme-catalyzed synthesis of poly[(*R*)-(-)-3-hydroxybutyrate]: formation of macroscopic granules *in vitro*. Proc Natl Acad Sci USA 92:6279–6283

Gerngross TU, Reilly P, Stubbe J, Sinskey AJ, Peoples OP (1993) Immunocytochemical analysis of poly-beta-hydroxybutyrate (PHB) synthase in *Alcaligenes eutrophus* H16: localization of the synthase enzyme at the surface of PHB granules. J Bacteriol 175:5289–5293

Gerngross TU, Snell KD, Peoples OP, Sinskey AJ, Csuhai E, Masamune S, Stubbe J (1994) Overexpression and purification of the soluble polyhydroxyalkanoate synthase from *Alcaligenes eutrophus*: evidence for a required posttranslational modification for catalytic activity. Biochemistry 33:9311–9320

Grage K, Jahns AC, Parlane N, Palanisamy R, Rasiah IA, Atwood JA, Rehm BHA (2009) Bacterial polyhydroxyalkanoate granules: biogenesis, structure, and potential use as nano-/micro-beads in biotechnological and biomedical applications. Biomacromolecules 10:660–669. https://doi.org/10.1021/bm801394s

Griebel R, Smith Z, Merrick JM (1968) Metabolism of poly-beta-hydroxybutyrate. I. Purification, composition, and properties of native poly-beta-hydroxybutyrate granules from *Bacillus megaterium*. Biochemistry 7:3676–3681

Handrick R, Reinhardt S, Jendrossek D (2000) Mobilization of poly(3-hydroxybutyrate) in *Ralstonia eutropha*. J Bacteriol 182:5916–5918

Handrick R, Reinhardt S, Schultheiss D, Reichart T, Schüler D, Jendrossek V, Jendrossek D (2004a) Unraveling the function of the *Rhodospirillum rubrum* activator of polyhydroxybutyrate (PHB) degradation: the activator is a PHB-granule-bound protein (phasin). J Bacteriol 186:2466–2475. https://doi.org/10.1128/JB.186.8.2466-2475.2004

Handrick R, Technow U, Reichart T, Reinhardt S, Sander T, Jendrossek D (2004b) The activator of the *Rhodospirillum rubrum* PHB depolymerase is a polypeptide that is extremely resistant to high temperature (121 degrees C) and other physical or chemical stresses. FEMS Microbiol Lett 230:265–274. https://doi.org/10.1016/S0378-1097(03)00919-4

Hauf W, Watzer B, Roos N, Klotz A, Forchhammer K (2015) Photoautotrophic polyhydroxybutyrate granule formation is regulated by cyanobacterial phasin PhaP in *Synechocystis* sp. strain PCC 6803. Appl Environ Microbiol 81:4411–4422. https://doi.org/10.1128/AEM.00604-15

Horowitz DM, Sanders JKM (1994) Amorphous, biomimetic granules of polyhydroxybutyrate: preparation, characterization, and biological implications. J Am Chem Soc 116:2695–2702. https://doi.org/10.1021/ja00086a001

Huijberts GN, Eggink G, de Waard P, Huisman GW, Witholt B (1992) *Pseudomonas putida* KT2442 cultivated on glucose accumulates poly(3-hydroxyalkanoates) consisting of saturated and unsaturated monomers. Appl Environ Microbiol 58:536–544

Huisman GW, de Leeuw O, Eggink G, Witholt B (1989) Synthesis of poly-3-hydroxyalkanoates is a common feature of fluorescent pseudomonads. Appl Environ Microbiol 55:1949–1954

Huisman GW, Wonink E, Meima R, Kazemier B, Terpstra P, Witholt B (1991) Metabolism of poly (3-hydroxyalkanoates) (PHAs) by *Pseudomonas oleovorans*. Identification and sequences of genes and function of the encoded proteins in the synthesis and degradation of PHA. J Biol Chem 266:2191–2198

Isabel de Eugenio L, Escapa IF, Morales V, Dinjaski N, Galan B, Luis Garcia J, Prieto MA (2010a) The turnover of medium-chain-length polyhydroxyalkanoates in *Pseudomonas putida* KT2442 and the fundamental role of PhaZ depolymerase for the metabolic balance. Environ Microbiol 12:207–221. https://doi.org/10.1111/j.1462-2920.2009.02061.x

Isabel de Eugenio L, Galan B, Escapa IF, Maestro B, Sanz JM, Luis Garcia J, Prieto MA (2010b) The PhaD regulator controls the simultaneous expression of the pha genes involved in polyhydroxyalkanoate metabolism and turnover in *Pseudomonas putida* KT2442. Environ Microbiol 12:1591–1603. https://doi.org/10.1111/j.1462-2920.2010.02199.x

Jahns AC, Rehm BHA (2009) Tolerance of the *Ralstonia eutropha* class I polyhydroxyalkanoate synthase for translational fusions to its C terminus reveals a new mode of functional display. Appl Environ Microbiol 75:5461–5466. https://doi.org/10.1128/AEM.01072-09

James BW, Mauchline WS, Dennis PJ, Keevil CW, Wait R (1999) Poly-3-hydroxybutyrate in *Legionella pneumophila*, an energy source for survival in low-nutrient environments. Appl Environ Microbiol 65:822–827

Jendrossek D (2001) Microbial degradation of polyesters. Adv Biochem Eng Biotechnol 71:293–325

Jendrossek D (2005) Fluorescence microscopical investigation of poly(3-hydroxybutyrate) granule formation in bacteria. Biomacromolecules 6:598–603. https://doi.org/10.1021/bm049441r

Jendrossek D (2007) Peculiarities of PHA granules preparation and PHA depolymerase activity determination. Appl Microbiol Biotechnol 74:1186–1196. https://doi.org/10.1007/s00253-007-0860-9

Jendrossek D (2009) Polyhydroxyalkanoate granules are complex subcellular organelles (carbonosomes). J Bacteriol 191:3195–3202. https://doi.org/10.1128/JB.01723-08

Jendrossek D, Handrick R (2002) Microbial degradation of polyhydroxyalkanoates. Annu Rev Microbiol 56:403–432. https://doi.org/10.1146/annurev.micro.56.012302.160838

Jendrossek D, Pfeiffer D (2014) New insights in the formation of polyhydroxyalkanoate granules (carbonosomes) and novel functions of poly(3-hydroxybutyrate). Environ Microbiol 16:2357–2373. https://doi.org/10.1111/1462-2920.12356

Jendrossek D, Selchow O, Hoppert M (2007) Poly(3-hydroxybutyrate) granules at the early stages of formation are localized close to the cytoplasmic membrane in *Caryophanon latum*. Appl Environ Microbiol 73:586–593. https://doi.org/10.1128/AEM.01839-06

Jiang Y, Ye J, Wu H, Zhang H (2004) Cloning and expression of the polyhydroxyalkanote depolymerase gene from *Pseudomonas putida*, and characterization of the gene product. Biotechnol Lett 26:1585–1588. https://doi.org/10.1023/B:BILE.0000045657.93818.18

Jüngert JR, Borisova M, Mayer C, Wolz C, Brigham CJ, Sinskey AJ, Jendrossek D (2017) Absence of ppGpp Leads to Increased Mobilization of Intermediately Accumulated Poly (3-hydroxybutyrate) (PHB) in *Ralstonia eutropha* H16. Appl Environ Microbiol AEM.:00755–00717. https://doi.org/10.1128/AEM.00755-17

Kadouri D, Jurkevitch E, Okon Y (2003) Involvement of the reserve material poly-beta-hydroxybutyrate in *Azospirillum brasilense* stress endurance and root colonization. Appl Environ Microbiol 69:3244–3250. https://doi.org/10.1128/aem.69.6.3244-3250.2003

Kawaguchi Y, Doi Y (1992) Kinetics and Mechanism of Synthesis and Degradation of Poly (3-Hydroxybutyrate) in *Alcaligenes eutrophus*. Macromolecules 25:2324–2329. https://doi.org/10.1021/ma00035a007

Kim T-K, Jung Y-M, Vo MT, Shioya S, Lee Y-H (2006) Metabolic engineering and characterization of phaC1 and phaC2 genes from *Pseudomonas putida* KCTC1639 for overproduction of medium-chain-length polyhydroxyalkanoate. Biotechnol Prog 22:1541–1546. https://doi.org/10.1021/bp0601746

Kim J, Kim Y-J, Choi SY, Lee SY, Kim K-J (2017a) Crystal structure of *Ralstonia eutropha* polyhydroxyalkanoate synthase C-terminal domain and reaction mechanisms. Biotechnol J. https://doi.org/10.1002/biot.201600648

Kim Y-J, Choi SY, Kim J, Jin KS, Lee SY, Kim K-J (2017b) Structure and function of the N-terminal domain of *Ralstonia eutropha* polyhydroxyalkanoate synthase, and the proposed structure and mechanisms of the whole enzyme. Biotechnol J. https://doi.org/10.1002/biot.201600649

Kobayashi T, Saito T (2003) Catalytic triad of intracellular poly(3-hydroxybutyrate) depolymerase (PhaZ1) in *Ralstonia eutropha* H16. J Biosci Bioeng 96:487–492. https://doi.org/10.1016/S1389-1723(03)70136-4

Kobayashi T, Shiraki M, Abe T, Sugiyama A, Saito T (2003) Purification and properties of an intracellular 3-hydroxybutyrate-oligomer hydrolase (PhaZ2) in *Ralstonia eutropha* H16 and its identification as a novel intracellular poly(3-hydroxybutyrate) depolymerase. J Bacteriol 185:3485–3490. https://doi.org/10.1128/JB.185.12.3485-3490.2003

Kobayashi T, Uchino K, Abe T, Yamazaki Y, Saito T (2005) Novel intracellular 3-hydroxybutyrate-oligomer hydrolase in *Wautersia eutropha* H16. J Bacteriol 187:5129–5135. https://doi.org/10.1128/JB.187.15.5129-5135.2005

Koch M, Doello S, Gutekunst K, Forchhammer K (2019) PHB is Produced from Glycogen Turnover during Nitrogen Starvation in *Synechocystis* sp. PCC 6803. Int J Mol Sci 20:1942. https://doi.org/10.3390/ijms20081942

Koskimäki JJ, Kajula M, Hokkanen J, Ihantola E-L, Kim JH, Hautajärvi H, Hankala E, Suokas M, Pohjanen J, Podolich O, Kozyrovska N, Turpeinen A, Pääkkönen M, Mattila S, Campbell BC, Pirttilä AM (2016) Methyl-esterified 3-hydroxybutyrate oligomers protect bacteria from hydroxyl radicals. Nat Chem Biol 12:332–338. https://doi.org/10.1038/nchembio.2043

Kuchta K, Chi L, Fuchs H, Pötter M, Steinbüchel A (2007) Studies on the influence of phasins on accumulation and degradation of PHB and nanostructure of PHB granules in *Ralstonia eutropha* H16. Biomacromolecules 8:657–662. https://doi.org/10.1021/bm060912e

Lageveen RG, Huisman GW, Preusting H, Ketelaar P, Eggink G, Witholt B (1988) Formation of polyesters by *Pseudomonas oleovorans*: effect of substrates on formation and composition of poly-(*R*)-3-hydroxyalkanoates and poly-(*R*)-3-hydroxyalkenoates. Appl Environ Microbiol 54:2924–2932

Lawrence AG, Choi J, Rha C, Stubbe J, Sinskey AJ (2005) In vitro analysis of the chain termination reaction in the synthesis of poly-(R)-beta-hydroxybutyrate by the class III synthase from *Allochromatium vinosum*. Biomacromolecules 6:2113–2119. https://doi.org/10.1021/bm0501048

Leuprecht D (2012) Untersuchung des intrazellulären PHB Abbaus von *Ralstonia eutropha* H16 durch Fluoreszenzmikroskopie und Two-Hybrid-Analyse. Diploma thesis, University Stuttgart

Liebergesell M, Schmidt B, Steinbüchel A (1992) Isolation and identification of granule-associated proteins relevant for poly(3-hydroxyalkanoic acid) biosynthesis in *Chromatium vinosum* D. FEMS Microbiol Lett 78:227–232. https://doi.org/10.1016/0378-1097(92)90031-i

Liebergesell M, Sonomoto K, Madkour M, Mayer F, Steinbüchel A (1994) Purification and characterization of the poly(hydroxyalkanoic acid) synthase from *Chromatium vinosum* and localization of the enzyme at the surface of poly(hydroxyalkanoic acid) granules. Eur J Biochem 226:71–80. https://doi.org/10.1111/j.1432-1033.1994.tb20027.x

Liu Q, Luo G, Zhou XR, Chen G-Q (2011) Biosynthesis of poly(3-hydroxydecanoate) and 3-hydroxydodecanoate dominating polyhydroxyalkanoates by β-oxidation pathway inhibited *Pseudomonas putida*. Metab Eng 13:11–17. https://doi.org/10.1016/j.ymben.2010.10.004

Liu G, Hou J, Cai S, Zhao D, Cai L, Han J, Zhou J, Xiang H (2015) A patatin-like protein associated with the polyhydroxyalkanoate (PHA) granules of *Haloferax mediterranei* acts as an efficient depolymerase in the degradation of native PHA. Appl Environ Microbiol 81:3029–3038. https://doi.org/10.1128/AEM.04269-14

Liu G, Cai S, Hou J, Zhao D, Han J, Zhou J, Xiang H (2016) Enoyl-CoA hydratase mediates polyhydroxyalkanoate mobilization in *Haloferax mediterranei*. Sci Rep. https://doi.org/10.1038/srep24015

Lütke-Eversloh T, Bergander K, Luftmann H, Steinbüchel A (2001) Identification of a new class of biopolymer: bacterial synthesis of a sulfur-containing polymer with thioester linkages. Microbiology (Reading, Engl) 147:11–19. https://doi.org/10.1099/00221287-147-1-11

Madison LL, Huisman GW (1999) Metabolic engineering of poly(3-hydroxyalkanoates): from DNA to plastic. Microbiol Mol Biol Rev 63:21–53

Maestro B, Galan B, Alfonso C, Rivas G, Prieto MA, Sanz JM (2013) A new family of intrinsically disordered proteins: structural characterization of the major phasin PhaF from *Pseudomonas putida* KT2440. PLoS One. https://doi.org/10.1371/journal.pone.0056904
Mandon K, Michel-Reydellet N, Encarnación S, Kaminski PA, Leija A, Cevallos MA, Elmerich C, Mora J (1998) Poly-beta-hydroxybutyrate turnover in *Azorhizobium caulinodans* is required for growth and affects *nifA* expression. J Bacteriol 180:5070–5076
Matin A, Veldhuis C, Stegeman V, Veenhuis M (1979) Selective advantage of a *Spirillum* sp. in a carbon-limited environment. Accumulation of poly-beta-hydroxybutyric acid and its role in starvation. J Gen Microbiol 112:349–355. https://doi.org/10.1099/00221287-112-2-349
Mato A, Tarazona NA, Hidalgo A, Cruz A, Jiménez M, Pérez-Gil J, Prieto MA (2019) Interfacial activity of phasin PhaF from *Pseudomonas putida* KT2440 at hydrophobic-hydrophilic biointerfaces. Langmuir 35:678–686. https://doi.org/10.1021/acs.langmuir.8b03036
Mayer F, Hoppert M (1997) Determination of the thickness of the boundary layer surrounding bacterial PHA inclusion bodies, and implications for models describing the molecular architecture of this layer. J Bas Microbiol 37:45–52
McCool GJ, Cannon MC (1999) Polyhydroxyalkanoate inclusion body-associated proteins and coding region in *Bacillus megaterium*. J Bacteriol 181:585–592
Meng D-C, Chen G-Q (2018) Synthetic Biology of Polyhydroxyalkanoates (PHA). Adv Biochem Eng Biotechnol 162:147–174. https://doi.org/10.1007/10_2017_3
Merrick JM, Doudoroff M (1964) Depolymerization of Poly-beta-hydroxybutyrate by intracellular enzyme system. J Bacteriol 88:60–71
Mezzina MP, Pettinari MJ (2016) Phasins, multifaceted polyhydroxyalkanoate granule-associated proteins. Appl Environ Microbiol 82:5060–5067. https://doi.org/10.1128/AEM.01161-16
Mezzolla V, D'Urso OF, Poltronieri P (2018) Role of PhaC type I and type II enzymes during PHA biosynthesis. Polymers (Basel) 10:910. https://doi.org/10.3390/polym10080910
Moldes C, Garcia P, García JL, Prieto MA (2004) In vivo immobilization of fusion proteins on bioplastics by the novel tag BioF. Appl Environ Microbiol 70:3205–3212. https://doi.org/10.1128/AEM.70.6.3205-3212.2004
Moreno S, Castellanos M, Bedoya-Pérez LP, Canales-Herrerías P, Espín G, Muriel-Millán LF (2019) Outer membrane protein I is associated with poly-β-hydroxybutyrate granules and is necessary for optimal polymer accumulation in *Azotobacter vinelandii* on solid medium. Microbiology (Reading, Engl) 193:265. https://doi.org/10.1099/mic.0.000837
Myshkina VL, Nikolaeva DA, Makhina TK, Bonartsev AP, Bonartseva GA (2008) Effect of growth conditions on the molecular weight of poly-3-hydroxybutyrate produced by *Azotobacter chroococcum* 7B. Appl Biochem Microbiol 44:482–486. https://doi.org/10.1134/S0003683808050050
Nelson KE, Weinel C, Paulsen IT, Dodson RJ, Hilbert H, Santos dos V, Fouts DE, Gill SR, Pop M, Holmes M, Brinkac L, Beanan M, DeBoy RT, Daugherty S, Kolonay J, Madupu R, Nelson W, White O, Peterson J, Khouri H, Hance I, Lee PC, Holtzapple E, Scanlan D, Tran K, Moazzez A, Utterback T, Rizzo M, Lee K, Kosack D, Moestl D, Wedler H, Lauber J, Stjepandic D, Hoheisel J, Straetz M, Heim S, Kiewitz C, Eisen J, Timmis KN, Dusterhoft A, Tummler B, Fraser CM (2002) Complete genome sequence and comparative analysis of the metabolically versatile *Pseudomonas putida* KT2440. Environ Microbiol 4:799–808
Neumann L, Spinozzi F, Sinibaldi R, Rustichelli F, Pötter M, Steinbüchel A (2008) Binding of the major phasin, PhaP1, from *Ralstonia eutropha* H16 to poly(3-hydroxybutyrate) granules. J Bacteriol 190:2911–2919. https://doi.org/10.1128/JB.01486-07
Nowroth V, Marquart L, Jendrossek D (2016) Low temperature-induced viable but not culturable state of *Ralstonia eutropha* and its relationship to accumulated polyhydroxybutyrate. FEMS Microbiol Lett 363:fnw249. https://doi.org/10.1093/femsle/fnw249
Obruca S, Sedlacek P, Mravec F, Samek O, Marova I (2016) Evaluation of 3-hydroxybutyrate as an enzyme-protective agent against heating and oxidative damage and its potential role in stress response of poly(3-hydroxybutyrate) accumulating cells. Appl Microbiol Biotechnol 100:1365–1376. https://doi.org/10.1007/s00253-015-7162-4

Obruca S, Sedlacek P, Mravec F, Krzyzanek V, Nebesarova J, Samek O, Kucera D, Benesova P, Hrubanova K, Milerova M, Marova I (2017) The presence of PHB granules in cytoplasm protects non-halophilic bacterial cells against the harmful impact of hypertonic environments. New Biotechnol 39:68–80. https://doi.org/10.1016/j.nbt.2017.07.008

Olivera ER, Arcos M, Naharro G, Luengo JM (2010) Unusual PHA biosynthesis. Microbiol Monogr 14:133–186

Peoples OP, Sinskey AJ (1989) Poly-beta-hydroxybutyrate (Phb) biosynthesis in *Alcaligenes eutrophus* H16—Identification and characterization of the PHB polymerase gene (PhbC). J Biol Chem 264:15298–15303. https://doi.org/10.1016/0014-5793%2888%2980693-8

Peplinski K, Ehrenreich A, Doering C, Boemeke M, Reinecke F, Hutmacher C, Steinbüchel A (2010) Genome-wide transcriptome analyses of the "Knallgas" bacterium *Ralstonia eutropha* H16 with regard to polyhydroxyalkanoate metabolism. Microbiology (Reading, Engl) 156:2136–2152. https://doi.org/10.1099/mic.0.038380-0

Peters V, Rehm BHA (2005) In vivo monitoring of PHA granule formation using GFP-labeled PHA synthases. FEMS Microbiol Lett 248:93–100. https://doi.org/10.1016/j.femsle.2005.05.027

Pfeiffer D, Jendrossek D (2011) Interaction between poly(3-hydroxybutyrate) granule-associated proteins as revealed by two-hybrid analysis and identification of a new phasin in *Ralstonia eutropha* H16. Microbiology (Reading, Engl) 157:2795–2807. https://doi.org/10.1099/mic.0.051508-0

Pfeiffer D, Jendrossek D (2012) Localization of poly(3-hydroxybutyrate) (PHB) granule-associated proteins during PHB granule formation and identification of two new phasins, PhaP6 and PhaP7, in *Ralstonia eutropha* H16. J Bacteriol 194:5909–5921. https://doi.org/10.1128/JB.00779-12

Pfeiffer D, Jendrossek D (2014) PhaM is the physiological activator of poly(3-hydroxybutyrate) (PHB) synthase (PhaC1) in *Ralstonia eutropha*. Appl Environ Microbiol 80:555–563. https://doi.org/10.1128/AEM.02935-13

Pfeiffer D, Wahl A, Jendrossek D (2011) Identification of a multifunctional protein, PhaM, that determines number, surface to volume ratio, subcellular localization and distribution to daughter cells of poly(3-hydroxybutyrate), PHB, granules in *Ralstonia eutropha* H16. Mol Microbiol 82:936–951. https://doi.org/10.1111/j.1365-2958.2011.07869.x

Pieper-Furst U, Madkour MH, Mayer F, Steinbüchel A (1994) Purification and characterization of a 14-kilodalton protein that is bound to the surface of polyhydroxyalkanoic acid granules in *Rhodococcus ruber*. J Bacteriol 176:4328–4337

Pohlmann A, Fricke WF, Reinecke F, Kusian B, Liesegang H, Cramm R, Eitinger T, Ewering C, Pötter M, Schwartz E, Strittmatter A, Voss I, Gottschalk G, Steinbüchel A, Friedrich B, Bowien B (2006) Genome sequence of the bioplastic-producing "Knallgas" bacterium *Ralstonia eutropha* H16. Nat Biotechnol 24:1257–1262. https://doi.org/10.1038/nbt1244

Pötter M, Steinbüchel A (2006) Biogenesis and structure of polyhydroxyalkanoate granules. Microbiol Monogr 1:1–28

Pötter M, Madkour MH, Mayer F, Steinbüchel A (2002) Regulation of phasin expression and polyhydroxyalkanoate (PHA) granule formation in *Ralstonia eutropha* H16. Microbiology (Reading, Engl) 148:2413–2426. https://doi.org/10.1099/00221287-148-8-2413

Pötter M, Müller H, Reinecke F, Wieczorek R, Fricke F, Bowien B, Friedrich B, Steinbüchel A (2004) The complex structure of polyhydroxybutyrate (PHB) granules: four orthologous and paralogous phasins occur in *Ralstonia eutropha*. Microbiology (Reading, Engl) 150:2301–2311. https://doi.org/10.1099/mic.0.26970-0

Pötter M, Müller H, Steinbüchel A (2005) Influence of homologous phasins (PhaP) on PHA accumulation and regulation of their expression by the transcriptional repressor PhaR in *Ralstonia eutropha* H16. Microbiology (Reading, Engl) 151:825–833. https://doi.org/10.1099/mic.0.27613-0

Prieto MA, Bühler B, Jung K, Witholt B, Kessler B (1999) PhaF, a polyhydroxyalkanoate-granule-associated protein of *Pseudomonas oleovorans* GPo1 involved in the regulatory expression system for pha genes. J Bacteriol 181:858–868

Prieto A, Escapa IF, Martinez V, Dinjaski N, Herencias C, la Pena de F, Tarazona N, Revelles O (2016) A holistic view of polyhydroxyalkanoate metabolism in *Pseudomonas putida*. Environ Microbiol 18:341–357. https://doi.org/10.1111/1462-2920.12760

Qi Q, Steinbüchel A, Rehm BH (2000) *In vitro* synthesis of poly(3-hydroxydecanoate): purification and enzymatic characterization of type II polyhydroxyalkanoate synthases PhaC1 and PhaC2 from *Pseudomonas aeruginosa*. Appl Microbiol Biotechnol 54:37–43

Ratcliff WC, Kadam SV, Denison RF (2008) Poly-3-hydroxybutyrate (PHB) supports survival and reproduction in starving rhizobia. FEMS Microbiol Ecol 65:391–399. https://doi.org/10.1111/j.1574-6941.2008.00544.x

Rehm BHA (2003) Polyester synthases: natural catalysts for plastics. Biochem J 376:15–33. https://doi.org/10.1042/BJ20031254

Rehm BHA (2006) Genetics and biochemistry of polyhydroxyalkanoate granule self-assembly: the key role of polyester synthases. Biotechnol Lett 28:207–213. https://doi.org/10.1007/s10529-005-5521-4

Rehm BHA (2010) Bacterial polymers: biosynthesis, modifications and applications. Nat Rev Microbiol 8:578–592. https://doi.org/10.1038/nrmicro2354

Rehm BH, Krüger N, Steinbüchel A (1998) A new metabolic link between fatty acid de novo synthesis and polyhydroxyalkanoic acid synthesis. The PHAG gene from *Pseudomonas putida* KT2440 encodes a 3-hydroxyacyl-acyl carrier protein-coenzyme a transferase. J Biol Chem 273:24044–24051

Ren Q, de Roo G, Witholt B, Zinn M, Thöny-Meyer L (2009) Overexpression and characterization of medium-chain-length polyhydroxyalkanoate granule bound polymerases from *Pseudomonas putida* GPo1. Microb Cell Fact 8:60–69. https://doi.org/10.1186/1475-2859-8-60

Ruth K, de Roo G, Egli T, Ren Q (2008) Identification of two acyl-CoA synthetases from *Pseudomonas putida* GPo1: one is located at the surface of polyhydroxyalkanoates granules. Biomacromolecules 9:1652–1659. https://doi.org/10.1021/bm8001655

Saegusa H, Shiraki M, Kanai C, Saito T (2001) Cloning of an intracellular Poly[D(-)-3-Hydroxybutyrate] depolymerase gene from *Ralstonia eutropha* H16 and characterization of the gene product. J Bacteriol 183:94–100. https://doi.org/10.1128/JB.183.1.94-100.2001

Schubert P, Steinbüchel A, Schlegel HG (1988) Cloning of the *Alcaligenes eutrophus* genes for synthesis of poly-beta-hydroxybutyric acid (PHB) and synthesis of PHB in *Escherichia coli*. J Bacteriol 170:5837–5847

Schultheiss D, Handrick R, Jendrossek D, Hanzlik M, Schüler D (2005) The presumptive magnetosome protein Mms16 is a poly(3-hydroxybutyrate) granule-bound protein (phasin) in *Magnetospirillum gryphiswaldense*. J Bacteriol 187:2416–2425. https://doi.org/10.1128/JB.187.7.2416-2425.2005

Schulze S (2018) Neue Einlicke in die Funktionalität PHB-Granula-assoziierter Proteine in *Ralstonia eutropha* H16. PhD thesis, University Stuttgart

Sedlacek P, Slaninova E, Koller M, Nebesarova J, Marova I, Krzyzanek V, Obruca S (2019) PHA granules help bacterial cells to preserve cell integrity when exposed to sudden osmotic imbalances. New Biotechnol 49:129–136. https://doi.org/10.1016/j.nbt.2018.10.005

Senior PJ, Dawes EA (1971) Poly-β-hydroxybutyrate biosynthesis and the regulation of glucose metabolism in *Azotobacter beijerinckii*. Biochem J 125:55–66. https://doi.org/10.1042/bj1250055

Solaiman DKY, Ashby RD, Foglia TA (2003) Effect of inactivation of poly(hydroxyalkanoates) depolymerase gene on the properties of poly(hydroxyalkanoates) in *Pseudomonas resinovorans*. Appl Microbiol Biotechnol 62:536–543. https://doi.org/10.1007/s00253-003-1317-4

Stam H, van Verseveld HW, de Vries W, Stouthamer AH (1986) Utilization of poly-β-hydroxybutyrate in free-living cultures of *Rhizobium* ORS571. FEMS Microbiol Lett 35:215–220

Steinbüchel A, Valentin HE (1995) Diversity of bacterial polyhydroxyalkanoic acids. FEMS Microbiol Lett 128:219–228

Steinbüchel A, El-Mehdi D, Marchessault RH, Timm A (1993a) Synthesis and production of poly (3-hydroxyvaleric acid) homopolyester by *Chromobacterium violaceum*. Appl Biochem Biotechnol 39:443–449

Steinbüchel A et al (1993b) Molecular basis for biosynthesis and accumulation of polyhydroxyalkanoic acids in bacteria. FEMS Microbiol Lett 104:347–350

Steinbüchel A, Aerts K, Babel W, Follner C, Liebergesell M, Madkour MH, Mayer F, Pieper-Furst U, Pries A, Valentin HE (1995) Considerations on the structure and biochemistry of bacterial polyhydroxyalkanoic acid inclusions. Can J Microbiol 41(Suppl 1):94–105

Stuart ES, Foster LJ, Lenz RW, Fuller RC (1996) Intracellular depolymerase functionality and location in *Pseudomonas oleovorans* inclusions containing polyhydroxyoctanoate. Int J Biol Macromol 19:171–176

Stubbe J, Tian J (2003) Polyhydroxyalkanoate (PHA) hemeostasis: the role of PHA synthase. Nat Prod Rep 20:445–457

Stubbe J, Tian J, He A, Sinskey AJ, Lawrence AG, Liu P (2005) Nontemplate-dependent polymerization processes: polyhydroxyalkanoate synthases as a paradigm. Annu Rev Biochem 74:433–480. https://doi.org/10.1146/annurev.biochem.74.082803.133013

Sun Y-W, Li Y, Hu Y, Chen W-X, Tian C-F (2019) Coordinated regulation of size and number of PHB granules by core and accessory phasins in the facultative microsymbiont *Sinorhizobium fredii* NGR234. Appl Environ Microbiol AEM:00717–00719. https://doi.org/10.1128/AEM.00717-19

Suzuki T, Deguchi H, Yamane T, Shimizu S, Gekko K (1988) Control of molecular weight of poly-p-hydroxybutyric acid produced in fet-batch culture of *Protomonas extorquens*. Appl Microbiol Biotechnol 27:487–491

Sznajder A, Jendrossek D (2011) Biochemical characterization of a new type of intracellular PHB depolymerase from *Rhodospirillum rubrum* with high hydrolytic activity on native PHB granules. Appl Microbiol Biotechnol 89:1487–1495. https://doi.org/10.1007/s00253-011-3096-7

Sznajder A, Jendrossek D (2014) To be or not to be a poly(3-hydroxybutyrate) (PHB) depolymerase: PhaZd1 (PhaZ6) and PhaZd2 (PhaZ7) of *Ralstonia eutropha*, highly active PHB depolymerases with no detectable role in mobilization of accumulated PHB. Appl Environ Microbiol 80:4936–4946. https://doi.org/10.1128/AEM.01056-14

Sznajder A, Pfeiffer D, Jendrossek D (2015) Comparative proteome analysis reveals four novel polyhydroxybutyrate (PHB) granule-associated proteins in *Ralstonia eutropha* H16. Appl Environ Microbiol 81:1847–1858. https://doi.org/10.1128/AEM.03791-14

Tarazona NA, Maestro B, Revelles O, Sanz JM, Prieto MA (2019) Role of leucine zipper-like motifs in the oligomerization of *Pseudomonas putida* phasins. Biochimica et Biophysica Acta (BBA) - General Subjects 1863:362–370. https://doi.org/10.1016/j.bbagen.2018.11.002

Thakor N, Lütke-Eversloh T, Steinbüchel A (2005) Application of the BPEC pathway for large-scale biotechnological production of poly(3-mercaptopropionate) by recombinant *Escherichia coli*, including a novel in situ isolation method. Appl Environ Microbiol 71:835–841. https://doi.org/10.1128/AEM.71.2.835-841.2005

Tian J, He A, Lawrence AG, Liu P, Watson N, Sinskey AJ, Stubbe J (2005a) Analysis of transient polyhydroxybutyrate production in *Wautersia eutropha* H16 by quantitative Western analysis and transmission electron microscopy. J Bacteriol 187:3825–3832. https://doi.org/10.1128/JB.187.11.3825-3832.2005

Tian J, Sinskey AJ, Stubbe J (2005b) Kinetic studies of polyhydroxybutyrate granule formation in *Wautersia eutropha* H16 by transmission electron microscopy. J Bacteriol 187:3814–3824. https://doi.org/10.1128/JB.187.11.3814-3824.2005

Timm A, Steinbüchel A (1990) Formation of polyesters consisting of medium-chain-length 3-hydroxyalkanoic acids from gluconate by *Pseudomonas aeruginosa* and other fluorescent pseudomonads. Appl Environ Microbiol 56:3360–3367

Tirapelle EF, Müller-Santos M, Tadra-Sfeir MZ, Kadowaki MAS, Steffens MBR, Monteiro RA, Souza EM, Pedrosa FO, Chubatsu LS (2013) Identification of proteins associated with

polyhydroxybutyrate granules from *Herbaspirillum seropedicae* SmR1—old partners, new players. PLoS One 8:e75066. https://doi.org/10.1371/journal.pone.0075066
Uchino K, Saito T (2006) Thiolysis of poly(3-hydroxybutyrate) with polyhydroxyalkanoate synthase from *Ralstonia eutropha*. J Biochem 139:615–621. https://doi.org/10.1093/jb/mvj069
Uchino K, Saito T, Gebauer B, Jendrossek D (2007) Isolated poly(3-hydroxybutyrate) (PHB) granules are complex bacterial organelles catalyzing formation of PHB from acetyl coenzyme A (CoA) and degradation of PHB to acetyl-CoA. J Bacteriol 189:8250–8256. https://doi.org/10.1128/JB.00752-07
Uchino K, Saito T, Jendrossek D (2008) Poly(3-hydroxybutyrate) (PHB) depolymerase PhaZa1 is involved in mobilization of accumulated PHB in *Ralstonia eutropha* H16. Appl Environ Microbiol 74:1058–1063. https://doi.org/10.1128/AEM.02342-07
Ushimaru K, Tsuge T (2016) Characterization of binding preference of polyhydroxyalkanoate biosynthesis-related multifunctional protein PhaM from *Ralstonia eutropha*. Appl Microbiol Biotechnol 100:4413–4421. https://doi.org/10.1007/s00253-015-7225-6
Wahl A, Schuth N, Pfeiffer D, Nussberger S, Jendrossek D (2012) PHB granules are attached to the nucleoid via PhaM in *Ralstonia eutropha*. BMC Microbiol 12:262. https://doi.org/10.1186/1471-2180-12-262
Wang Q, Yu H, Xia Y, Kang Z, Qi Q (2009) Complete PHB mobilization in *Escherichia coli* enhances the stress tolerance: a potential biotechnological application. Microb Cell Fact 8:47–49. https://doi.org/10.1186/1475-2859-8-47
Wieczorek R, Pries A, Steinbüchel A, Mayer F (1995) Analysis of a 24-kilodalton protein associated with the polyhydroxyalkanoic acid granules in *Alcaligenes eutrophus*. J Bacteriol 177:2425–2435
Wilde E (1962) Untersuchungen über Wachstum und Speicherstoffsynthese von *Hydrogenomonas*. Arch Mikrobiol 43:109. https://doi.org/10.1007/BF00406429
Wittenborn EC, Jost M, Wei Y, Stubbe J, Drennan CL (2016) Structure of the catalytic domain of the class I polyhydroxybutyrate synthase from *Cupriavidus necator*. J Biol Chem 48:25264–25277. https://doi.org/10.1074/jbc.M116.756833
Wodzinska J, Snell KD, Rhomberg A, Sinskey AJ, Biemann K, Stubbe J (1996) Polyhydroxybutyrate synthase: evidence for covalent catalysis. J Am Chem Soc 118:6319–6320. https://doi.org/10.1021/ja961108a
Yeung T, Gilbert GE, Shi J, Silvius J, Kapus A, Grinstein S (2008) Membrane phosphatidylserine regulates surface charge and protein localization. Science 319:210–213. https://doi.org/10.1126/science.1152066
York GM, Stubbe J, Sinskey AJ (2002) The *Ralstonia eutropha* PhaR protein couples synthesis of the PhaP phasin to the presence of polyhydroxybutyrate in cells and promotes polyhydroxybutyrate production. J Bacteriol 184:59–66. https://doi.org/10.1128/JB.184.1.59-66.2002
York GM, Lupberger J, Tian JM, Lawrence AG, Stubbe J, Sinskey AJ (2003) *Ralstonia eutropha* H16 encodes two and possibly three intracellular poly[D-(-)-3-hydroxybutyrate] depolymerase genes. J Bacteriol 185:3788–3794. https://doi.org/10.1128/JB.13.3788.3794.2003
Zhao YH, Li HM, Qin LF, Wang HH, Chen G-Q (2007) Disruption of the polyhydroxyalkanoate synthase gene in *Aeromonas hydrophila* reduces its survival ability under stress conditions. FEMS Microbiol Lett 276:34–41. https://doi.org/10.1111/j.1574-6968.2007.00904.x

The manufacturer's authorised representative in the EU is Springer Nature Customer Service Centre GmbH, Europaplatz 3, 69115 Heidelberg, Germany. If you have any concerns regarding our products, please contact ProductSafety@springernature.com

Printed and bound by CPI Group (UK) Ltd, Croydon, CR0 4YY
30/09/2026
02225254-0013